LASER PHYSICS
AND
LASER INSTABILITIES

LASER PHYSICS
AND
LASER INSTABILITIES

LORENZO M. NARDUCCI
Drexel University

NEAL B. ABRAHAM
Bryn Mawr College

World Scientific
Singapore • New Jersey • Hong Kong

Published by

World Scientific Publishing Co. Pte. Ltd.

5 Toh Tuck Link, Singapore 596224

USA office: 27 Warren Street, Suite 401-402, Hackensack, NJ 07601

UK office: 57 Shelton Street, Covent Garden, London WC2H 9HE

British Library Cataloguing-in-Publication Data
A catalogue record for this book is available from the British Library.

LASER PHYSICS AND LASER INSTABILITIES

ISBN-13 978-9971-5-0062-7
ISBN-10 9971-5-0062-0
ISBN-13 978-9971-5-0063-4 (pbk)
ISBN-10 9971-5-0063-9 (pbk)

Preface

The 1985 Summer School for Laser Physics was the result of the efforts of many committed scientists and educators, who donated freely of their time and energy for the successful completion of this project.

Professor Gao Jin Yue, the Director of the School and a recognized laser physicist in his own right, conceived the notion of a Laser Institute that would encourage the interaction of young Chinese researchers with one another and with selected foreign representatives, and that would address some of the most active issues in this field. The realization of this enterprise and its scientific success are largely the result of many months of preparation and careful planning by Professor Gao and the members of his group.

The School was organized under the auspices of Jilin University in Changchun, People's Republic of China, and with the direct support and encouragement of the Physics Department Chairman, Professor Liu Yue Zuo. The lectures were delivered by Lorenzo M. Narducci of Drexel University in Philadelphia, Pennsylvania, and by Neal B. Abraham of Bryn Mawr College in Bryn Mawr, Pennsylvania.

The goal of their presentations was to expose the participants, some of them relative newcomers to the laser field, to the basic theoretical and experimental facts, with an emphasis on the problem of instabilities. The lectures were delivered in English, and were organized in an alternating pattern that would maximize the logical connections between theory and experiments. For the purpose of this volume, the authors have decided to prepare two separate sets of lecture notes, feeling that this would be the best mode of presentation in written form. The style, however, retains much of the spontaneity of the verbal presentation, so that, apart from some modifications, improvements and completions, the following articles are indeed faithful reproductions of the original lectures notes.

The lecturers are grateful to Professors Gao and Liu for their invitation, and to the Administration of Jilin University for creating the ideal environment for the development of the free and spirited dialogue that was to be one of the highlights of the school. They are also indebted to the students and researchers for their active participation and for their enthusiasm. Their joy of learning and their dedication was an inspiration for the entire staff.

The lecturers owe a special debt of gratitude to Debbie D. Hughes, LeAnn Davis and Ann Daudert for their professional preparation of the first version of the notes and to Liu Yan Ling, Xie Qun, and Su Dang Sheng for working extra hard in making copies of notes and transparencies available to all the attendees. Professor Narducci's travel and living expenses were provided through a grant of the World Bank, while Professor Abraham's support was contributed in part by Jilin University.

List of Participants

Gao Jin-Yue	Physics Department, Jilin University Changchun, PRC
Lorenzo M. Narducci	Department of Physics, Drexel University Philadelphia, Pa. 19104, USA
Neal B. Abraham	Department of Physics, Bryn Mawr College Bryn Mawr, Pa. 19010, USA
Bai Xiao-Ming	Physics Department, Jilin University Changchun, PRC
Cai Yong-Qang	Physics Department, South China Institute of Technology, Guang Zhou, PRC
Cui Jie	Changchun Institute of Optics and Fine Mechanics, Changchun, PRC
Dong Jiu-Zhao	Physics Department, Lan Zhuo University Lan Zhuo, PRC
Fan Bin	Physics Department, Northwestern University Xian, Shanxi, PRC
Feng JianMin	Changchun Institute of Optics and Fine Mechanics, Changchun, PRC
Gao JunJie	Changchung Institute of Optics and Fine Mechanics, Changchun, PRC
Guo Xiu Zhen	Physics Department, Jilin University Changchun, PRC
Guan Shi-Rong	Physics Department, Fujian Teachers University, Fuzhou, Fujian, PRC
He Lin-Sheng	P.O. Box 25, He Fei, PRC
Jin Guang-Shu	Physics Department, Jilin University Changchun, PRC
Li Je-Fei	Department of Applied Physics, Harbin Institute of Technology, Harbin, PRC
Li Xi-Zeng	Physics Department, Tian Jin University Tian Jin, PRC
Li Yong-Da	Changchun Institute of Optics and Fine Mechanics, Changchun, PRC

Lin Shuai-Bin	The Institute of Modern Optical Science, Nankai University, Tian Jin, PRC
Ling Hong-Yan	Xian Institute of Optics and Fine Mechanics, Xian, PRC
Liu Ya-Bo	Physics Department, Jilin University Changchun, PRC
Liu Yi	Changchun Institute of Optics and Fine Mechanics, Changchun, PRC
Liu Yong	Physics Department, Jilin University Changchun, PRC
Luan Shu Wen	Physics Department, Jilin University Changchun, PRC
Lu Deren	Changchun Institute of Optics and Fine Mechanics, Changchun, PRC
Ma Tie	Physics Department, Ning Xia University, Yin Chan, Ning Xia, PRC
Miao Xin-Lin	Changchun Institute of Optics and Fine Mechanics, Changchun, PRC
Qin Zi-Xiong	Physics Department, South China Institute of Technology, Guang Zhou, PRC
Shan Ying	Physics Department, Tian Jin University, Tian Jin, PRC
Shen Ke	Changchun Institute of Optics and Fine Mechanics, Changchun, PRC
Shun Jian-Wei	Physics Department, Jilin University Changchun, PRC
Wang Cheng-Fei	Physics Department, Jilin University Changchun, PRC
Wang Xi-Jun	Changchun Institute of Optics and Fine Mechanics, Changchun, PRC
Wang En-Yuan	Physics Department, Jilin University Changchun, PRC
Wang Xi-Po	Changchun Institute of Applied Chemistry, Academia Sinica, Changchun, PRC
Wang Xu-Hua	Physics Department, Jilin University Changchun, PRC

Wang Zhao-Hui	Physics Department, Northwestern University, Xian, Shanxi, PRC
Wu Ting-Wan	Physics Department, South China Institute of Technology, Guang Zhou, PRC
Xue Fuzeng	Changchun Institute of Optics and Fine Mechanics, Changchun, PRC
Yu Liang	Department of Physics, Shandong College of Oceanography, Quin Dao, PRC
Zhang Lei	Department of Applied Physics, Harbin Institute of Technology, Harbin, PRC
Zhang Peixun	Changchun Institute of Optics and Fine Mechanics, Changchun, PRC
Zhang Xiaodong	Physics Department, South China Institute of Technology, Guang Zhou, PRC
Zhao Ming-Di	Physics Department, Jilin University Changchun, PRC
Zhao Xiao-Ping	Physics Department, Jilin University Changchun, PRC
Zhang Zhi-Ren	Physics Department, Jilin University Changchun, PRC

CONTENTS

LASER PHYSICS
AND
LASER INSTABILITIES

A Theoretical Outline of Laser Physics and Laser Instabilities

by

Lorenzo M. Narducci

Department of Physics and Atmospheric Science
Drexel University
Philadelphia, Pa. 19104 USA

Introductory Comments

The objective of these notes is to provide an overview of some basic aspects of laser theory for the participants of the Jilin University Summer School in Laser Physics. Because the presentation of a complete theoretical development is not feasible during the limited time available to this School, I will focus my attention on one of the most popular and, to a large extent, successful models of laser action, the so called plane-wave, Maxwell-Bloch model. Many of the essential physical aspects of laser operation are represented well by this theoretical framework. After careful review of these notes, the students should be able to pursue their own research interests independently or with the help of specialized literature.

In spite of the rather idealized nature of the Maxwell-Bloch model, a large number of important results will emerge from our discussion. In general, one should keep in mind that gas lasers tend to be more accurately described by this version of the theory than their solid-state counterparts, mainly because our simple picture of the active medium cannot reflect the complications that are typical, for example, of ruby, neodymium, or semiconductor lasers . In addition, for the most part, we shall focus on laser systems that can be designed to operate in a ring cavity configuration, and which can be forced to function in a unidirectional mode. The reason is mainly one of convenience because the presence of counter-propagating fields inside the cavity introduces considerable additional complications due to the interference of the optical fields and the small-scale modulation that this imposes on the active medium. The reader, however, will be able to form an idea of how the unidirectional theory can be extended to include bidirectional propagation with the help of Appendix C at the end of these notes.

In constructing a laser theory we discover that certain operating configurations in which the output intensity is stationary in time (here we consider lasers that are pumped steadily, for example by a well stabilized discharge mechanism) can become unstable, and develop pulsations in the form of either regular or irregular pulses. These pulse-trains are also characterized by a stability domain of their own; outside this range of stability, new and different kinds of temporal oscillations emerge. Often, also very irregular pulsations can develop out of the nonlinear dynamics of the system. There is strong evidence that internal or external noise sources are not the basic physical cause of this randomness, although they may be responsible for triggering these phenomena. The term deterministic chaos has been introduced in laser physics, and in fact in many other disciplines as well, to indicate that noise-free equations of motion are capable of producing completely aperiodic temporal behaviors.

The field of laser instabilities has grown to such an extent, as a major area of research in its own right, that it is quite appropriate to devote a major fraction of these lectures to this subject. In addition to its intrinsic theoretical interest, there are also practical reasons why a laser physicist should become knowledgeable about these aspects of the laser phenomenology. Instabilities represent disruptions in existing patterns of output radiation that cannot be corrected by working more diligently with the basic support instrumentation (e.g., power supplies, mechanical mounts, etc.); they can be avoided only by varying the parameters in such a way that the laser is forced to occupy a stable domain of operation for the given configuration. On the other hand, there may be practical reasons why a particular unstable pattern is desirable; for example, a periodically modulated output intensity may be useful for specific applications. In this case also, it is desirable to understand the physics of the system as completely as possible in order to maintain stable control of these oscillations.

Pulsations in lasers came unexpectedly from the very beginning. One would expect that a cw-pumped laser, or a laser excited by a slowly-varying pump, should produce a time-independent (or slowly-varying) output that follows the pump profile. Most early lasers and, in fact, also some masers did not match this expectation. The inability of the early models to describe this behavior provided a strong motivation for the development of more accurate theories. At this time we have learned to recognize the importance of atomic coherence in the context of laser dynamics, and we have explained a number of interesting unstable behaviors, at least qualitatively, with the help of theories that have gone beyond the elementary rate equation approach. Much more research is still needed to clarify the open issues, particularly because quantitative agreement with the experiments is still only a dream in the eyes of the theorists.

It is our hope that, stimulated by these discussions, many of the participants of this Summer School will feel challenged enough to join the search for a more complete theoretical understanding.

4

1. An introduction to the Theory of the Laser: the Rate Equations

The word laser is an acronym for Light Amplification by Stimulated Emission of Radiation. This name was introduced in the early days to stress the difference between the new devices, which operated in the visible part of the electromagnetic spectrum, and the existing microwave systems (masers) whose output frequency was of the order of several tens of GHz (the ammonia maser, for example, operates at a frequency of 23.8 GHz corresponding to a wavelength of about 1 cm).

In this course we shall learn that the development of a theory of the laser is not a very difficult task, but that some of the ideas that are useful in describing ordinary light sources, are no longer appropriate for dealing with the essential aspects of laser light: coherence, unidirectionality, spectral purity and similar attributes of a laser can be discussed most effectively in theoretical terms that depart significantly from those that are appropriate to the description of blackbody radiation.

The conventional laser, as a device, consists of three essential components:

i)   An active medium whose elementary units are either atoms or molecules;
ii)   A resonant electromagnetic structure, or cavity;
iii)   A pump source that delivers the energy needed to initiate and to maintain the laser process.

We use the term "conventional laser" because a new device has been added recently to the laser family; this is the so-called free electron laser, whose principle of operation, in spite of its name, is entirely different from that of an ordinary laser.

With the inclusion of the free electron laser, laser systems cover almost eight orders of magnitude in their output frequency, from the NMR laser (radiowaves, 1 MHz) to the soft x-ray laser produced by beams of relativistic electrons. Practically every conceivable atomic of molecular unit can be made to function as a laser, as witnessed by the hundreds of laser lines that have been tabulated in specialized handbooks [Beck, Englisch and Gurs (1980)].

The physical principles of light emission and absorption were laid out in the well known paper by Einstein (1917) entitled "On the Quantum Theory of Radiation". Einstein proposed a heuristic view of the processes of absorption and emission of radiation, and demonstrated how the Boltzmann population formula for systems in thermodynamic equilibrium could be used to "derive" Planck's blackbody radiation law.

In simple terms, Einstein's argument suggests the following picture of the energy exchanges that take place between light and matter. We consider an arbitrary pair of energy levels in a molecular or atomic system. If an electron is placed in a higher energy state (which we label m, for definiteness), it can decay spontaneously to a lower state (n) by a process which Einstein equated to radioactive decay. Through the mechanism of spontaneous emission, the excess energy stored in the medium is converted into electromagnetic radiation. The requirement of energy conservation forces the energy of the emitted photon, hv, to be the same as the energy difference, E_m-E_n between the higher and lower states of the medium.

By a reverse process, the emitted quantum may interact with a similar atom or molecule in its ground state and be absorbed. In this case, the system is left in its excited state, and the electromagnetic field loses an amount of energy equal to hv. Significantly, Einstein proposed a third energy exchange mechanism, the so-called stimulated emission process. In this case, a photon of energy hv interacts with an excited atom or molecule, and forces it to decay to the ground state. At the end of the process, the atom is left unexcited, and the electromagnetic field carries away two units of energy hv.

More quantitatively, the elementary probabilities that each of the three processes occur in a time dt are given by

$$dW_{sp.\ em.} = A_{m \to n} dt \qquad \text{spontaneous emission} \qquad (1.1a)$$

$$dW_{st.\ em.} = \rho B_{m \to n} dt \qquad \text{simulated emission} \qquad (1.1b)$$

$$dW_{abs.} = \rho B_{n \to m} dt \qquad \text{absorption} \qquad (1.1c)$$

where ρ denotes the electromagnetic energy density.

If the molecules of the medium are in canonical equilibrium with the radiation, the probability of finding a molecule in the n-th excited state is proportional to the Boltzmann factor

$$W_n = \alpha_n \exp(-\frac{E_n}{kT}) \qquad (1.2)$$

where α_n is a number that characterizes the molecular state, but does not depend on the temperature T of the environment. The thermodynamic equilibrium condition can be summarized as follows

number of absorption events per unit time = number of emission events per unit time   (1.3a)

or, more explicitly,

$$\alpha_n e^{-E_n/kT} B_{n \to m} \rho = \alpha_m e^{-E_m/kT} (B_{m \to n} \rho + A_{m \to n}) \qquad (1.3b)$$

If ρ increases monotonically with increasing temperature, then, at a sufficiently high temperature, it must be true that

$$\alpha_n B_{n \to m} \cong \alpha_m B_{m \to n} \qquad (1.4)$$

As a consequence of Eqs. (1.4) and (1.3b), the energy density of the field must take the form

$$\rho = \frac{\alpha_m e^{-\frac{E_m}{kT}} A_{m \to n}}{\alpha_n e^{-\frac{E_n}{kT}} B_{n \to m} - \alpha_m e^{-\frac{E_m}{kT}} B_{m \to n}}$$

$$= \frac{A_{m \to n}}{B_{m \to n}} \frac{1}{\exp[\,(E_m - E_n)/kT\,] - 1} \qquad (1.5)$$

From Wien's displacement law, it follows at once that

$$\frac{A_{m \to n}}{B_{m \to n}} \sim \nu^3 \qquad (1.6)$$

and from the second Bohr hypothesis we have that

$$E_m - E_n = h\nu \qquad (1.7)$$

We must stress an essential point in this argument: the atomic or molecular systems are assumed to be in a state of equilibrium with the radiation. In this case, then, the radiation density acquires the familiar functional dependence on ν and T, as proposed by Planck.

A very simplified view of the process of amplification of light can be formulated on the basis of the above ideas. One begins with a collection of identical atoms or molecules in an enclosure having two highly reflecting parallel walls. Some atoms begin to decay spontaneously; the photons that happen to be emitted in a direction perpendicular to the reflecting surfaces have a high chance of being reflected and, therefore, have a high probability of causing subsequent stimulated emission processes or to be reabsorbed. Of course, if we want to produce amplification, the former processes must occur with a higher probability than the latter ones, and this implies that the excited state population must be larger than the population in the ground state. Under these conditions, it is easy to see how a large increase of electromagnetic energy density can be produced in the active volume. If the mirrors are partially transparent, some of the radiation escapes from the active volume. As long as new excited atoms are steadily supplied, the process can continue indefinitely. The unidirectionality of the emerging radiation can be explained, roughly, on the basis of the geometric arrangement, while the monochromatic character of the emission can be connected, superficially, to the fact that only two levels are involved in the process.

We must insist, however, that this simple view of the laser process is just that--a simple picture. If one should attempt to extract finer details from it, one would quickly find out that a deeper understanding of the underlying physics is required. This picture, for example, does not explain why the linewidth of the laser light is so much narrower than that of the active atomic transition, nor, at a deeper level, why the statistical character of laser radiation is so distinctively different from that of thermal light. This picture of laser operation is not really wrong, in the same sense as models of reality are not wrong, if they provide a reasonable description at some level. However, we must keep in mind that the range of validity of this description is limited, and confusion can arise if one tries to apply it indiscriminately.

Yet, there is a pedagogical value in pursuing this line of reasoning at a more quantitative level because it will give us a chance to introduce some useful notions in a simple way. At a later time, we shall show explicitly under what conditions these results can be obtained as an approximation to a more refined theory.

We consider an idealized active medium, consisting of only two energy levels, and we denote with $N_1(t)$ and $N_2(t)$ the populations densities of the lower and higher levels, respectively. If we ignore, for the moment, the effects of spontaneous decay, the rate equations that govern the evolution of the two populations are

$$\frac{dN_1}{dt} = - W_{1 \to 2} \, nN_1 + W_{2 \to 1} \, nN_2 \qquad\qquad (1.8a)$$

$$\frac{dN_2}{dt} = W_{1 \to 2} \, nN_1 - W_{2 \to 1} \, nN_2 \qquad\qquad (1.8b)$$

where n is the number of photons in the active volume and $W_{i \to j}$ are transition rates from state i to state j. The population difference $D = N_2 - N_1$ obeys the equation

$$\frac{dD}{dt} = - 2WnD \qquad\qquad (1.9)$$

where we have set $W_{1 \to 2} = W_{2 \to 1} = W$, following Einstein's argument.

Now we introduce, phenomenologically, the combined action of the spontaneous decay and of the pump, with the result

$$\frac{dD}{dt} = - 2WnD - \frac{1}{T_1} (D - D_0) \qquad\qquad (1.10)$$

where T_1 is the lifetime of the equilibration process of the population with the pump, and D_0 is the population difference in the absence of electromagnetic energy in the cavity (n=0).

The intensity of the radiation is governed by a rate equation of its own. For every atomic transition of the type 2→1, one photon is added to the field, while for every transition 1→2, one photon is removed. On the basis of this argument, it is easy to see that the number of photons satisfies the rate equation

$$\frac{dn}{dt} = WnN_2 - WnN_1 = WnD \qquad\qquad (1.11)$$

Note, that if we neglect the pump term in Eq. (1.10) we have

$$\frac{d}{dt} (2n+D) = 0 \quad \text{or} \quad 2(\delta n) = - \delta D \qquad\qquad (1.12)$$

as it must be, because for every individual act of absorption or creation of photons, the population difference changes by two units. To include the effects of the losses of radiation out of the active volume we modify Eq. (1.11) as follows

$$\frac{dn}{dt} = WnD - \frac{1}{T_c} n \qquad\qquad (1.13)$$

where T_c denotes the lifetime of a photon in the cavity. The two coupled equations

$$\frac{dD}{dt} = - 2WnD - \frac{1}{T_1} (D - D_0) \qquad\qquad (1.14a)$$

$$\frac{dn}{dt} = WnD - \frac{1}{T_c} n \qquad\qquad (1.14b)$$

are called the laser rate equations [a more precise derivation of the rate equations, useful for quantitative studies of the laser can be found, for example in Svelto (1982)]. Even in this idealized form, these equations allow us to make some useful statements.

(a)   If, initially, very few photons exist in the cavity, the rate of change dn/dt is positive, i.e. light amplification will occur, if

$$WD(0) - \frac{1}{T_c} > 0$$

or

$$D(0) > \frac{1}{WT_c} \qquad (1.15)$$

For obvious reasons, the quantity

$$D_{thr} \equiv \frac{1}{WT_c} \qquad (1.16)$$

is called the population threshold value for laser action. Because the initial value of D is, usually, the equilibrium population D_0, a necessary condition for laser action is

$$D_0 > D_{thr} \equiv \frac{1}{WT_c}$$

or

$$D_0 > \frac{1}{WT_c} \qquad (1.17)$$

Clearly, laser action requires population inversion, i.e., $N_2 > N_1$. In addition, the population inversion must be sufficiently large to overcome the losses. From Eq. (1.17) it is clear that a given equilibrium population difference can be made to provide enough energy for laser action if T_c or W (or both) are made sufficiently large. The condition "T_c large" implies a high quality optical cavity, while "W large" implies the selection of a pair of levels with a large transition rate. As we shall see more clearly in future lectures, this also implies a large induced dipole moment between the chosen levels.

(b)   The rate equations are nonlinear equations and cannot be solved by elementary techniques. A common strategy to extract useful physical information from a set of nonlinear equations is to begin with an analysis of their possible steady states. As a first step we set all the time derivatives equal to zero. In our case, we have

$$2WnD + \frac{1}{T_1}(D - D_0) = 0 \qquad (1.18a)$$

$$n\left(WD - \frac{1}{T_c}\right) = 0 \qquad (1.18b)$$

Thus, two possible solutions exist:

i) $n = 0,$ $D = D_0$ $\qquad (1.19)$

ii) $D = \dfrac{1}{WT_c} \equiv D_{thr}$,

$$n = \frac{D_0 - D_{thr}}{2WD_{thr}T_1} = \frac{T_c}{2T_1}(D_0 - D_{thr}) \qquad (1.21)$$

Equation (1.19) corresponds to the so-called trivial state of equilibrium with the pump and no photons in the cavity; Equations (1.20) and (1.21) correspond to a nontrivial operating condition. Now we discuss both cases in some detail.

Case 1

On physical grounds, there are two ways in which this solution can be realized: either the laser is below threshold, and nothing is expected to happen (of course), or the laser is above threshold and the slightest deviation from this state will cause amplification. This is a good place to stress an important shortcoming of the rate equation model: the absence of a mechanism that can initiate laser action [if we select $n(0)=0$ as the initial condition for Eq. (1.14b), the only possible solution is $n(t)=0$ for all time, regardless of the initial inversion]. Laser action is primed by spontaneous emission, which is missing from our description, unless we insert it in the theory "by hand". A properly developed quantum theory displays this noise term automatically [See, for example, Haken (1970), Sargent, Scully and Lamb (1979)].

Returning now to our discussion, we label the two possibilities that may develop when the system is initially in the state (i) as being stable and unstable configurations, respectively. There are techniques by which we can probe the stability of a given steady state in a more sophisticated way than by a qualitative analysis, as we shall see shortly.

Case 2

In this case, also, there are two possibilities. From Eq. (1.21) we see that if $D_0 < D_{thr}$ not enough population inversion is available to overcome the losses, and this solution is physically meaningless. If, on the other hand, $D_0 > D_{thr}$ the solution represents a possible laser steady state. Our expectation is that this steady state will be stable.

This problem is simple enough that the stability or instability of the steady state can be argued on physical grounds without much mathematical help. In general, this will not be the case, especially with more sophisticated classes of laser models. For this reason, we introduce some elementary notions of linear stability analysis, using this simple situation as a test case, in order to gain some experience before undertaking more elaborate studies.

(c)   The physical realizability (or stability) of a solution can be studied with the help of a technique called linear stability analysis. It is likely that this method will be already known to the readers of these notes. In simple terms one can summarize the procedure as follows: we expand the unknown solution of a problem in the neighborhood of a stationary state and retain only the leading corrections; we solve the resulting linearized equations for the correction variables, or fluctuations as we often call them, in terms of trial exponential functions. The complex rate constants of the exponential solutions produce convergence back to the steady state or divergence away from it, depending on the sign of their real parts. If all the rate constants of the problem have a negative real part, the steady state is stable; if even one of the real parts is positive, one is faced with an instability.

An important comment is appropriate at this point: an implicit requirement of the linear stability method is that the steady state should be perturbed by an infinitesimal amount. Thus, this procedure can only provide meaningful answers in the local neighborhood of each steady state configuration. Often, stable, time-dependent solutions exist in different regions of the parameter space. These solutions are difficult to find analytically (they are certainly not accessible by linearization methods), because only a finite perturbation from steady state can drive the system into their domain of attraction. Thus, it takes much care before one can claim knowledge of all the possible solutions of a nonlinear system of equations.

We now return to our problem and denote a steady state solution of Eqs. (1.14) by n_∞ and D_∞; furthermore, we label with ε_n and ε_D two arbitrarily small perturbations around this solution. In the neighborhood of the steady state the time-dependent solutions of Eqs. (1.14) take the form

$$n(t) = n_\infty + \varepsilon_n(t) \tag{1.22a}$$

$$D(t) = D_\infty + \varepsilon_D(t) \tag{1.22b}$$

Because of the smallness of the perturbations, the equations of motion can be linearized by neglecting terms that contain quadratic combinations of the fluctuation variables. The result of this simple calculation is

$$\frac{d\varepsilon_D}{dt} = -2W\,(n_\infty \varepsilon_D + D_\infty \varepsilon_n) - \frac{1}{T_1}\,\varepsilon_D \tag{1.23a}$$

$$\frac{d\varepsilon_n}{dt} = W\,(n_\infty \varepsilon_D + D_\infty \varepsilon_n) - \frac{1}{T_c}\,\varepsilon_n \tag{1.23b}$$

We seek elementary solutions of the form

$$\begin{pmatrix} \varepsilon_D(t) \\ \varepsilon_n(t) \end{pmatrix} = e^{\lambda t} \begin{pmatrix} \varepsilon_D(0) \\ \varepsilon_n(0) \end{pmatrix} \tag{1.24}$$

and obtain the two coupled linear algebraic equations

$$(\lambda + 2Wn_\infty + \frac{1}{T_1})\,\varepsilon_D(0) + 2WD_\infty \varepsilon_n(0) = 0 \tag{1.25a}$$

$$-Wn_\infty \varepsilon_D(0) + (\lambda - WD_\infty + \frac{1}{T_c})\,\varepsilon_n(0) = 0 \tag{1.25b}$$

Nontrivial solutions for this system will exists if and only if the determinant of the coefficients vanishes identically, i.e. if

$$(\lambda + 2Wn_\infty + \frac{1}{T_1}) \, (\lambda - WD_\infty + \frac{1}{T_c}) + 2W^2 n_\infty D_\infty = 0 \qquad (1.26)$$

This is the so-called secular, or characteristic, equation for the rate constants λ. The stability requirement is given by the following necessary and sufficient condition

$$\text{Re } \lambda_i < 0 \qquad \text{for } i=1,2 \qquad (1.27)$$

Consider now the steady state of case (i) with n_∞ and D_∞ given by Eq. (1.19). The characteristic equation takes the form

$$(\lambda + \frac{1}{T_1}) \, (\lambda - WD_0 + \frac{1}{T_c}) = 0 \qquad (1.28)$$

and the roots are

$$\lambda_1 = -\frac{1}{T_1} \qquad (1.29a)$$

$$\lambda_2 = WD_0 - \frac{1}{T_c} \qquad (1.29b)$$

If $WD_0 - 1/T_c < 0$ (laser below threshold), this stationary state is stable. If, instead, $WD_0 - 1/T_c > 0$ (laser above threshold), the steady state is unstable. In the latter case, the system under the action of an arbitrarily small perturbation departs exponentially from the stationary state. After some time, the linearized solution is no longer useful to describe the evolution of the system, and the full machinery of the nonlinear differential equations will have to be used if we want to obtain information about the actual time dependence of the system.

Consider now case (ii). Here the secular equation can be put into the form

$$\lambda \, (\lambda + \frac{D_0}{D_{thr}T_1}) + \frac{W \, (D_0 - D_{thr})}{T_1} = 0 \qquad (1.30)$$

The roots are

$$\lambda_{1,2} = \frac{1}{2} \left\{ -\frac{D_0}{D_{thr}T_1} \pm \left[\left(\frac{D_0}{D_{thr}T_1}\right)^2 - \frac{4W \, (D_0 - D_{thr})}{T_1} \right]^{1/2} \right\} \qquad (1.31)$$

The only physically meaningful solution in this case corresponds to $D_0 > D_{thr}$, i.e., the laser must be above threshold, otherwise n_∞ is negative. Thus, the eigenvalues have a negative real part, as one can confirm with a a quick analysis of Eq. (1.31).

The approach to steady state is monotonic or oscillatory depending on the sign of the quantity

$$\Delta = \left(\frac{D_0}{D_{thr}T_1}\right)^2 - \frac{4W \, (D_0 - D_{thr})}{T_1 T_c D_{thr}}$$

If $\Delta < 0$, $|\Delta|^{1/2}$ represents the frequency of the relaxation oscillations. The parameter Δ is negative, in lasers near threshold, if T_1 is sufficiently larger than T_c, as it happens to be true in many solid state lasers. Thus, when $T_1 >> T_c$, the transient approach to steady state involves a damped train of spikes whose damping rate is of the order of $1/T_1$ while the pulsing rate is of the order of $(T_1 T_c)^{-1/2}$ and the width of the largest spike may be as short as T_c.

The relaxation oscillations play a special role when a laser, operating in the rate equation regime, is driven externally by a periodic forcing function. If the frequency of the external modulation approaches that of a damped resonance of the system, the transfer of energy is greatly enhanced, as with all resonantly driven systems, and interesting types of unstable behavior can be excited. However, free running lasers in the rate equation regime are always stable according to this prescription, so that the rate equations are unable to describe the sustained spiking behavior of many of the early solid state devices. A detailed proof of the global stability of the rate equations was given by Hofelich-Abate and Hofelich (1968a,b,c,d).

In conclusion, the steady state root (i) is stable below threshold and unstable above threshold; root (ii) is stable above threshold (and meaningless below). These results can be summarized by the so-called bifurcation graph shown in Fig. 1.1 which displays the dependence of the steady state intensity upon the population difference D_0 viewed as a control parameter.

(d)   In the low intensity limit, the population difference is essentially constant and equal to its initial value (for example, D_0). In this case, the rate equation for the photon number can be interpreted as a balance equation between linear gain and loss mechanisms

$$\frac{dn}{dt} = \underbrace{W D_0 n}_{\text{gain}} - \underbrace{\frac{1}{T_c} n}_{\text{loss}} \qquad (1.32)$$

The quantity WD_0 is proportional to the so-called unsaturated gain, which has the dimensions of reciprocal length and is related to WD_0 by a simple scale factor. In order to clarify the reason why this gain constant is called "unsaturated", consider the behavior of the population difference when the photon flux becomes very large. Using the population equation, we can estimate D by calculating its value in steady state. The result is

$$D = \frac{D_0}{1 + 2WT_1 n} \qquad (1.33)$$

It is clear that as n becomes larger, D becomes smaller, until eventually, for extremely large intensity levels, the population difference approaches zero (i.e., $N_1 = N_2$). Under these conditions, the quantity WD, which plays the role of a gain "constant" in Eq. (1.14a) becomes progressively smaller. This implies that the gain "seen" by an intense flux of photons is smaller than in the low intensity limit, or that the gain becomes "saturated" at high intensity levels. Thus, by definition, the saturated gain is always smaller than the unsaturated gain, if everything else is held fixed.

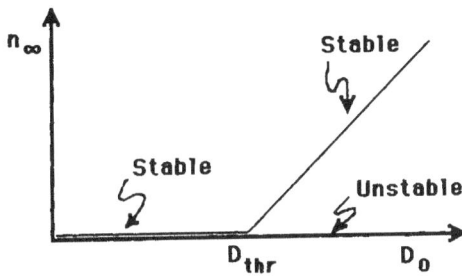

1.1   Behavior of the steady state intensity as a function of the equilibrium population difference D_0. For $D_0 < D_{thr}$ the output intensity is zero; for $D_0 > D_{thr}$ the output intensity can take two values. The lower branch above the bifurcation point is unstable.

In this preliminary discussion, we have introduced a number of ideas that will reappear at a later stage in a more complete and rigorous fashion. These notions are useful, in spite of the fact that the rate equation picture is only a first-cut at a laser theory. For example, from a fundamental point of view, one would expect that absorption (or its counterpart, amplification) should be accompanied by dispersive effects. These have not been mentioned in this section, and will be discussed at length only in the context of the more precise formulation of the following chapters. Other aspects that are missing from this picture involve the spectral character of the amplified radiation, and the possible build-up of atomic coherence among the individual radiators. These questions, also, are best posed within the framework of more detailed models. It will be best to develop our theory through a series of progressively more sophisticated steps: for this reason, we will begin with an analysis of the basic interaction mechanism between radiation and matter which is entirely classical. This will be followed by a description that borrows simultaneously from classical and quantum physics, and, for this reason, will be called the semiclassical approach. The end point of this study will be the full set of Maxwell-Bloch equations for the laser.

This machinery is all we need to explore some of the interesting aspects of laser dynamics. In fact, at present, the dynamical behavior of lasers and their instabilities have become again a focal point of attention in many laboratories. One of the main reasons for the current revival of interest in laser dynamics is the realization that lasers are not unique in their tendency to undergo transitions from time-independent to time-dependent behaviors and from pulsations of one kind to oscillations of entirely different form.

Spontaneous oscillations have long been recognized as common occurrences in quite unrelated scientific disciplines. The numerous similarities that have emerged between the behavior of lasers and, for example, chemical, hydrodynamical and plasma instabilities, or the formation of spatial and temporal patterns in living and man-made systems have brought foward suggestions and arguments in favor of the existence of underlying general principles [See, for example Nicholis and Prigogine (1877); Swinney and Gollub (1981); Haken (1982), (1983a,b); Frehland (1984), Abraham, Gollub and Swinney (1984)]. In addition, recent mathematical advances [Sparrow (1982)] have stimulated new conceptual, analytical and numerical methods for the study of nonlinear dynamical phenomena. Within the wide range of activities that have flourished under the conceptual framework of synergetics, the laser has offered a nearly ideal setting for both theoretical and experimental investigations [JOSAB (1985); Boyd, Raymer, Narducci, (1986); Chrostowski and Abraham (1986)].

Comprehensive reviews of this field have already appeared in the form of articles and monographs [Oraevskii (1964); Oraevskii and Uspenskii (1968); Oraevskii (1981); Abraham (1983); Englund, Snapp and Schieve (1984); Abraham, Lugiato and Narducci (1985); Ackerhalt, Milonni and Shih (1985) and Harrison and Biswas (1985)] The intent of the following chapters is to guide the newcomer through some useful background information, and to review the basic issues that are currently at the cutting edge of research.

2. The Interaction of Light and Matter

2.1 An overview - The early history

There is sometimes an incorrect perception surrounding the semiclassical theory of the laser: that it was developed to overcome the limitations of the rate equation approach in handling such issues as the coherence and spectral purity of the output radiation and, of course the spiking phenomenon. In fact, attempts to create a first-principle theory of the laser were already well under way from the very early 1960's [See, for example, Fain (1958); Oraevskii (1959)], and undoubtedly had a significant influence on Lamb's pioneering developments [Lamb (1964)] and the subsequent quantum mechanical formulations [Haken and Sauermann (1963a,b), (1964); Lax (1968); Scully and Lamb (1967), (1968 a,b) and Haken (1970)].

Fain was one of the first to formalize the idea that the coherent coupling between the electromagnetic radiation and the active medium was at the heart of the maser operation and who developed the basic theoretical tools starting from this premise. An important advance was made by Tang (1963) with his discovery that the rate equations are the natural limit of the coherent laser equations when the relaxation time of the atomic polarization is much smaller than all the other characteristic times of the system. Tang's method in deriving the rate equations has since evolved into a well established procedure known as the adiabatic elimination process [Haken (1986) and additional references therein; see also Lugiato, Mandel and Narducci (1984)].

The earliest studies of the coherent laser equations appear to originate with the Russian school. Uspenski (1963), (1964), for example, showed that the single-mode laser model predicts the existence of a transition from stable to unstable operation and the development of spiking in the output intensity. This type of behavior requires simultaneously a sufficiently high gain and the so-called "bad cavity condition" which, in modern terminology, denotes a situation where the decay rate of the cavity field exceeds the sum of the polarization and population decay rates.

A related important advance was recorded independently by Grasyuk and Orsevkii (1964 a,b) and by Buley and Cummings (1964) who confirmed the existence of undamped pulsations in their numerical solutions of the laser equations of motion and provided evidence for the existence of irregular spiking. The discovery of what is known, in modern language, as deterministic chaos received only limited attention by the laser community for quite some time. Subsequent investigations based on the more general set of Maxwell-Bloch equations also revealed a stability breakdown and the development of time-dependent behavior under appropriate conditions [Haken (1966), Risken, Schmidt and Weidlich (1966), Risken and Nummedal (1968 a,b), and Graham and Haken (1968)].

By a remarkable coincidence, at about the same time as the laser equations were yielding irregular pulsations, Lorenz (1963), derived a highly simplified version of the Navier-Stokes equations as a description of a class of hydrodynamical instabilities. The numerical solutions of the Lorenz model displayed the same kind of erratic pulsing behavior as observed from the single-mode laser equations. A few years later, Haken (1975) showed that the single-mode laser and the Lorenz equations were actually isomorphic to one another, a discovery that in a single stroke enriched the theory of the laser with an extensive chaotic phenomenology of its own.

The basic difference between the rate equations and the coherent laser equations is that the former describe only the energetic aspects of the interaction between radiation and matter; the coherent laser equations, instead, are rooted in the fundamental point of view that electromagnetic radiation is generated by macroscopic polarization sources. In fact, according to traditional electrodynamics, the coupling between radiation and matter resides in the mutual action of charges and fields. If the charges are bound to atomic or molecular units, their displacement from equilibrium generates an induced macroscopic polarization which acts as the source of re-radiated fields.

We now explore this point of view in some detail. Our goal is to derive appropriate equations for the description of a laser within the simplifying plane-wave approximation. This is done most easily in the context of a classical description, which is also pedagogically the best starting point for an appreciation of the conceptual foundation of the theory.

2.2 The classical picture of the interaction of radiation and matter

The passage of light through matter is accompanied by the simultaneous occurence of absorption (or amplification in an inverted medium) and dispersion. In vacuum, the wave equation is

$$\nabla^2 \vec{E}(\vec{r},t) = \frac{1}{c^2} \frac{\partial^2}{\partial t^2} \vec{E}(\vec{r},t)$$

(2.1)

and arbitrary field configurations can be built from linear superpositions of elementary solutions of the form

$$\vec{E}_{k,\omega}(\vec{r},t) = \vec{E}_{k,\omega} \, e^{i(\vec{k}\cdot\vec{r} - \omega t)}$$

(2.2)

where $|k| = \omega/c$. Note that the phase velocity of each monochromatic component is the same, in vacuum, so that no dispersion is possible (and of course, there is no absorption, either). In a medium, the wave equation is modified in an essential way. A rather general model to describe the propagation of an electric field through a medium is based on the wave equation

$$\vec{\nabla} \times (\vec{\nabla} \times \vec{E}) + \frac{1}{c^2} \frac{\partial^2 \vec{E}}{\partial t^2} = -\mu_0 \frac{\partial^2 \vec{P}}{\partial t^2} - \mu_0 \frac{\partial \vec{J}}{\partial t}$$

(2.3)

where P represents the macroscopic polarization per unit volume, induced in the medium by the passage of light or permanently present in the medium itself, and J denotes possible current densities that might be present as well.

In the Coulomb gauge we have:

$$\vec{\nabla} \times (\vec{\nabla} \times \vec{E}) = -\nabla^2 \vec{E}$$

(2.4)

(because $\nabla \cdot E = 0$ in the absence of free charges). Furthermore, if the free current density obeys Ohm's law in the medium, we have

$$\vec{J} = \sigma \vec{E} \tag{2.5}$$

where σ is the conductivity of the medium. Thus, the wave equation takes the form

$$\nabla^2\vec{E} - \frac{1}{c^2}\frac{\partial^2\vec{E}}{\partial t^2} = \mu_0\frac{\partial^2\vec{P}}{\partial t^2} + \mu_0\sigma\frac{\partial\vec{E}}{\partial t} \tag{2.6}$$

It is clear that, in this form, Eq. (2.6) is not sufficient to describe the propagation of an electromagnetic wave. The response of the medium, which is embodied in the macroscopic polarization, must be specified. The simplest statement that we can make is to assume that P is proportional to E, according to a relation of the type

$$\vec{P} = \varepsilon_0 \chi \vec{E} \tag{2.7}$$

where ε_0 is the vacuum electric permeability, and χ is the susceptibility, a scalar quantity if the medium is isotropic, and a second rank tensor if the medium is anisotropic. Of course, this statement puts strong restrictions on the behavior of the medium, as we can easily appreciate from the fact that, according to Eq. (2.7), its reaction is completely characterized by a single field-independent quantity χ, for any field E. If this model is not accurate enough for our purposes, we have no other choice but to study the behavior of the medium through suitable equations of motion which will tell us how P varies as a function of the applied electric field.

The one-dimensional scalar version of the wave equation (2.6), in a physical setting where Eq. (2.7) provides an adequate description of the medium response, supports the propagation of elementary solutions of the type:

$$E_{k,\omega}(z,t) = E_{k,\omega}\,e^{i(kz-\omega t)} \tag{2.8}$$

as we can readily verify by direct substitution. It is clear, however, that for every selected monochromatic component ω, the appropriate wave number |k| is no longer given by the simple vacuum dispersion relation. In this case we have

$$k^2 = \frac{\omega^2}{c^2}(1+\chi) + i\mu_0\sigma\omega \tag{2.9}$$

Hence, the wave number k is a complex quantity which we may write quite generally in the form

$$k = k_R + i\,k_I \tag{2.10}$$

where k_R and k_I are both functions of ω (note that χ is usually also a complex number). Thus, the elementary solutions can be written explicitly as follows

$$E_{k,\omega}(z,t) = E_{k,\omega}\,e^{i(k_R z-\omega t)}\,e^{-k_I z} \tag{2.11}$$

Two comments may be useful:

1) $k_R(\omega)$ sets the index of refraction of the medium as seen by the monochromatic component ω. Because of its dependence on the carrier frequency, the index of refraction imposes a different phase velocity to each monochromatic component. This causes a light packet to spread during the propagation.

18

2) $k_i(\omega)$ is responsible for the attenuation or the amplification of the wave, depending on its sign. The origin of k_i can be traced back to the presence of Ohmic dissipation and to the imaginary part of the susceptibility. The latter is a natural feature of the medium response that will be clarified at greater length in following lectures.

An obvious limitation of this description, in which we set "by fiat" the nature of the response of the medium through the susceptibility, is that this may be a sensitive function of the strength of the propagating field. This effect may not be noticeable at low light fluxes in most solid and gaseous media, but it becomes very apparent at large levels of illumination. Thus, in the presence of strong fields, one must allow the medium to adjust itself in response to the applied electric field of the incident wave.

A self-consistent scheme that incorporates this requirement was formulated by Lamb (1964) in connection with his theory of the gas laser. Schematically, the argument evolves along the following lines. An incident electromagnetic field creates a macroscopic polarization in the medium by acting on the individual microscopic dipoles (the field may be viewed as the ordering agent that forces the displaced electric charges to oscillate coherently with the driving force). The macroscopic polarization, which acts as the source term in Maxwell's wave equation, generates a new field in the medium which sets up a new polarization, and so on. In a graphical way, this concept can be illustrated as shown in Fig. 2.1. The arrows show the logical sequence of events and reveal the existence of a feedback mechanism. They also show pictorially the self-consistent nature of the argument.

Now we want to illustrate this idea in a classical context. As a first step, we focus on a specific model of the microscopic units of interest. The best known classical picture is the Lorentz oscillating dipole model. We consider a plane wave propagating along the z-axis and polarized along the x-axis. This wave forces the bound atomic electrons to oscillate along the x-axis. The microscopic polarization induced by the field on each atom is given by

$$p(t) = e\, x(t) \tag{2.12}$$

where e represent the electronic charge; the macroscopic polarization of the medium is

$$P(t) = N\, e\, x(t) \tag{2.13}$$

where N is the number of oscillating charges per unit volume. This is the macroscopic source of Maxwell's equations. The self-consistent loop is shown schematically in Fig. 2.2

Suppose now that the dipole oscillations are governed by the damped and driven harmonic equation

$$\frac{d^2 x}{dt^2} + \frac{2}{\tau_0}\frac{dx}{dt} + \omega_A^2 x = \frac{e}{m} E \tag{2.14}$$

where ω_A is the resonant frequency of the oscillating charge and

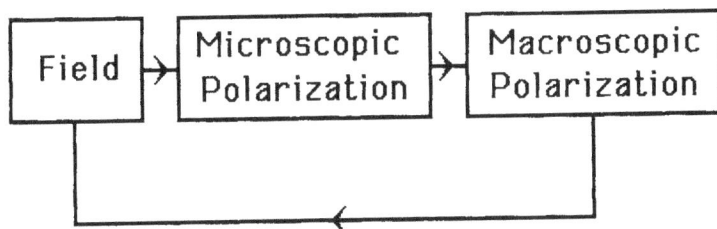

2.1   Schematic representation of the self-consistent approach for the description of the interaction of light and matter. An incident field interacts with a collection of microscopic dipoles and creates a macroscopic polarization; the bulk polarization acts as a source for the electric field which further interacts with the microscopic dipoles, etc.

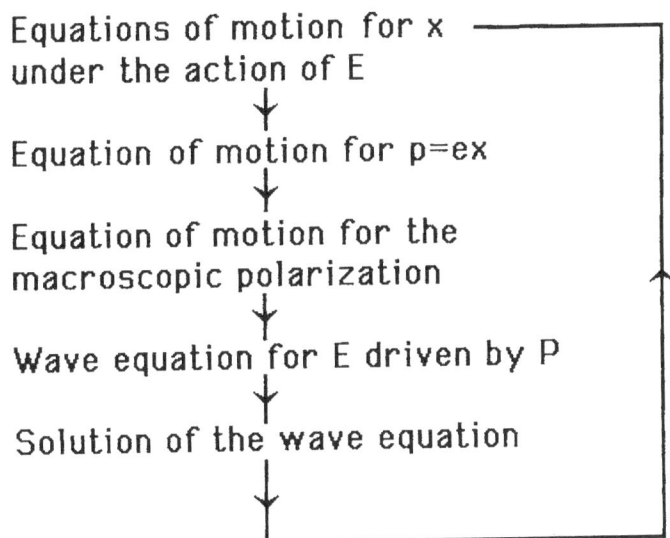

2.2   Schematic representation of the self-consistent approach to the interaction of a field and a collection of Lorentz oscillators.

$$\frac{2}{\tau_0} = \frac{2e^2 \omega_A^2}{3mc^3} \qquad (2.15)$$

is the decay rate of the oscillator due to the radiation reaction. The macroscopic polarization is governed by the equation of motion:

$$\frac{d^2 P}{dt^2} + \frac{2}{\tau_0} \frac{dP}{dt} + \omega_A^2 P = \frac{Ne^2}{m} E \qquad (2.16)$$

Note that P is also a function of the longitudinal coordinate z, in addition to time, because of the space-time dependence of the electric field. The coupled equations (2.6) and (2.16) form a closed set that can be solved upon specification of appropriate initial and boundary conditions.

As an illustration of these ideas, we consider the propagation of a wave in a semi-infinite medium. Here the physical setting is such that a wave with a prescribed known frequency ω_E enters the medium, for example, in the form of a pulse

$$E(z=0,t) = E_0(t) e^{-i \omega_E t} \qquad (2.17)$$

where $E_0(t)$ is an assigned envelope function at the left boundary. Given that, at the arbitrary position z and time t, the field is represented in the form

$$E(z,t) = E_0(z,t) e^{i(kz - \omega_E t)} \qquad (2.18)$$

we want to calculate $E_0(z,t)$ and the wave number k. The solution of this problem is complicated by the transient evolution when the charges and the field try to "come to terms" with one another. For simplicity, we consider a related but simpler problem. We assume to be positioned far away from the left boundary of the medium, and in addition we allow enough time for all the transients to die away. The incident field, of course, is no longer a pulse but it has infinite temporal duration. In this case we look for a solution of the form

$$E(z,t) = E_0 e^{i(kz - \omega_E t)} \qquad (2.19a)$$

$$P(z,t) = P_0 e^{i(kz - \omega_E t)} \qquad (2.19b)$$

where k is unknown, and E_0 and P_0 are time independent. At this point, the one-dimensional wave equation and the equation for P reduce to a pair of coupled linear equations for the unknown amplitudes E_0 and P_0

$$(-k^2 + \frac{\omega_E^2}{c^2} + \mu_0 \sigma i \omega_E) E_0 + \mu_0 \omega_E^2 P_0 = 0 \qquad (2.20a)$$

$$-\frac{Ne^2}{m} E_0 + (-\omega_E^2 - \frac{2i\omega_E}{\tau_0} + \omega_A^2) P_0 = 0 \qquad (2.20b)$$

Nontrivial solutions of Eqs. (2.20) exist if the determinant of the coefficients is identically equal to zero. This condition provides an equation for k^2 that can be solved very easily. The polarization amplitude can be calculated at once from Eq. (2.20b) with the result

$$P_0 = \frac{Ne^2}{m} \frac{E_0}{-\omega_E^2 - \dfrac{2i\omega_E}{\tau_0} + \omega_A^2} \tag{2.21}$$

The constant of proportionality between P_0 and E_0 is related directly to the usual susceptibility χ, which is now given explicitly in terms of the microscopic parameters of the model.

It is instructive to take a closer look at the results of this analysis. The wave vector is given by the rather complicated-looking expression

$$k^2 = \frac{1}{\omega_A^2 - \omega_E^2 - 2i\,\dfrac{\omega_E}{\tau_0}} \left\{ \left(\frac{\omega_E^2}{c^2} + i\,\mu_0\sigma\,\omega_E \right) \left(\omega_A^2 - \omega_E^2 - \frac{2i\,\omega_E}{\tau_0} \right) + \right.$$
$$\left. + \mu_0\,\omega_E^2\,\frac{Ne^2}{m} \right\} \tag{2.22}$$

Because we are more interested in the interaction of the field with the atomic oscillators than with its interaction with the background medium, which is the source of the Ohmic losses, we set $\sigma = 0$. In this case the wave number becomes

$$k^2 = \frac{\omega_E^2}{c^2} \left\{ 1 + \mu_0 c^2 \frac{Ne^2}{m} \frac{1}{\omega_A^2 - \omega_E^2 - 2i\dfrac{\omega_E}{\tau_0}} \right\} \tag{2.23}$$

Thus, k^2 differs from the vacuum value $(\omega_E/c)^2$ because of an additive contribution which is proportional to N, the number density of oscillators. As a further simplification, we assume that the medium is sufficiently dilute so that the second term in the curly brackets of Eq. (2.23) is much smaller than unity. With the help of the binomial expansion, we obtain

$$k \cong \frac{\omega_E}{c} \left\{ 1 + \frac{\mu_0 c^2 Ne^2}{2m} \frac{1}{\omega_A^2 - \omega_E^2 - 2i\dfrac{\omega_E}{\tau_0}} \right\} \tag{2.24}$$

The real and imaginary parts of the wave number are given by

$$k_R = \frac{\omega_E}{c^2}\left\{1 + \frac{\mu_0 c^2 N e^2}{2m} \frac{\omega_A^2 - \omega_E^2}{(\omega_A^2 - \omega_E^2)^2 + (2\frac{\omega_E}{\tau_0})^2}\right\} \qquad (2.25a)$$

$$k_I = \frac{2\omega_E}{c\tau_0} \frac{\mu_0 c^2 N e^2}{2m} \frac{1}{(\omega_A^2 - \omega_E^2)^2 + (2\frac{\omega_E}{\tau_0})^2} \qquad (2.25b)$$

As we have already mentioned, the real part of k is related to the index of refraction, while its imaginary part provides the attenuation constant. In the frequency range that is typical of optical phenomena, one normally has

$$|\omega_A - \omega_E| \ll \omega_A + \omega_E$$

so that k_R and k_I can also be expressed in the following way, to an excellent approximation,

$$k_R = \frac{\omega_E}{c}\left\{1 + \frac{\mu_0 c^2 N e^2}{2m} \frac{1}{2\omega_A} \frac{\omega_A - \omega_E}{(\omega_A - \omega_E)^2 + (\frac{1}{\tau_0})^2}\right\} \qquad (2.26a)$$

$$k_I = \frac{\mu_0 c^2 N e^2}{4mc\tau_0} \frac{1}{(\omega_A - \omega_E)^2 + (\frac{1}{\tau_0})^2} \qquad (2.26b)$$

The frequency dependence of the index of refraction and of the attenuation constant are sketched in Figs. 2.3a,b. These results should be already familiar from elementary studies in optics. Note that the width of the attenuation curve is inversely proportional to the relaxation time of the dipole oscillators, a result that will play a useful role even in the semiclassical theory developed in the next chapter.

a

b

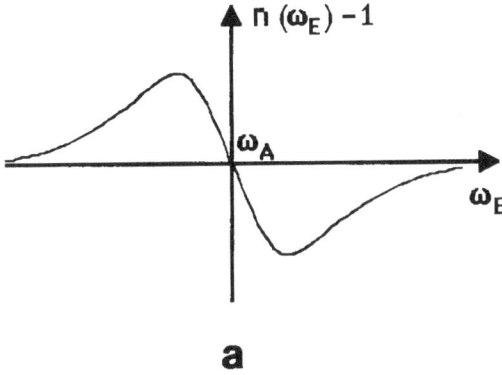

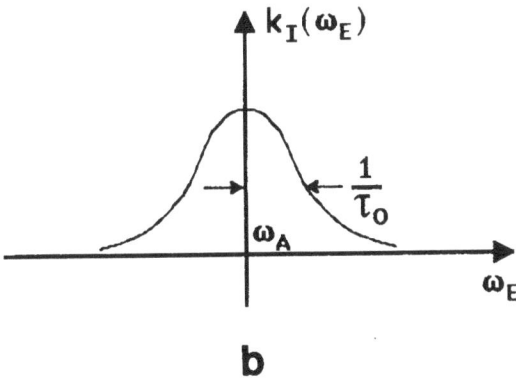

2.3   (a) Schematic representation of the behavior of the index of refraction in the neighborhood of the resonance frequency of a dipole; (b) the behavior of the attenuation constant in the neighborhood of the same resonance.

3.   The Basic Laser Equation - Homogeneously
      Broadened Systems

A complete description of the operation of a laser is an ambitious undertaking, even in the context of the so-called semi-classical approach (classical wave equation and quantized active atoms). The finite transverse cross-section of the resonator requires consideration of the full three-dimensional wave equation if one wants to describe the cavity field exactly. Transverse and longitudinal modes can be excited simultaneously, and can couple to one another through geometry-dependent coefficients. Each cavity mode requires specification of an electromagnetic field variable (usually complex), so that as more modes become exited, more dynamical variables become involved in the evolution.

The active medium, even in its simplest conception, involves more than just the two levels that are coupled by the laser transition. The laser transition itself may be broadened by collisional and Doppler effects in gases, or by local field inhomogeneities in solid systems. The large organic molecules that make up a dye laser have very complicated internal structures, with intricate pathways between the relevant levels. Semiconductor lasers involve transitions between states that are distributed in energy, and the typical band structures may be significantly nonparabolic in shape, and therefore difficult to represent in theoretical terms.

All these complications need to be addressed in the practical design of a laser system; they are, in fact, the focus of a large segment of the technical laser literature. Our goal, however, is to formulate a description of laser action that retains the essential features of the process without including complicated but (hopefully) nonessential aspects. In these lectures we focus on a simplified model and explore its most significant predictions. In line with this philosophy, we assume the laser field to behave as a classical variable governed by Maxwell's wave equation. The atoms are modelled as a collection of two-level systems with a transition frequency ω_A between the chosen levels (Fig 3-1).

For the moment, we assume the atoms to be identical to one another (homogeneous broadening) and to be governed by the Schroedinger equation. The active medium is enclosed in a volume V which can be visualized as a cylinder with a uniform cross-sectional area A and length L. The atoms are placed inside an optical resonator whose effect on the field will be discussed later in detail. It is clear that the possible presence of right and left travelling waves is a serious complication. Even in the absence of the medium, counter-propagating waves interfere with one another and create a modulated intensity pattern on a scale of the order of the wavelegth. In the presence of a medium, the fine-scale structure of the interfering waves requires a special handling, as we will discuss briefly in Appendix C (In essence, the atoms exposed to different intensity levels acquire a modulated pattern of inversion which acts as a grating, and is capable of scattering the forward propagating wave in the backward direction, and viceversa; this effect is responsible for dynamical coupling of the two waves).

In order to avoid this type of complication, we focus on a cavity configuration that allows only one direction of propagation; in practice this can be accomplished with the help of a ring resonator and a non-reciprocal device, such as a Faraday isolator, to suppress one of the two propagating waves (ring cavity model). Even under these

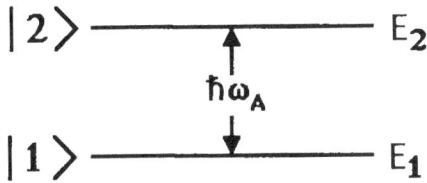

3.1   Schematic representation of the energy levels of the active atoms; the frequency spacing between the upper and lower levels is ω_A.

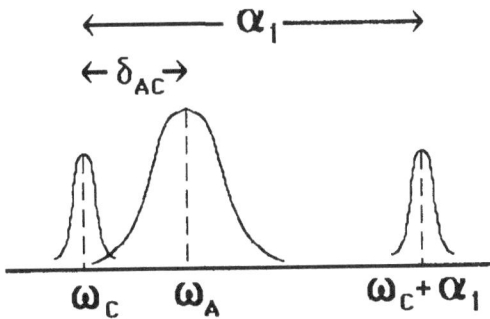

3.2   The broadened lines at frequencies ω_C and $\omega_C+\alpha_1$ represent two adjacent cavity resonances separated by the intermode spacing α_1. The line at frequency ω_A represents the atomic gain line; δ_{AC} is the detuning parameter between the center of the atomic line and the cavity resonance selected as a reference.

restrictive conditions, typical cavity modes may have a rather complicated transverse structure [Siegman (1971); Yariv (1985); Svelto (1982)]. We avoid this by insisting that the cavity be designed to support only one TEM_{00} mode whose transverse profile is Gaussian. The inclusion of a variable transverse field profile is also a source of complications of its own which are presently under active investigation. We then approximate the Gaussian beam with a plane wave, assuming that the beam waist is sufficiently large. It is important to stress that the plane-wave approximations may be responsible for some of the most obvious shortcomings of this model, particularly in connection with the unstable behaviors. This conjecture needs to be verified but, in any event, the accuracy of the plane-wave approximation should not be taken for granted.

In summary, the cavity field is assumed to propagate along one direction only, and to resemble a plane wave with a constant intensity profile along the transverse direction. In this way we have removed diffraction effects, but left the possibility open that multiple longitudinal modes can be excited. We note that the notion of cavity modes in the presence of transmission losses offers some additional difficulties that need to be discussed with care. We postpone this issue until needed for our development.

With the the final stipulation that the cavity field is linearly polarized in a direction perpendicular to the axis of the active medium (z-axis), the scalar wave equation in the MKS system of units is

$$\frac{\partial^2 E}{\partial z^2} - \frac{1}{c^2}\frac{\partial^2 E}{\partial t^2} = \mu_0 \frac{\partial^2 P}{\partial t^2} \qquad (3.1)$$

At this point we inject an important assumption concerning the cavity field and the atomic polarization; we require that these observales be described accurately by expressions of the type

$$E(z,t) = \frac{1}{2}\left\{ E_0(z,t)\, e^{\,i\,(k_c z\,-\,\omega_c t)} + c.c \right\} \qquad (3.2)$$

and

$$P(z,t) = \frac{1}{2}\, N\mu i \left\{ P_0(z,t)\, e^{\,i\,(k_c z\,-\,\omega_c t)} - c.c. \right\} \qquad (3.3)$$

where N is the number of atoms per unit volume, μ is the modulus of the atomic transition dipole moment and ω_c is one of the empty cavity eigenfrequencies which we select as a reference along the frequency axis. The specific choice of ω_c is entirely arbitrary; for convenience, however, we choose as our reference marker the mode that lies closest to the atomic transition frequency. The selected resonance will be called the resonant mode; all the others will be generally referred to as off-resonant modes (Fig. 3.2).

An essential point of the ansatz expressed by Eqs. (3.2) and (3.3) is that the complex envelope functions E_0 and P_0 should be slowly varying both in space and in time. Possible exceptions may arise in lasers that produce ultrashort pulses with durations of the order of a few femtoseconds (10^{-15} sec) at optical frequencies. This type of laser operation will not be considered in these notes, although it represents an extremely interesting problem of its own.

In the so-called slowly varying envelope approximations, i.e. under the assumption that

$$| \frac{\partial E_0}{\partial z} | \ll k_c | E_0 | , \quad | \frac{\partial E_0}{\partial t} | \ll \omega_c | E_0 | \qquad (3.4)$$

the equation for the field envelope can be put in the form of the transport equation

$$c \frac{\partial E_0}{\partial z} + \frac{\partial E_0}{\partial t} = - \frac{N \mu \omega_c}{2 \varepsilon_0} P_0 \qquad (3.5)$$

A similar result holds for the complex conjugate variable E_0^*.

The evolution of the atomic variables is governed by the Schroedinger equation

$$ih \frac{\partial}{\partial t} | \psi(t) \rangle = H | \psi(t) \rangle \qquad (3.6)$$

for the state vector $| \psi(t) \rangle$ of a single two-level atom. The Hamiltonian operator H is the sum of an unperturbed part, which sets the atomic energy spectrum, and of a coupling contribution describing the interaction of the atoms with their common radiation field. If we model the atom as a two-level system with unperturbed eigenstates $|1\rangle$ and $|2\rangle$ and unperturbed energies $E_1 < E_2$, the time-dependent state vector $| \psi(t) \rangle$ can be represented by the linear superposition

$$| \psi(t) \rangle = a_1 | 1 \rangle + a_2 | 2 \rangle \qquad (3.7)$$

where a_1 and a_2 are the time-dependent complex probability amplitudes, and where

$$H_0 | i \rangle = E_i | i \rangle , \quad i = 1,2 \qquad (3.8)$$

The interaction Hamiltonian in the usual dipole approximation has the form [for a particularly lucid description of the dipole approximation in problems involving optical frequencies, see Cohen-Tannoudji, Diu and Laloe (1977), Vol. II, p. 1306]

$$H_1 = - \wp E \qquad (3.9)$$

where $\wp$ is the projection of the dipole moment operator along the direction of polarization of the field, and E is the electric field given by Eq. (3.2). Our strategy is to construct an equation of motion for the atomic polarization using the Schroedinger equation as the starting point. In this sense, this procedure is conceptually identical to the one developed in Section 2 along classical lines where the equation for the Lorentz oscillator was given by Newton's second law.

First of all, we construct the equations of motion for the amplitudes a_1 and a_2 of the state vector with the result

$$\frac{da_1}{dt} = - i \frac{E_1}{\hbar} a_1 + i \frac{\mu}{\hbar} E a_2 \qquad (3.10a)$$

$$\frac{da_2}{dt} = - i \frac{E_2}{\hbar} a_2 + i \frac{\mu}{\hbar} E a_1 \qquad (3.10b)$$

where the dipole matrix elements $\mu \equiv \langle 1 | \wp | 2 \rangle = \langle 2 | \wp | 1 \rangle$ are assumed to be real without loss of generality. Next, we identify the macroscopic polarization with the expression

$$P = N \langle \psi(t) \mid \wp \mid \psi(t) \rangle = N\mu (a_1^*(t) a_2(t) + a_2^*(t) a_1(t)) \qquad (3.11)$$

This must be compared with the ansatz (3.3) used in making the slowly varying amplitude approximation, i.e.,

$$P = \frac{1}{2} N\mu i \left(P_0 e^{i(k_c z - \omega_c t)} - c.c \right) \qquad (3.12)$$

Thus, it follows that the slowly varying amplitude of the atomic polarization P_0 is connected to the Schroedinger amplitudes a_1 and a_2 by the relation

$$P_0 = - 2ia_1^* a_2 e^{-i(k_c z - \omega_c t)} \qquad (3.13)$$

It is clear that, in order to assign a specific forcing function to the right hand side of Maxwell's equation (3.5), we need to construct an equation of motion for P_0. This is done very easily with the help of Eqs. (3.10). The details of the construction are left as an exercise; here we only point out that

$$\frac{\partial P_0}{\partial t} = - 2i \left\{ \frac{\partial a_1^*}{\partial t} a_2 e^{-i(k_c z - \omega_c t)} + a_1^* \frac{\partial a_2}{\partial t} e^{-i(k_c z - \omega_c t)} \right.$$

$$\left. + a_1^* a_2 i \omega_c e^{-i(k_c z - \omega_c t)} \right\} \qquad (3.14)$$

and that the time rates of change of the atomic amplitudes are available from Eqs. (3.10). Note that, on the left hand side of Eq. (3.14), we have used the partial derivative symbol as a reminder that the atomic variables are also functions of the spatial coordinate z.

At this point an important difference emerges relative to the classical description of Section 2. In fact, the classical equations for the field and the atomic polarization form a closed linear system of equations. Here the equation of motion for P_0 involves a new variable

$$D_0 = a_2^* a_2 - a_1^* a_1 \qquad (3.15)$$

which can be identified immediately as the population difference per atom. Obviously, we need to construct an equation of motion for D_0, as well. In general, there is no guarantee that a procedure of this type will lead eventually to a closed system of equations because, a priori, more and more new variables could arise at every turn. The two-level model is kind, in this respect, because the equations of motion do close in terms of P_0, P_0^* and D_0. In fact, after some elementary algebra, we can show that

$$\frac{\partial P_0}{\partial t} = - i \delta_{AC} P_0 - \frac{\mu E_0}{h} D_0 \qquad (3.16a)$$

$$\frac{\partial P_0^*}{\partial t} = i \delta_{AC} P_0^* - \frac{\mu E_0}{h} D_0 \qquad (3.16b)$$

$$\frac{\partial D_0}{\partial t} = \frac{\mu}{2\hbar}\left(P_0 E_0^* + P_0^* E_0\right) \tag{3.16c}$$

where $\delta_{AC} = \omega_A - \omega_C$ represents the detuning parameter between the center of the atomic line and the reference cavity mode.

Thus, the basic equations of motion for two-level systems interacting with radiation are

$$c\frac{\partial E_0}{\partial z} + \frac{\partial E_0}{\partial t} = -\frac{N\mu\omega_c}{2\varepsilon_0} P_0 \tag{3.17a}$$

$$\frac{\partial P_0}{\partial t} = -i\,\delta_{AC}P_0 - \frac{\mu E_0}{\hbar} D_0 \tag{3.17b}$$

$$\frac{\partial D_0}{\partial t} = \frac{\mu}{2\hbar}\left(P_0 E_0^* + P_0^* E_0\right) \tag{3.17c}$$

Two additional equations hold for E_0^* and P_0^*. Note that, in this form, the atomic equations are consistent with the conservation law

$$\frac{d}{dt}\left(P_0^* P_0 + D_0^2\right) = 0$$

or

$$P_0^* P_0 + D_0^2 = \text{const} \tag{3.18}$$

which is a consequence of the probability conservation built into the Schroedinger equation.

Three additional items are needed before Eqs. (3.17) can be used as an operating laser model: (i) we must simulate the irreversible processes that cause the decay of the polarization and population difference; (ii) we need to simulate the external pump mechanism and, finally, (iii) we must specify the boundary conditions for the ring resonator.

The incoherent decay and pumping processes can be included easily by adding the following irreversible terms to the atomic equations

$$\frac{\partial P_0}{\partial t} = \ldots\ldots - \frac{1}{T_2} P_0 \tag{3.19a}$$

$$\frac{\partial D_0}{\partial t} = \ldots\ldots - \frac{1}{T_1}(D_0 - 1) \tag{3.19b}$$

In fact, when the cavity field is turned off, the population difference relaxes to the value

$$D_0 = 1 \tag{3.20}$$

which implies $|a_2|^2 = 1$ and $|a_1|^2 = 0$.

The resonant structure that is consistent with the unidirectional model has the form of a ring, as shown schematically in Fig 3-3. Mirrors 1 and 2 have the same amplitude reflectivity coefficient $\sqrt{R}$; mirrors 3 and 4 are assumed to be perfect reflectors (R=1). The active medium occupies a segment of length L; while the full length of the ring resonator is Λ. The boundary conditions for the Maxwell field are

$$E(0,t) = R E (L, t - \frac{\Lambda - L}{c})$$ (3.21)

or, with the help of Eq. (3.2),

$$E_0(0,t) e^{-i\omega_c t} = R E_0(L, t - \frac{\Lambda - L}{c}) \exp\left\{i [k_c L - \omega_c (t - \frac{\Lambda - L}{c})]\right\}$$ (3.22)

(We talk about "boundary conditions", using the plural form, to indicate that Eq. (3.21) imposes two constraints, one for the modulus and another for the phase of the field). If we take into account that $k_c = \omega_c/c$, and that $\omega_c(\Lambda/c) = k_c\Lambda = 2\pi m$, where m is an arbitrary integer number, we arrive at the required result

$$E_0(0,t) = R E_0(L, t - \frac{\Lambda - L}{c})$$ (3.23)

These boundary conditions are not of the usual periodicity type because of the presence of the time delay and of the scale factor R on the right hand side of Eq. (3.23). Before closing this section we introduce a new set of dimensionless variables

$$\bar{F}(z,t) = \frac{\mu E_0(z,t)}{\hbar\sqrt{\gamma_\perp \gamma_\parallel}},$$

$$\bar{P}(z,t) = \sqrt{\frac{\gamma_\perp}{\gamma_\parallel}} P_0$$ (3.24)

$$\bar{D}(z,t) = D_0(z,t)$$

where $\gamma_\perp = 1/T_2$ and $\gamma_\parallel = 1/T_1$ denote the polarization and population relaxation rates. At this point, the full set of Maxwell-Bloch equations and boundary conditions take the form

$$\frac{\partial \bar{F}}{\partial z} + \frac{1}{c}\frac{\partial \bar{F}}{\partial t} = -\alpha\bar{P}$$ (3.25a)

$$\frac{\partial \bar{P}}{\partial t} = -(\gamma_\perp + i\delta_{A0})\bar{P} - \gamma_\perp \bar{F}\bar{D}$$ (3.25b)

$$\frac{\partial \bar{D}}{\partial t} = \frac{1}{2}\gamma_\parallel(\bar{P}^*\bar{F} + \bar{P}\bar{F}^*) - \gamma_\parallel(\bar{D} - 1)$$ (3.25c)

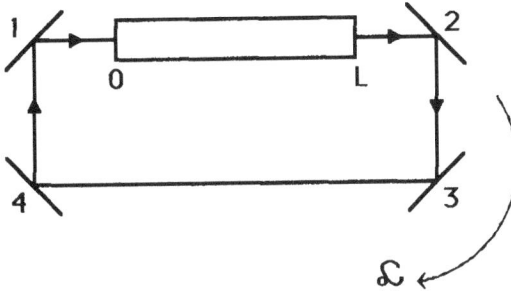

3.3 Schematic representation of a ring cavity. The mirrors labelled 1 and 2 have a power reflectivity R; mirrors 3 and 4 are ideal reflectors. The active medium is confined within the region 0<z<L.

$$\bar{F}(0,t) = R\,\bar{F}(L, t - \frac{\Lambda - L}{c})$$

(3.26)

where

$$\alpha = \frac{N\mu^2\omega_c}{2\hbar\epsilon_0\gamma_\perp c}$$

(3.27)

denotes the gain constant per unit length of the active medium.

4. Steady State Behavior of a Homogeneously Broadened Ring Laser

We now address the question of the long-time behavior of our model, when the active medium is detuned by an arbitrary amount $\delta_{AC} = \omega_A - \omega_C$ from the resonant cavity mode [The procedure adopted in this and following sections was originally developed by L.A. Lugiato and collaborators for the description of Optical Bistability; for details see L.A. Lugiato (1984). It has also been used in connection with the theory of the laser; see, Lugiato, Narducci, Eschenazi, Bandy and Abraham (1985); Narducci, Tredicce, Lugiato, Abraham and Bandy (1986)]. Under these conditions, the output field is expected to oscillate with a carrier frequency ω_L which is neither equal to ω_C or ω_A, but to some intermediate value determined by the cavity and atomic parameters. For this reason we look for stationary solutions of the type

$$\bar{F}(z,t) = \bar{F}_{st}(z)\, e^{-i\delta\omega t} \tag{4.1a}$$

$$\bar{P}(z,t) = \bar{P}_{st}(z)\, e^{-i\delta\omega t} \tag{4.1b}$$

$$\bar{D}(z,t) = \bar{D}_{st}(z) \tag{4.1c}$$

where $\delta\omega$ is the frequency offset of the operating laser line from the resonant mode (i.e. $\delta\omega = \omega_L - \omega_C$). Of course, $\delta\omega$ is unknown and must be calculated. The space-dependent steady state variables are solutions of the equations

$$\frac{d}{dz}\bar{F}_{st}(z) - i\frac{\delta\omega}{c}\bar{F}_{st}(z) = -\alpha\bar{P}_{st}(z) \tag{4.2a}$$

$$0 = \bar{F}_{st}(z)\,\bar{D}_{st}(z) + (1+i\bar{\Delta})\,\bar{P}_{st}(z) \tag{4.2b}$$

$$0 = -\frac{1}{2}\left(\bar{F}_{st}^{*}\bar{P}_{st} + \bar{F}_{st}\bar{P}_{st}^{*}\right) + \bar{D}_{st} - 1 \tag{4.2c}$$

where

$$\bar{\Delta} \equiv \bar{\delta}_{AC} - \bar{\delta}\omega = \frac{\delta_{AC} - \delta\omega}{\gamma_{\perp}} = \frac{\omega_A - \omega_L}{\gamma_{\perp}} \tag{4.3}$$

The atomic longitudinal profiles can be calculated at once as functions of the stationary field envelope from Eqs. (4.2b) and (4.2c) with the result

$$\bar{P}_{st}(z) = -\bar{F}_{st}(z)\,\frac{1 - i\bar{\Delta}}{1 + \bar{\Delta}^2 + |\bar{F}_{st}(z)|^2} \tag{4.4a}$$

$$\bar{D}_{st}(z) = \frac{1 + \bar{\Delta}^2}{1 + \bar{\Delta}^2 + |\bar{F}_{st}(z)|^2} \tag{4.4b}$$

Several facts should be immediately obvious. The steady state polarization and the field envelope are generally out of phase from one another by an amount that depends on the detuning δ_{AC} and the position of the operating laser line. In resonance, however, P_{st} and F_{st} have the same phase. The steady state population difference saturates at high intensity levels in the sense that $D_{st}(z) \to 0$ as $|F_{st}(z)| \to \infty$.

Our next step is to determine the value of the output field and the form of its longitudinal profile in steady state. For this purpose, we substitute Eq. (4.4a) into Eq. (4.2a) and, at the same time, we decompose the laser field into its modulus and phase

$$\bar{F}_{st}(z) = \rho(z)\, e^{i\,\theta(z)} \tag{4.5}$$

In this way, Eq. (4.2a) takes the form

$$\frac{d\rho}{dz} = \alpha \frac{\rho}{1 + \bar{\Delta}^2 + \rho^2} \tag{4.6a}$$

$$\frac{d\theta}{dz} = \frac{\delta\omega}{c} - \alpha \frac{\bar{\Delta}}{1 + \bar{\Delta}^2 + \rho^2} \tag{4.6b}$$

The two coupled equations (4.6) can be combined to yield the first integral of the problem

$$\ln\left(\frac{\rho(z)}{\rho(0)}\right) = -\frac{1}{\bar{\Delta}}\left(\theta(z) - \theta(0) - \frac{\delta\omega}{c}z\right) \tag{4.7}$$

while Eq. (4.6a) can be integrated at once to give

$$(1 + \bar{\Delta}^2)\ln\left(\frac{\rho(z)}{\rho(0)}\right) + \frac{1}{2}\left(\rho^2(z) - \rho^2(0)\right) = \alpha z \tag{4.8}$$

The boundary conditions (3.26), expressed in terms of the field modulus and phase, provide the two constraining relations

$$\rho(0) = R\,\rho(L) \tag{4.9a}$$

$$\theta(L) - \theta(0) = -\delta\omega \frac{\Lambda\text{-}L}{c} + 2\pi j\,; \quad j = 0, \pm 1, \pm 2, \dots\dots \tag{4.9b}$$

which show that, in principle, the boundary conditions can be satisfied by more that one solution. This is not surprising in view of the resonant nature of the cavity; the result is important, however, because it suggests the possibility of coexisting steady states and mode-mode interactions, as we shall confirm shortly.

The output laser intensity can be calculated at once from Eq. (4.8) after selecting $z = L$, and using the boundary condition (4.9a). The result is

$$\rho^2(L) = \frac{2}{1 - R^2}\left(\alpha L - (1 + \bar{\Delta}^2)\,|\ln R\,|\right) \tag{4.10}$$

The operating frequency follows after setting $z = L$ in Eq. (4.7) and using the boundary condition (4.9b). The required result is

$$\ln\left(\frac{1}{R}\right) = -\frac{1}{\Delta}\left(-\delta\omega\frac{\Lambda-L}{c} + 2\pi j - \frac{\delta\omega}{c}L\right)$$

or

$$\delta\omega\left(1 + \frac{c\,|\ln R|}{\Lambda\gamma_\perp}\right) = \frac{c\,|\ln R|}{\Lambda\gamma_\perp}\delta_{AC} + \frac{2\pi c}{\Lambda}j \qquad (4.11)$$

The quantity $c|\ln R|/\Lambda\gamma_\perp$ represents the decay rate of the cavity field; $2\pi c/\Lambda$ is the spacing between adjacent cavity resonances. After introducing the symbols

$$\kappa = \frac{c\,|\ln R|}{\Lambda} \qquad (4.12a)$$

$$\alpha_1 = \frac{2\pi c}{\Lambda} \qquad (4.12b)$$

we obtain the required result

$$\delta\omega_j = \omega_L - \omega_C = \frac{\kappa\delta_{AC} + \alpha_1\gamma_\perp j}{\gamma_\perp + \kappa} \quad ; \quad j = 0, \pm1, \pm2, \dots \qquad (4.13)$$

where the sub-index j reminds us of the possible existence of multiple solutions. This is the well known mode-pulling formula written in a slightly unconventional way. In most textbooks, Eq. (4.13) is usually given as

$$\omega_L^{(j)} = \frac{(\omega_C + \alpha_1 j)\,\gamma_\perp + \omega_A\kappa}{\gamma_\perp + \kappa} \qquad (4.14)$$

This result shows that the laser operating frequency is a weighted average of the atomic resonant frequency and the frequency of one of the cavity modes. Obviously, this steady state analysis cannot decide on which the cavity modes will be operating under given conditions. It is possible, however, to identify the range of allowed values of j by using Eqs. (4.10), (4.3) and (4.13) and by requiring that $\rho^2(L)$ remain greater than zero for all allowed values of the steady state index j.

For a fixed value of detuning parameter δ_{AC} and for a given steady state index, Eq. (4.10) sets the threshold gain that is needed to bring the system into action. This is given by

$$(\alpha L)_{thr, j} = (1 + \overline{\Delta}_j^2)\,|\ln R| \qquad (4.15)$$

It is also interesting to inquire into the longitudinal profile of the field under given steady state conditions. This can be done using Eq. (4.8) with $\rho(0) = R\,\rho(L)$ and $\rho(L)$ given by Eq. (4.10). The solution of this transcendental equation can be obtained easily by numerical means. Selected sample solutions are shown in Fig. (4.1,2). In both cases shown here, the field longitudinal profile undergoes a fairly large variation (about 10% in Fig. 1, and 65% in Fig. 2). The details of the spatial variation of the laser intensity inside the medium are governed by the degree of saturation and by the mirror reflectivity.

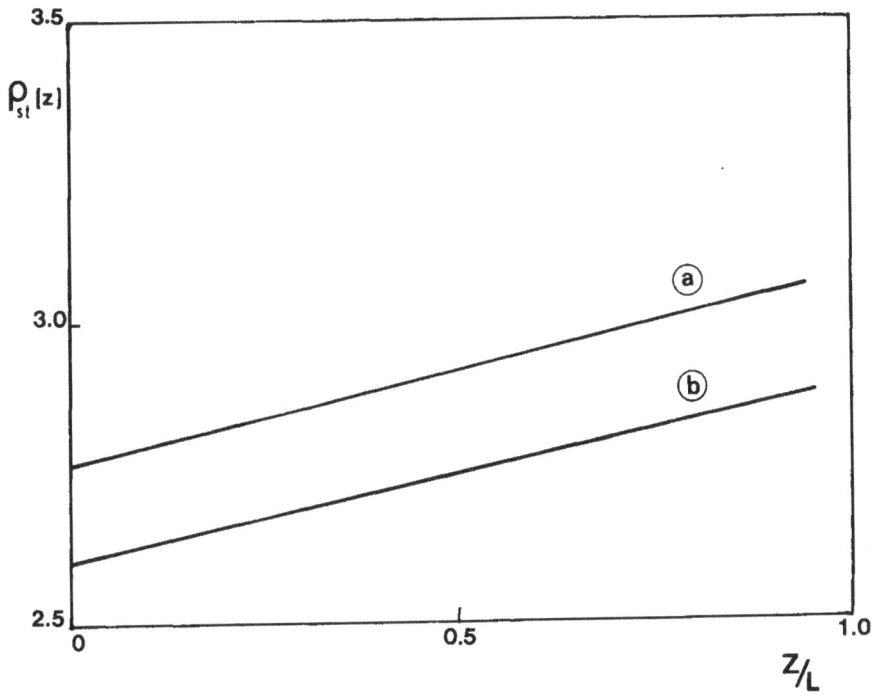

4.1   Longitudinal variation of the steady state modulus of the field for $\alpha L = 1.0$, $R = 0.9$, $\bar{\alpha}_1 = 2.0$, and (a) $\bar{\delta}_{AC} = 0$, (b) $\bar{\delta}_{AC} = 1.0$. The overall variation of the field modulus is 10%.

36

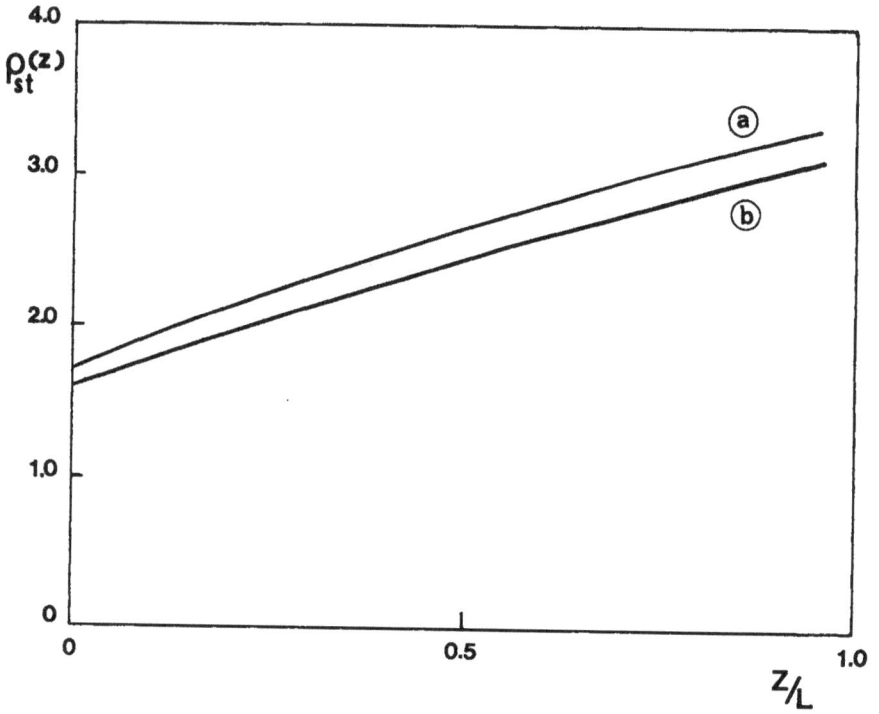

4.2   Longitudinal variation of the steady state modulus of the field for $\alpha l = 5.0$, R = 0.5, $\bar{\alpha}_1 = 2.0$, and (a) $\bar{\delta}_{AC} = 0$, (b) $\delta_{AC} = 1.0$. The overall variation of the field modulus is 65%.

A convenient mathematical limit [Bonifacio, Lugiato (1978)] that has received a considerable amount of attention corresponds to a situation where

$$\alpha L \to 0, \quad R \to 1, \quad (\text{or } T \to 0) \tag{4.16a}$$

in such a way that

$$\frac{\alpha L}{T} \equiv 2C = \text{arbitrary} \tag{4.16b}$$

where $T = 1-R$ is the mirror power transmittivity. This is not a trivial limit because, even if the gain per pass is made progressively smaller, the transmittivity of the mirrors is simultaneously reduced. Thus, on decreasing αL the field grows less and less on every round trip, but it remains trapped within the cavity for gradually longer times, as T is made smaller. In this limit, which we call the uniform field limit, the field profile becomes uniform along the longitudinal direction and acquires the value

$$\rho_j^2 = 2C - (1 + \bar{\Delta}_j^2) \tag{4.17}$$

The threshold gain, instead, is given by

$$(2C)_{thr, j} = 1 + \bar{\Delta}_j^2 \tag{4.18}$$

5.   Linear Stability Analysis: The Resonant Case

The general stability analysis of the Maxwell-Bloch equations (3.25) is a rather difficult problem that has been solved exactly only a short time ago [Lugiato, Narducci, Squicciarini(1986)].   The main source of complications originates from the spatial dependence of the field and of the atomic variables.   In an attempt to get around this problem, most linear stability analyses have been carried out within the uniform field limit.   While this may not appear to be a very realistic approach, there are good reasons, in fact, why useful information can be extracted even from this limiting case:

(i)   we can reformulate the Maxwell-Bloch problem in terms of a new set of atomic and field variables that are not very sensitive to limited departures from the ideal limit. For this reason it is not necessary to operate with unrealistically low values of the gain or the mirror transmittivity;

(ii)   the mean field limit is a good indicator of instabilities and functions well as a rough diagnostic tool.   This is fortunate because the numerical solutions of the time-dependent   Maxwell-Bloch equations require considerable efforts and some guidance can produce significant saving of time.

The resonant case, instead, is not very complicated and can be studied exactly with limited effort.   For this reason, in this chapter we limit ourselves to the exact analysis of the resonant laser problem, without any restrictions on the gain of the active medium or the reflectivity of the mirrors [Lugiato, Narducci, Eschenazi, Bandy and Abraham (1985)]. Our starting point is the full set of Maxwell-Bloch equations (3.25) with $\delta_{AC}=0$. Because the phase of the stationary field is undetermined, it is possible to select $\bar{F}_{st}(z)$ as a real quantity. In principle, a random fluctuation of the cavity field could force the growth of an imaginary part through a process called phase instability (See Section 7 of these notes).   In this section we simply assume that no phase instabilities can develop, so that both the field and polarization variables remain real during the linearized evolution. In later lectures we will give convincing arguments in support of this assumption, in the case of a resonant homogeneously broadened laser, by showing that phase instabilities can emerge only when the detuning parameter is sufficiently large. We substitute

$$\bar{F}(z,t) = \bar{F}_{st}(z) + \delta F(z,t) \tag{5.1a}$$

$$\bar{P}(z,t) = \bar{P}_{st}(z) + \delta P(z,t) \tag{5.1b}$$

$$\bar{D}(z,t) = \bar{D}_{st}(z) + \delta D(z,t) \tag{5.1c}$$

into Eqs. (0.25) and, upon neglecting fluctuation terms of order higher than one, we obtain the linearized equations

$$\frac{\partial}{\partial z} \delta F + \frac{1}{c}\frac{\partial}{\partial t} \delta F = -\alpha\, \delta P \tag{5.2a}$$

$$\frac{\partial}{\partial t} \delta P = -\gamma_\perp \left(\delta F\, \bar{D}_{st} + \bar{F}_{st}\delta D + \delta P\right) \tag{5.2b}$$

$$\frac{\partial}{\partial t}\delta D = -\gamma_\parallel\left(-\delta F\,\bar{P}_{st} - \bar{F}_{st}\delta P + \delta D\right)$$ (5.2c)

We seek solutions of the type:

$$\begin{pmatrix}\delta F(z,t)\\ \delta P(z,t)\\ \delta D(z,t)\end{pmatrix} = e^{\lambda t}\begin{pmatrix}\delta f(z)\\ \delta p(z)\\ \delta d(z)\end{pmatrix}$$ (5.3)

and derive the following expression for the polarization fluctuation $\delta p(z)$

$$\delta p(z) = -\frac{\gamma_\perp}{1+\bar{F}_{st}^2}\,\frac{\lambda+\gamma_\parallel-\gamma_\parallel\,\bar{F}_{st}^2(z)}{(\lambda+\gamma_\perp)(\lambda+\gamma_\parallel)+\gamma_\perp\gamma_\parallel\,\bar{F}_{st}^2(z)}\,\delta f(z)$$ (5.4)

After combining Eqs. (5.2a), (5.3) and (5.4), the field fluctuation equation takes the simple form

$$\frac{d}{dz}\delta f(z) = H(z)\,\delta f(z)$$ (5.5)

where

$$H(z) = -\frac{\lambda}{c}+\frac{\alpha\gamma_\perp}{1+\bar{F}_{st}^2(z)}\,\frac{\lambda+\gamma_\parallel-\gamma_\parallel\,\bar{F}_{st}^2(z)}{(\lambda+\gamma_\perp)(\lambda+\gamma_\parallel)+\gamma_\perp\gamma_\parallel\bar{F}_{st}^2(z)}$$ (5.6)

The formal solution of Eq. (5.5) is

$$\delta f(z) = \delta f(0)\exp\left(\int_0^z dz'\,H(z')\right) \equiv \delta f(0)\,e^{\psi(z)}$$ (5.7)

The problem is that $\bar{F}_{st}(z)$ is not known in closed analytic form, but only as the solution of the transcendental equation (4.8) (in resonance, $\bar{F}_{st}(z) = \rho(z)$). We can get around this difficulty with a change of independent variable from z to $\bar{F}_{st}(z)$, if we take advantage of the fact that

$$dz = \frac{1}{\dfrac{d\bar{F}_{st}}{dz}}\,d\bar{F}_{st}$$ (5.8)

and that $d\bar{F}_{st}/dz$ (or $d\rho/dz$) is known explicitly from Eq. (4.6a) [Carmichael (1983)]. Thus, the integral leading to $\psi(z)$ can be put into the form

$$\psi(z) = -\frac{\lambda z}{c}+\gamma_\perp\int_{\bar{F}_{st}(0)}^{\bar{F}_{st}(z)}\frac{d\bar{F}_{st}}{\bar{F}_{st}}\,\frac{\lambda+\gamma_\parallel-\gamma_\parallel\,\bar{F}_{st}^2}{(\lambda+\gamma_\perp)(\lambda+\gamma_\parallel)+\gamma_\parallel\gamma_\perp\,\bar{F}_{st}^2}$$ (5.9)

which can be solved by elementary techniques. In summary, the field fluctuation variable $\delta F(z,t)$ is given by the simple formula

$$\delta F(z,t) = e^{\lambda t}\, \delta f(0)\, e^{\psi(z)} \tag{5.10}$$

while the boundary conditions are

$$\delta F(0,t) = R\, \delta F(L, t - \frac{\Lambda - L}{c}) \tag{5.11}$$

If we now substitute Eq. (5.10) into (5.11), we easily obtain the following transcendental equation for the eigenvalues of the linearized problem

$$\lambda_n = - i\alpha_n - \frac{c\,|\ln R|}{\Lambda}\,\frac{\lambda_n}{\lambda_n + \gamma_\perp} - \frac{c}{2\Lambda}\,\frac{\lambda_n + 2\gamma_\perp}{\lambda_n + \gamma_\perp}\, x$$

$$\ln\left(\frac{(\lambda_n + \gamma_\perp)\,(\lambda_n + \gamma_\parallel) + \gamma_\perp \gamma_\parallel\, \bar{F}_{st}^2(L)}{(\lambda_n + \gamma_\perp)\,(\lambda_n + \gamma_\parallel) + R\,\gamma_\perp \gamma_\parallel\, \bar{F}_{st}^2(L)}\right) \tag{5.12}$$

The origin of the term $- i\alpha_n$ in Eq. (5.12) can be traced back to the equality $\exp(0) = \exp(2\pi n i)$ for $n = 0, \pm 1, \pm 2 \ldots$. Note that setting $\exp(0) = 1$ would be a mistake because it would eliminate practically the entire spectrum of eigenvalues.

At this point we have reduced the linearized problem (5.2) to the solution of an infinite number of characteristic equations, one for each value of α_n. The existence of an infinite number of eigenvalues is not surprising in view of the space-time dependent nature of the field and atomic variables and of the boundary conditions of the laser resonator. One is reminded of the ordinary vibration problems, linear string, two-dimensional membrane, etc., except that here we are dealing simultaneously with three fluctuation variables (see Eq. (5.3)), and thus on physical grounds, one expects three characteristic roots $\lambda_n^{(1)}, \lambda_n^{(2)}, \lambda_n^{(3)}$ for each value of n. Because α_n represents the frequency separation between the n-th empty cavity resonance and the selected reference mode, it is easy to interpret the set of roots $\lambda_n^{(i)}$, $i = 1, 2, 3$, as descriptive of the growth or decay of an initial fluctuation that develops in correspondence to the n-th mode of the cavity.

This interpretation forms the basis for a classification of the possible unstable behaviors of the system. If Re λ_0 is positive and Re λ_n ($n \neq 0$) are all negative, an initial fluctuation of the resonant mode will grow exponentially and evolve with the same carrier frequency as the stationary state. Thus, the linearized dynamics of the laser can be described only in terms of the behavior of the resonant mode fluctuation (all the other fluctuations are damped because Re $\lambda_n < 0$, $n \neq 0$) and the instability will be called of the single-mode type.

If, on the other hand, Re $\lambda_0 < 0$ and, for some value of n, Re $\lambda_n > 0$, the n-th cavity mode will support the growth of a fluctuation whose carrier frequency is different from that of the stationary state. Actually, because $\lambda(\alpha_n) = \lambda^*(-\alpha_n) = \lambda^*(\alpha_{-n})$, the instability condition Re $\lambda_n > 0$ implies also Re $\lambda_{-n} > 0$. Thus, an off-resonance instability forces the exponential growth of a pair of sidebands and produces an intensity modulation of the laser output; this is the result of the superposition of the stationary state and the two sidebands that oscillate with different carrier frequencies. This type of instability is called multi-mode instability [Risken and Nummedal (1968); Graham and Haken (1986); for a modern discussion of single and multimode instabilities see Lugiato, Asquini and Narducci (1986)].

In this discussion, we have suggested the existence of a one-to-one correspondence between the index n, that appears in Eq. (5.12), and the longitudinal cavity modes. Our informal suggestion is founded on physical grounds. The main conceptual difficulty with this interpretation is that the notion of "mode" is not well defined when the resonator mirrors have a finite reflectivity, and the elementary cavity excitations have a finite lifetime. In fact, in solving the linearized problem, we have not even introduced resonator eigenfunctions, as one normally would in a standard boundary value problem. In Section 7, we show that it is possible to introduce a rigorous modal description that justifies our present interpretation of the results. For this reason, we continue to refer to $\lambda_n^{(l)}$ as the set of linearized eigenvalues of the n-th cavity resonance.

A complete analysis of Eq. (5.12), particulary with regard to the role played by the basic laser parameters, gain, intermode spacing, reflectivity and the atomic decay rates, has not been carried out. Two facts, however, can be established with confidence: Eq. (5.12) predicts both single and multimode unstable behavior, as we now show with a few examples.

We begin our analysis by scaling all the relevant rates of the problem to the linewidth $\gamma_\perp$ of the active medium. In this way, Eq. (5.12) takes the form

$$
\overline{\lambda}_n = -i\,\overline{\alpha}_n - \frac{c\,|\ln R|}{\Lambda\gamma_\perp}\,\frac{\overline{\lambda}_n}{\overline{\lambda}_n + 1} - \frac{c}{2\Lambda\gamma_\perp}\,\frac{\overline{\lambda}_n + 2}{\overline{\lambda}_n + 1}\,x
$$

$$
\ln\!\left(\frac{(\overline{\lambda}_n + 1)\,(\overline{\lambda}_n + \overline{\gamma}) + \overline{\gamma}\,\overline{F}_{st}^2}{(\overline{\lambda}_n + 1)\,(\overline{\lambda}_n + \overline{\gamma}) + R^2\,\overline{\gamma}\,\overline{F}_{st}^2}\right) \tag{5.13}
$$

where

$$
\overline{\lambda}_n = \frac{\lambda_n}{\gamma_\perp}, \quad \overline{\alpha}_n = \frac{\alpha_n}{\gamma_\perp}, \quad \text{and} \quad \overline{\gamma} = \frac{\gamma_\parallel}{\gamma_\perp}
$$

Furthermore, we remind ourselves of the following equalities (see Eqs. (4.12))

$$
\frac{c}{\Lambda\gamma_\perp} = \frac{\overline{\alpha}_1}{2\pi} = \frac{\overline{\kappa}}{|\ln R|} \tag{5.14}
$$

and of the form of the steady state output intensity

$$
\overline{F}_{st}^2 = \frac{2}{1 - R^2}\,(\alpha L - |\ln R|) \tag{5.15}
$$

The characteristic equation

$$\bar{\lambda}_n + i\,\bar{\alpha}_n + \bar{\kappa}\,\frac{\bar{\lambda}_n}{\bar{\lambda}_n + 1} + \frac{1}{2}\,\frac{\bar{\kappa}}{|\ln R|}\,\frac{\bar{\lambda}_n + 2}{\bar{\lambda}_n + 1}\,x$$

$$\ln\!\left(\frac{(\bar{\lambda}_n + 1)\,(\bar{\lambda}_n + \bar{\gamma}) + \bar{\gamma}\,\bar{F}_{st}^{\,2}}{(\bar{\lambda}_n + 1)\,(\bar{\lambda}_n + \bar{\gamma}) + R^2\,\bar{\gamma}\,\bar{F}_{st}^{\,2}}\right) = 0 \qquad (5.16)$$

depends on the scaled cavity linewidth $\bar{\kappa}$, the intermode spacing $\bar{\alpha}_n$, the scaled rate of decay of the population difference $\bar{\gamma}$, and the gain of the active medium through the output field intensity $\bar{F}_{st}^{\,2}$.

A numerical study of this problem shows that single-mode instabilities ($\bar{\alpha}_n = 0$) tend to be favored in the presence of high gain and large cavity losses ($\bar{\kappa} > 1$). These conditions are difficult to realize in a practical system, although, successful attempts to operate a laser in this situation have been made by Weiss and collaborators [see, for example, Klische, Weiss, Al-Soufi and Huttmann (1986); Weiss (1986)].

In the absence of general guidelines for the search of optimum instability conditions, the numerical search for unstable eigenvalues is always more or less a matter of trials and errors. In general, however, it appears from Eq. (5.16) that single-mode instabilities require a scaled cavity linewidth which is sufficiently larger than unity (i.e., $\bar{\kappa} > 3$) and gain to loss ratios $\alpha L/|\ln R|$ of the order of 10-20. Examples of unstable situations are shown in Figure 5.1 for several values of the relevant parameters.

Multi-mode instabilities are not bound by the high-loss requirement, but they still require large values of the gain to reach their threshold. Figure 5.2 gives an example of some typical real parts of the linearized eigenvalues for parameter values that lead to a multimode instability. As shown in this figure, the beat frequency due to the superposition of the stationary solution and of the unstable sidebands is sensitive to the value of $\bar{\gamma}$.

An important feature of this problem is the monotonic shift of the positive real parts of the eigenvalues towards higher and higher values of $\bar{\alpha}_n$ for increasing values of the gain. This phenomenon is illustrated in Fig. 5.3, and could be responsible, at least in part, for some of the behaviors observed in the recent experiments by Stroud and collaborators [see, for example, Hillman, Krasinski, Boyd and Stroud (1984); Stroud, Koch, Chakmakjian (1986); Stroud, Koch, Chakmakjian, Hillman (1986)].

The single-mode instability discussed in this section was discovered during some of the earliest investigations of single-mode laser models (see our introductory discussion in Section 2), and was recognized by Haken (1975) as being mathematically isomorphic to the so-called Lorenz hydrodynamic instability. The multimode instability, instead, was discovered by Risken and Nummedal (1968a,b)) and by Graham and Haken (1968), as already mentioned. Our present discussion generalizes these early studies which assumed implicity the uniform field limit ($\alpha L \to 0$, $R \to 1$ with $\alpha L/(1 - R)$ = arbitrary). Our analysis, instead, is valid for arbitrary values of the gain and reflectivity parameters.

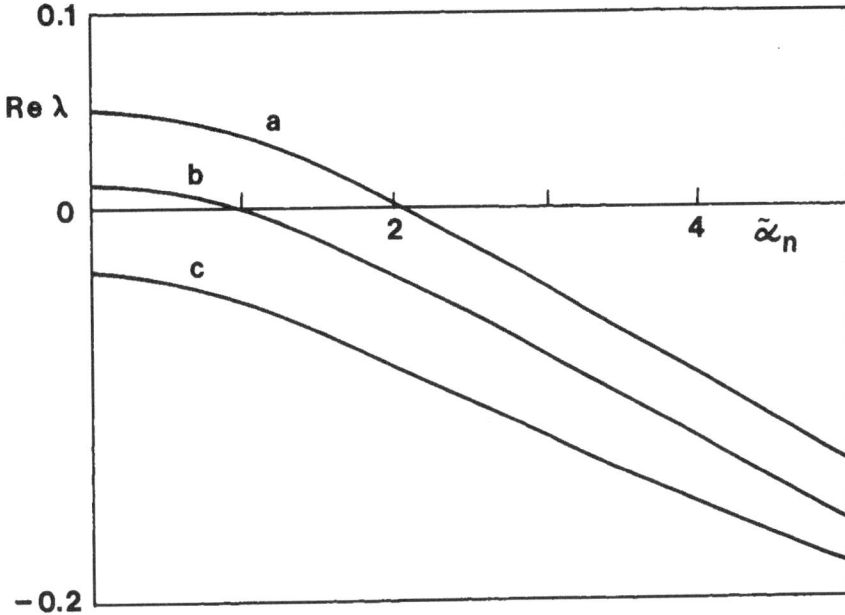

5.1   The largest real parts of the linearized eigenvalues are plotted as functions of $\tilde{\alpha}_n$ viewed as a continuous variable (remember that the only physically meaningful values of $\tilde{\alpha}_n$ are all the positive and negative multiples of the intermode spacing $\bar{\alpha}_1$). For all the curves displayed in the figure we have selected $R = 0.8$, $\bar{\alpha}_1 \approx 100$, $\bar{\gamma} = 0.1$; these parameters correspond to $\bar{\kappa} = 3.55$. (a) $\alpha L = 5.0$, (b) $\alpha L = 3.0$, (c) $\alpha L = 1.0$. The lasers corresponding to the parameters of curves (a) and (b) are subject to a single-mode instability.

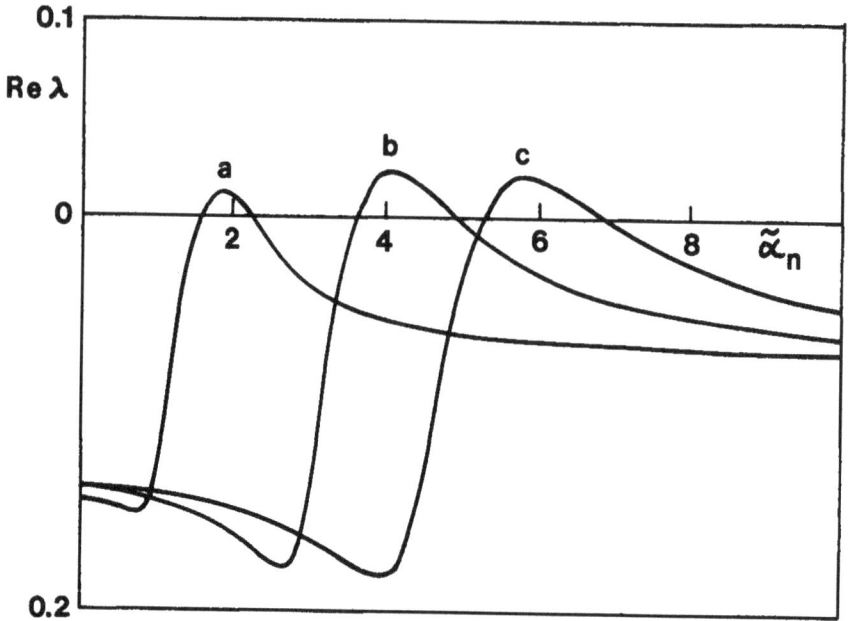

5.2   The largest real parts of the linearized eigenvalues are plotted as functions of $\tilde{\alpha}_n$ viewed as a continuous variable. For all the curves displayed in the figure we have selected $\alpha L = 6.0$ $R = 0.8$, $\bar{\alpha}_1 = 2.0$; these parameters correspond to $\bar{\kappa} = 0.071$. (a) $\bar{\gamma} = 0.1$, (b) $\bar{\gamma} = 0.5$, (c) $\bar{\gamma} = 1.0$. The lasers corresponding to the parameters of curves (a), (b) and (c) are all unstable and are expected to produce intensity pulsations at different frequencies.

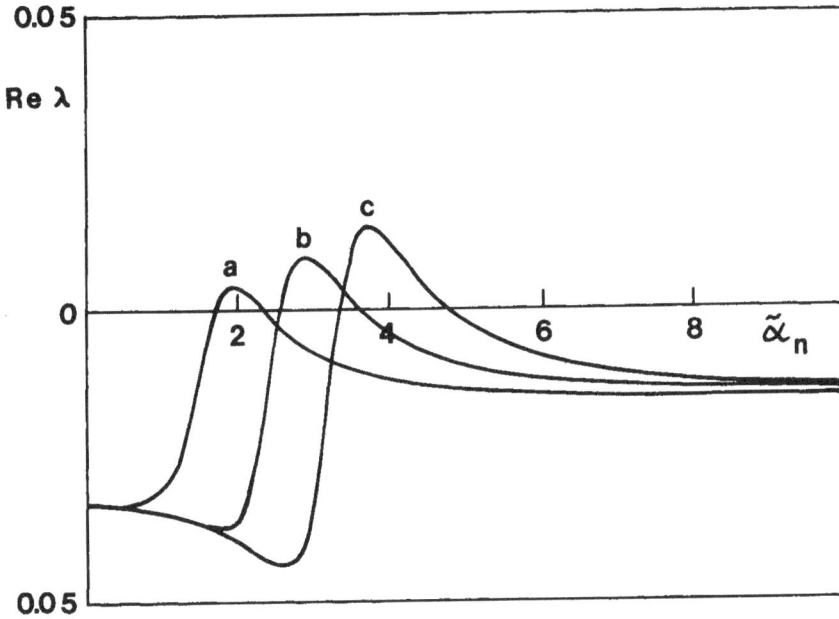

5.3   The largest real parts of the linearized eigenvalues are plotted as functions of
$\tilde{\alpha}_n$ viewed as a continuous variable. For all the curves displayed in the figure we have
selected $R = 0.9$, $\bar{\alpha}_1 = 1.0$, and $\bar{\gamma} = 0.1$; these parameters correspond to $\bar{\kappa} = 0.071$. (a)
$\alpha L = 3.0$, (b) $\alpha L = 7.0$, (c) $\alpha L = 12.0$. It is clear that, if the intermode spacing is
sufficiently small, a continuous increase of the gain will cause a progressively larger
beat frequency.

6.   The Adiabatic Elimination of the Fast Variables

In most laser systems the free relaxation rates of the basic physical variables tend to be quite different from one another, with one decay rate being much larger than the rest, in typical situations. Ruby, CO_2, Nd-YAG and semiconductor lasers, for example, are characterized by values of $\gamma_\perp$ which greatly exceed both $\gamma_\parallel$ and κ [see, Svelto (1982) for more specific information on these parameters). In a situation of this type, a mathematical process called adiabatic elimination is often useful to reduce the complexity of the problem. The limits of validity of the adiabatic elimination are not always easy to assess, as shown by Haken (1970), (1983a,b) and discussed also in a detailed study by Lugiato, Mandel and Narducci (1984). In particular, one needs great care in interpreting the results of this approximation because, often, important features of the exact mathematical equations disappear in the approximate version of the theory.

Here we discuss a situation that is common to laser physics, the case in which $\gamma_\perp$ is much larger than $\gamma_\parallel$ and κ We do so with the help of the adiabatic elimination process to offer a simple introduction to this technique. First, a few preliminary comments. Consider a dynamical system described by the two coupled equations of motion

$$\frac{dx}{dt} = -\gamma_x x + f(x,y) \tag{6.1a}$$

$$\frac{dy}{dt} = -\gamma_y y + g(x,y) \tag{6.1b}$$

The functions f and g are arbitrary functions of the dynamical variables x(t) and y(t) and the rates γ_x and γ_y are assumed to be such that $\gamma_x \ll \gamma_y$.

The first step in the adiabatic elimination process is to identify the variables that are characterized by the fastest relaxation processes and to set their time derivatives equal to zero. Next, we solve the resulting algebraic equations in such a way as to eliminate the fast varying variables from the problem; the result is a set of equations involving only variables whose relavation rates are comparable to one another. In our case, we set dy/dt = 0 and solve the algebraic equation

$$0 = -\gamma_y y + g(x,y) \tag{6.2}$$

whose solution we can denote formally by

$$y = G(x) \tag{6.3}$$

Next, we substitute Eq. (6.3) into (6.1a) and obtain a single differential equation of the form

$$\frac{dx}{dt} = -\gamma_x x + f(x, G(x)) \tag{6.4}$$

involving only the slower varying variable x. An intuitive justification of this process can be given with the following argument: we can integrate Eq. (6.1b) formally (and exactly) with the result

$$y(t) = y(0) e^{-\gamma_y t} + \int_0^t dt' \, e^{-\gamma_y (t-t')} \, g(x,y) \qquad (6.5)$$

If we agree that dynamical information contained over times of the order of $1/\gamma_y$, is not important, or is unresolvable by the detection process that measures the evolution of $x(t)$, we can approximate Eq. (6.5) with

$$y(t \gg 1/\gamma_y) \cong g(x,y) \int_0^t dt' \, e^{-\gamma_y (t-t')} \cong -\frac{1}{\gamma_y} g(x,y) \qquad (6.6)$$

The justification for this step is contained in the assumption that $g(x, y)$ is a slowly varying function of time as compared with $\exp(-\gamma_y(t-t'))$. The statement

$$y(t \gg 1/\gamma_y) \cong -\frac{1}{\gamma_y} g(x,y) \qquad (6.7)$$

is formally equivalent to setting $dy/dt \cong 0$, as indicated at the beginning.

We consider now the Maxwell-Bloch equation (3.25) under the assumption that $\gamma_\perp \gg \gamma_\parallel, \kappa$; in addition, we confine our attention to the resonant case $\delta_{AC} = 0$ (this is not a necessary requirement; it is made here only for simplicity). In this case we have

$$\bar{P} = -\bar{F} \bar{D} \qquad (6.8)$$

and

$$\frac{\partial \bar{F}}{\partial z} + \frac{1}{c} \frac{\partial \bar{F}}{\partial t} = \alpha \bar{F} \bar{D} \qquad (6.9a)$$

$$\frac{\partial \bar{D}}{\partial t} = -\gamma_\parallel \bar{F}^2 \bar{D} - \gamma_\parallel (\bar{D} - 1) \qquad (6.9b)$$

The steady state of this system of equations is identical to that described in Section 4 i.e.,

$$\bar{P}(z,\infty) = -\bar{F}_{st}(z) \frac{1}{1 + \bar{F}_{st}^2} \qquad (6.10a)$$

$$\bar{D}(z,\infty) = \frac{1}{1 + \bar{F}_{st}^2} \qquad (6.10b)$$

where $\bar{F}_{st}(z)$ is the solution of the transcendental equation

$$\ln\left(\frac{\bar{F}_{st}(z)}{\bar{F}_{st}(0)}\right) + \frac{1}{2} \left(\bar{F}_{st}^2(z) - \bar{F}_{st}^2(0)\right) = \alpha L \qquad (6.11a)$$

subject to the boundary condition

$$\bar{F}_{st}^2(0) = R^2 \, \bar{F}_{st}^2(L) \qquad (6.11b)$$

The output field is given by

$$\bar{F}^2_{st}(L) = \frac{2}{1-R^2}(\alpha L - |\ln R|) \tag{6.11c}$$

In resonance we ignore the possibility of multiple steady states which are usually irrelevant.

To study the stability of this steady state, we set

$$\bar{F}(z,t) = \bar{F}_{st} + \delta F(z,t) \tag{6.12a}$$

$$\bar{D}(z,t) = \bar{D}_{st} + \delta D(z,t) \tag{6.12b}$$

and obtain the linearized equations

$$\frac{\partial}{\partial z}\delta F + \frac{1}{c}\frac{\partial}{\partial t}\delta F = \alpha(\bar{F}_{st}\delta D + \bar{D}_{st}\delta F) \tag{6.13a}$$

$$\frac{\partial}{\partial t}\delta D = -\gamma_{\parallel}(2\bar{F}_{st}\bar{D}_{st}\delta F + \bar{F}^2_{st}\delta D + \delta D) \tag{6.13b}$$

We seek solutions of the type

$$\begin{pmatrix} \delta F(z,t) \\ \delta D(z,t) \end{pmatrix} = e^{\lambda t}\begin{pmatrix} \delta f(z) \\ \delta d(z) \end{pmatrix} \tag{6.14}$$

and obtain the following expression for the population fluctuation

$$\delta d(z) = -2\gamma_{\parallel}\frac{\bar{F}_{st}\bar{D}_{st}\delta f}{\lambda + \gamma_{\parallel} + \gamma_{\parallel}\bar{F}^2_{st}} \tag{6.15}$$

We now combine Eqs. (6.13a) and (6.15) with the result

$$\frac{\partial}{\partial z}\delta f + \frac{\lambda}{c}\delta f = \alpha\bar{D}_{st}\frac{\lambda + \gamma_{\parallel}(1 - \bar{F}^2_{st})}{\lambda + \gamma_{\parallel}(1 + \bar{F}^2_{st})}\delta f \tag{6.16}$$

As we have done in the previous section, we write Eq. (6.16) in the form

$$\frac{\partial}{\partial z}\delta f = H(z)\delta f \tag{6.17}$$

where

$$H(z) = -\frac{\lambda}{c} + \frac{\alpha}{1 + \bar{F}^2_{st}}\frac{\lambda + \gamma_{\parallel}(1 - \bar{F}^2_{st})}{\lambda + \gamma_{\parallel}(1 + \bar{F}^2_{st})} \tag{6.18}$$

The formal solution of Eq. (6.17) is

$$\delta f(z) = \delta f(0) \exp\left\{ \int_0^z dz' \ H(z') \right\} = \delta f(0) \ e^{\psi(z)} \tag{6.19}$$

The integral can be calculated following the same procedure adopted in handling Eq. (5.8); the result of this simple calculation is

$$\psi(z) = -\frac{\lambda}{c} z + \ln\left(\frac{\overline{F}_{st}^2(z)}{\overline{F}_{st}^2(0)}\right) + \ln\left(\frac{\lambda + \gamma_{\parallel} + \gamma_{\parallel} \overline{F}_{st}^2(z)}{\lambda + \gamma_{\parallel} + \gamma_{\parallel} \overline{F}_{st}^2(0)}\right) \tag{6.20}$$

and the field fluctuation takes the form

$$\delta F(z,t) = e^{\lambda t} \ \delta f(0) \ e^{\psi(z)} \tag{6.21}$$

Next, we impose the boundary condition

$$\delta F(0,t) = R \ \delta F(L, t - \frac{\Lambda - L}{c}) \tag{6.22}$$

we obtain the characteristic equation

$$\lambda_n = -i\alpha_n - \frac{c}{\Lambda} \ln\left(\frac{\lambda_n + \gamma_{\parallel} + \gamma_{\parallel} \overline{F}_{st}^2(L)}{\lambda_n + \gamma_{\parallel} + \gamma_{\parallel} R^2 \overline{F}_{st}^2(L)}\right) \tag{6.23}$$

Note that this result can also be derived, formally, from Eq. (5.12) in the limit $\gamma_{\perp} \to \infty$. No surprises emerge from the numerical study of this equation which predicts no instabilities for any values of the parameters.

We close now with a few remarks. Sections 4 and 5 detail a modern view of the steady state and stability properties of the resonant Maxwell-Bloch model. The result of this treatment indicates the existence of unstable behaviors for both single and multimode operation. The adiabatic elimination of the fast variables supresses these unstable behaviors as one can see from a detailed study of Eq. (6.23). There is really no basic inconsistency between the results of Sects. 4 and 5 and those presented in this Section, as long as we keep in mind that the adiabatic approximation remains meaningful only over a limited range in parameter space. The appearance of an instability involves faster temporal variations for all the coupled variables so that the basic premise [Eq. (6.7)] of the adiabatic elimination procedure fails.

One disappointing aspect of the exact analysis of the resonant problem is the large quantitive discrepancy between the experimentally observed thresholds for unstable behavior and the calculated values. It is well known, for example, that ruby lasers are notorious for developing a spiking output even at very low pumping levels, except under very carefully controlled conditions [Tang, Statz and DeMars (1963)]. One would think that such prominent instabilities would be predicted easily, even in a rather simplified model. Apparently, this is not the case for the plane-wave model.

One possible reason for this discrepancy is that, quite obviously, ruby lasers are very complicated systems, and that, perhaps, the Maxwell-Bloch model does not include enough essential physical features to provide a successful description. In addition, Casperson (1986) has suggested that, under hard excitation conditions, the system may be driven from its stable steady state into an isolated pulsing state. A hard excitation is a large perturbation, of the type that one is likely to encounter during the pumping process. Under these conditions, the linear stability analysis loses its meaning and only numerical methods would appear to be effective. Indeed, Casperson has shown by numerical integration of the single-mode laser equations that this model can be driven into a pulsing state, even well below the instability threshold predicted by the linear stability analysis. This is an appealing result. Unfortunately, his results indicate that even the instability boundary of the hard excitation terminates at a disappointingly high gain level (4-5 time above threshold). Thus, the issue remains still open.

7.   Linear Stability Analysis: The General Case

A common mode of operation for a laser involves a detuning between the center of the atomic line and one of the cavity resonances. In this case, the resonant analysis of Sect. 5 needs to be generalized. The general linearized problem is complicated because of the longitudinal variations of the field and of the atomic variables, and in fact it has been solved only recently by Lugiato, Narducci and Squicciarini (1986). This problem is much easier to handle in the uniform field limit, a situation where the field and the atomic variables are essentially uniform along the axis of the active medium [Narducci, Tredicce, Lugiato, Abraham and Bandy (1986)].

It is not difficult to visualize how one can reach this configuration, at least in principle. One needs to lower the unsaturated gain per pass to insure that the wave entering a given loop will experience a very small amplification. Of course, at this point, a laser normally stops functioning unless, at the same time, one adjusts the losses to a correspondingly low level. The point is that a uniform field profile requires a simultaneous decrease of both the unsaturated gain and the mirror transmittivity. In fact, in the mathematical limit [Bonifacio and Lugiato (1978)]

$$\alpha L \to 0, \quad T \to 0 \tag{7.1a}$$

with

$$\frac{\alpha L}{T} \equiv 2C = \text{arbitrary} \tag{7.1b}$$

the cavity field become, indeed, uniform. It is perhaps less obvious, but also not difficult to understand, that if the constant C is large enough, the laser will be able to operate above threshold and still maintain longitudinal uniformity for all the dynamical variables. Naturally, in this discussion we must assume that the only resonator losses arise from the finite transmittivity of the mirrors, otherwise any fixed residual losses will invalidate our premises. From a physical point of view, lowering the mirror transmittivity, increases the number of passes that the field makes through the active medium before leaking out of the cavity. Hence, if T is small enough, one can insure near field uniformity by selecting a small enough value of αL to keep the growth of the field envelope very small over one pass, and to allow the cumulative gain over many passes to be as large as needed for laser action. This mathematical model is often called "the mean field limit" in the literature; we prefer to label it as the "uniform field limit". From a practical viewpoint, the values of αL and T that would keep the variation of $\bar{F}_{st}(z)$ below, for example, 1% are unrealistically low (αL < 0.1, T < 0.01) so that this theoretical approach is very restrictive as it stands.

In a brilliant theoretical advance, Lugiato (1980a,b) succeded in getting around this difficulty and, at the same time, managed to introduce a suitable definition of cavity modes, even in the presence of reflectivity losses. We begin the review of Lugiato's approach with a restatement of the Maxwell-Bloch equations and their boundary conditions

$$\frac{\partial \bar{F}}{\partial z} + \frac{1}{c} \frac{\partial \bar{F}}{\partial t} = -\alpha \bar{P} \tag{7.2a}$$

$$\frac{\partial \bar{P}}{\partial t} = -\gamma_{\perp}\left(\bar{F}\,\bar{D} + (1 + i\,\bar{\delta}_{AC})\,\bar{P}\right) \tag{7.2b}$$

$$\frac{\partial \bar{D}}{\partial t} = -\gamma_{\parallel}\left\{-\frac{1}{2}\left(\bar{F}^*\,\bar{P} + \bar{F}\,\bar{P}^*\right) + \bar{D} - 1\right\} \tag{7.2c}$$

$$\bar{F}(0,t) = R\,\bar{F}(L, t - \frac{\Lambda\text{-}L}{c}) \tag{7.3}$$

As a first step, we define a new set of independent variables

$$z' = z \tag{7.4a}$$

$$t' = t + \frac{\Lambda\text{-}L}{c}\frac{z}{L} \tag{7.4b}$$

This transformation has the effect of mapping the two non-isochronous events $P_1 \equiv (0, t)$ and $P_2 \equiv (L, t - (\Lambda\text{-}L)/c)$ into two isochronous events in the new frame. In fact, according to the transformation (7.4) we have

$$P_1 \equiv (z=0, t=\bar{t}) \rightarrow (z'=0, t'=\bar{t})$$

$$P_2 \equiv (z=L, t'=\bar{t} - \frac{\Lambda\text{-}L}{c}) \rightarrow (z'=L, t'=\bar{t})$$

At this point, the transformed boundary conditions (7.3) begin to look very similar to the periodicity conditions that one encounters in the ordinary linear vibration problems. The only difference in our case is the multiplicative factor R. This can be eliminated by introducing the new set of field and atomic variables as follows

$$F(z',t') = \bar{F}(z',t')\,e^{\frac{z'}{L}\ln R} \tag{7.5a}$$

$$P(z',t') = \bar{P}(z',t')\,e^{\frac{z'}{L}\ln R} \tag{7.5b}$$

$$D(z',t') = \bar{D}(z',t') \tag{7.5c}$$

In fact, the boundary conditions for the new variables $F(z', t')$ are

$$F(0,t') = F(L,t') \tag{7.6}$$

The transformed Maxwell-Bloch equations take the form

$$\frac{\partial F}{\partial t'} + \frac{d}{\Lambda}\frac{\partial F}{\partial z'} = -\kappa\,(F + 2CP) \tag{7.7a}$$

$$\frac{\partial P}{\partial t'} = -\gamma_{\perp}\left(FD + (1+i\,\bar{\delta}_{AC})P\right) \tag{7.7b}$$

$$\frac{\partial D}{\partial t'} = -\gamma_{\parallel}\left\{-\frac{1}{2}(F^*P + FP^*)\exp[\frac{2z'}{L}|\ln R|] + D - 1\right\}$$ (7.7c)

where we have introduced the symbols

$$\kappa = \frac{c|\ln R|}{\Lambda}, \quad 2C = \frac{\alpha L}{|\ln R|}$$ (7.8)

The new equations of motion differ from the original set in two important respects:

i) the phase velocity of the new field amplitude is cL/Λ, instead of c; thus the transformation introduces an effective background dispersion that takes into account the field retardation due to the empty section of the cavity;

ii) the equations develop an explicit spatial dependence through the exponential factor $\exp(2z'|\ln R|/L)$.

The main advantage of the new picture is that the field amplitude obeys standard periodicity conditions (Eq. (7.6)) which make it possible to introduce an appropriate modal decomposition of the Fourier type and to define suitable modal amplitudes. An additional advantage is that the new field amplitude $F(z', t')$ maintains a large degree of uniformity, even under conditions that are removed significantly from the uniform field limit. Thus, in the new reference frame, it is possible to insist on the requirement of steady state field uniformity for $F(z', t')$ even without imposing unrealistic constraints on the laser parameters. Examples of this type of behavior are shown in Figs. (7.1) and (7.2) with the help of a direct comparison between the steady state longitudinal behavior of $|\bar{F}|$ and $|F|$.

We now introduce the following Fourier decomposition for the transformed variables:

$$\begin{pmatrix} F(z',t') \\ P(z',t') \end{pmatrix} = e^{-i\delta\Omega t'}\sum_{n=-\infty}^{+\infty} e^{ik_n z'}\begin{pmatrix} f_n(t') \\ p_n(t') \end{pmatrix}$$ (7.9a)

$$D(z',t') = \sum_{n=-\infty}^{+\infty} e^{ik_n z'}d_n(t')$$ (7.9b)

$$\begin{pmatrix} F^*(z',t') \\ P^*(z',t') \end{pmatrix} = e^{i\delta\Omega t'}\sum_{n=-\infty}^{+\infty} e^{-ik_n z'}\begin{pmatrix} f_n^*(t') \\ p_n^*(t') \end{pmatrix}$$ (7.9c)

where $\delta\Omega$ is an unknown frequency offset that measures the frequency separation between the selected cavity reference and the carrier frequency of the laser field. This parameter must be calculated from the steady state equations after imposing the boundary conditions. Note that $D(z', t')$ is a real variable, so that

54

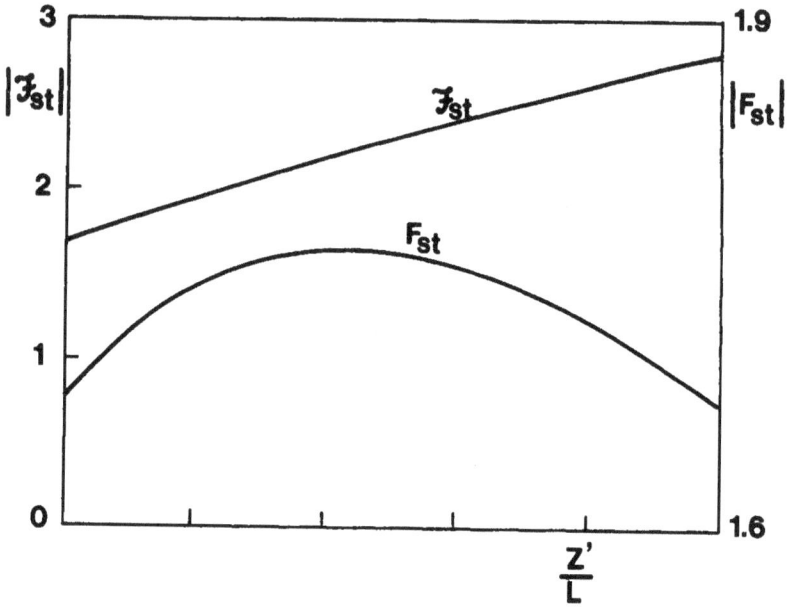

7.1   Steady state longitudinal profile of the original ($\overline{F}_{st}$) and transformed (F_{st}) field amplitudes for $\alpha L = 3.0$, $R = 0.6$, $\overline{\alpha}_1 = 10.0$ and $\overline{\delta}_{AC} = 0$. These parameters corresponds to $\overline{\kappa} = 0.813$ and $2C = 5.87$. The percentage variation of $|\overline{F}_{st}(z)|$ is 50%, while the percentage variation of $|F_{st}(z')|$ is 5%.

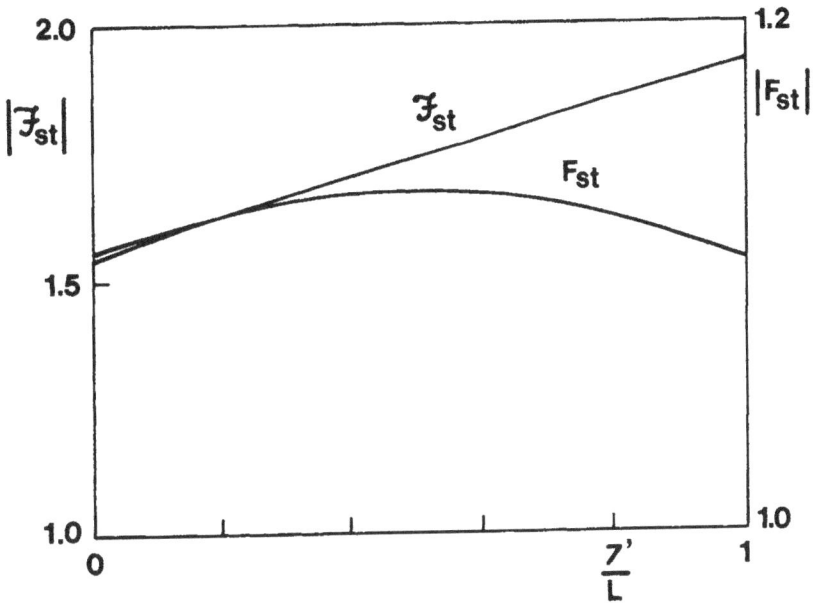

7.2   Steady state longitudinal profile of the original $(\overline{F}_{st})$ and transformed (F_{st}) field amplitudes for $\alpha L = 3.0$, $R = 0.6$, $\overline{\alpha}_1 = 10.0$ and $\overline{\delta}_{AC} = 3.0$. These parameters corresponds to $\overline{\kappa} = 0.813$ and $2C = 5.87$. The percentage variation of $|\overline{F}_{st}(z)|$ is 50%, while that of $|F_{st}(z')|$ is 2.3%.

$$d_n(t') = d_{-n}^*(t') \qquad (7.10)$$

This symmetry relation, of course, does not apply to the field and to the polarization.

The wave numbers k_n are selected in such a way that

$$k_n = \frac{2\pi}{L} n \quad (n = 0, \pm1, \pm2, ...) \qquad (7.11)$$

This choice insures that the boundary conditions (7.6) will be satisfied automatically. The time-dependent expansion coefficients $f_n(t')$ represent the amplitudes of the field modes. In the following analysis we show the connection between these exact modes of the ring resonator and the modes used in the traditional laser theory [see for example Risken and Nummedal (1968a,b), Graham and Haken (1968)]. The modal functions of the ring cavity in the new reference system are

$$u_n(z') = \frac{1}{\sqrt{L}} e^{ik_n z'} \qquad (7.12)$$

and satisfy the orthogonality relation

$$(u_n(z'), u_m(z')) \equiv \int_0^L dz'\, u_n^*(z')\, u_m(z') = \delta_{n,m} \qquad (7.13)$$

The above assignment of f_n and u_n as mode amplitudes and modal functions, respectively, could not be made in the original reference frame, except in the extreme limit $T \to 0$ where, in fact, the two pictures converge to one another. The steady state solutions for the field and the atomic variables can be found, as usual, by setting the time derivatives equal to zero, and by solving the resulting algebraic equations; in general, the solutions are linear combinations of the basis functions (7-12). This is clear for example from Fig (7-1) and (7-2) where the slight deviation of the exact steady state solution from perfect uniformity indicates the presence of additional harmonic Fourier components, beside the dominant constant term.

The infinite set of time-dependent functions $f_n(t')$, $p_n(t')$, $d_n(t')$ and the complex conjugate variables $f_n^*(t')$, $p_n^*(t')$ obey the following coupled equations of motion:

$$\frac{d}{dt'} f_n = i\, \delta\Omega f_n - \kappa\, (f_n + 2C\, p_n) - ick_n \frac{L}{\Lambda} f_n \qquad (7.14a)$$

$$\frac{d}{dt'} f_n^* = -i\, \delta\Omega f_n^* - \kappa\, (f_n^* + 2Cp_n^*) + ick_n \frac{L}{\Lambda} f_n^* \qquad (7.14b)$$

$$\frac{d}{dt'} p_n = \gamma_\perp \left[\sum_{n'} f_{n'} u_{n-n'} + [1 + i\, (\bar{\delta}_{AC} - \overline{\delta\Omega})]\, p_n \right\} \qquad (7.14c)$$

$$\frac{d}{dt'} p_n^* = -\gamma_\perp \left\{ \sum_{n'} f_{n'}^* d_{n-n'}^* + [1 - i\, (\bar{\delta}_{AC} - \overline{\delta\Omega})]\, p_n^* \right\} \qquad (7.14d)$$

$$\frac{d}{dt'} d_n = -\gamma_{\parallel} \left\{ -\frac{1}{2} \sum_{n' \, n''} \left(f_{n'}^* p_{n''} \, \Gamma_{n''-n'-n} + f_{n'} p_{n''}^* \, \Gamma_{n''-n'-n} \right) + d_n - \delta_{n,0} \right\} \qquad (7.14e)$$

where the complex quantities

$$\Gamma_p = \frac{1}{L} \int_0^L dz' \, e^{ik_p z'} \, e^{\frac{2z'}{L} |\ln R|} = \frac{1-R^2}{R^2} \frac{1}{2 |\ln R| + ik_p L} \qquad (7.15)$$

play the role of mode-mode coupling coefficients. This infinite set of exact modal equations can be studied by numerical techniques, but their complexity is such that the time dependent behavior of the ring laser model is more easily handled by direct numerical integration of Eqs. (7.2). An approximate treatment of this problem for the case of a single frequency laser was provided by Lugiato, Narducci, Bandy, and Tredicce (1986). Their analysis is summarized in Appendix B under the heading of "improved single-mode approximation". Here we mention that this procedure shows a way to produce much more accurate solutions in the single frequency case that can be obtained by making the single-mode approximations [i.e. by retaining only the n=0 modal amplitudes in Eqs. (7.14)]. The results is a less restrictive description of laser operation.

The exact modal equation (7.14) are especially useful in making contact with the earlier studies by Risken and Nummedal (1968a,b) and Graham and Haken (1968); in addition, they provide a convenient starting point for the derivation of the single-mode laser equations (See Section 8) and for the study of their limit of validity. To make these connections we must go to the uniform field limit. Thus, we consider a ring laser system in the limit in which $\alpha L \to 0$ and $T \to 0$ with $\alpha L/T \equiv 2C =$ arbitrary. Under these conditions, the mode-mode coupling coefficients take the form

$$\Gamma_p \to \frac{2T}{2T + ik_p L} \to \delta_{p,0} \qquad (7.16)$$

and the modal equations become

$$\frac{d}{dt'} f_n = i \, \delta\Omega \, f_n - \kappa \, (f_n + 2C p_n) \qquad (7.17a)$$

$$\frac{d}{dt'} f_n^* = -i \, \delta\Omega \, f_n^* - \kappa \, (f_n^* + 2C p_n^*) \qquad (7.17b)$$

$$\frac{d}{dt'} p_n = -\gamma_{\perp} \left\{ \sum_{n'} f_{n'} d_{n-n'} + [1 + i \, (\bar{\delta}_{AC} - \bar{\delta\Omega} - \bar{\alpha}_n)] \, p_n \right\} \qquad (7.17c)$$

$$\frac{d}{dt'} p_n^* = -\gamma_{\perp} \left\{ \sum_{n'} f_{n'}^* d_{n-n'}^* + [1 - i \, (\bar{\delta}_{AC} - \bar{\delta\Omega} - \bar{\alpha}_n)] \, p_n^* \right\} \qquad (7.17d)$$

$$\frac{d}{dt'} d_n = i \, \alpha_n d_n - \gamma_{\parallel} \left\{ -\frac{1}{2} \sum_{n'} \left(f_{n'} p_{n+n'} + f_{n'} p_{n'-n}^* \right) + d_n - \delta_{n,0} \right\} \qquad (7.17e)$$

where the new field and atomic amplitudes have been defined according to relations of type

$$x_n(t') = x_n(t') \exp(-ick_n t') \equiv x_n(t') \exp(-i\alpha_n t')$$

This representation is appropriate only in the uniform field limit because, in general, single-frequency field solutions are linear combinations of different spacial modes. Thus, in the uniform field limit, we recover the conventional travelling-wave expansion used by Risken and Nummedal, Graham and Haken, and reviewed by Risken (1986).

The stationary states of equations (7.17) correspond to the possible solutions of the algebraic system obtained by setting all the time derivatives equal to zero. It must be stressed that each steady state configuration corresponds to five strings of complex numbers of the type

$$f_0(\infty), \quad f_1(\infty), \quad f_{-1}(\infty). \quad ....$$

$$p_0(\infty), \quad p_1(\infty), \quad p_{-1}(\infty), \quad .....$$

etc.

Possible multiple solutions must be labelled with an additional index to avoid confusion. Thus, for example, one has:

$$f_0^{(j)}(\infty), \quad f_1^{(j)}(\infty), \quad f_{-1}^{(j)}(\infty), \quad .....$$

$$p_0^{(j)}(\infty), \quad p_1^{(j)}(\infty), \quad p_{-1}^{(j)}(\infty), \quad ....$$

etc., $\quad j = 0, 1, -1, ...$

The problem of finding all the possible steady states looks hopelessly complicated. In fact, close inspection of Eqs. (7.17), after setting the derivatives equal to zero, shows that the possible steady state solutions can be represented in the simple form

$$f_n^{(j)} = [2C - (1 + \bar{\Delta}_j^2)]^{1/2} \delta_{n,j} \tag{7.18a}$$

$$p_n^{(j)} = -f_n^{(j)} \frac{1 - i\bar{\Delta}_j}{1 + \bar{\Delta}_j^2 + |f_n^{(j)}|^2} \tag{7.18b}$$

$$d_n^{(j)} = \frac{1 + \bar{\Delta}_j^2}{1 + \bar{\Delta}_j^2 + |f_n^{(j)}|^2} \delta_{n,0} \tag{7 18c}$$

where

$$\bar{\Delta}_j = \frac{\bar{\delta}_{AC} - j\,\bar{\alpha}_1}{1 + \bar{\kappa}} \tag{7.19}$$

and

$$\overline{\delta\Omega} \equiv \overline{\delta\Omega}_j = \frac{\overline{\kappa}}{1+\overline{\kappa}}(\overline{\delta}_{AC} - j\,\overline{\alpha}_1) \tag{7.20}$$

Multiple steady state exist if the threshold condition

$$2C > 1 + \overline{\Delta}_j^2 \tag{7.21}$$

is satisfied for more than one value of j.

It is not difficult to advance a simple physical interpretation of these results. With reference to Fig. 7.3, consider a string of empty cavity resonances and a medium whose transition frequency is located at ω_A (vertical dashed line). A possible steady state solution corresponds to the laser line $\omega_L^{(0)}$ which is mode-pulled away from ω_C, as expected. The corresponding field amplitude oscillates at a frequency that is removed by an amount

$$\overline{\delta\Omega}_0 = \frac{\overline{\kappa}}{1+\overline{\kappa}}\overline{\delta}_{AC}$$

away from ω_C and is characterized by the modal amplitudes

$$f_0^{(0)} = [2C - (1 + \overline{\Delta}_0^2)]^{1/2}$$
$$f_1^{(0)} = f_{-1}^{(0)} = f_2^{(0)} = \ldots = 0$$

with

$$\overline{\Delta}_0 = \frac{\overline{\delta}_{AC}}{1+\overline{\kappa}}$$

The atomic variables have also Fourier components which are all zero, except for:

$$p_0^{(0)} = -f_0^{(0)}\frac{1 - i\overline{\Delta}_0}{1 + \overline{\Delta}_0^2 + |f_0^{(0)}|^2}$$

$$d_0^{(0)} = \frac{1 + \overline{\Delta}_0^2}{1 + \overline{\Delta}_0^2 + |f_0^{(0)}|^2}$$

If the medium has enough gain and/or the intermode spacing is sufficiently small, a second (or more) possible steady state(s) can coexist, such as the ones labelled schematically by $\omega_L^{(1)}$ and $\omega_L^{(-1)}$ in Fig. 7.3. These steady states are also characterized by Fourier amplitudes that are all zero, except for $f_1^{(1)}$, $p_1^{(1)}$, and $d_0^{(1)}$, in the case of $\omega_L^{(1)}$, as given by Eqs. (7.18). Each of these possible steady states is characterized by uniform field and atomic profiles, and by a purely harmonic structure, both in space and time.

Upon varying the detuning δ_{AC}, the output intensity varies smoothly, as expected.

$$\omega_L^{(-1)} \qquad \omega_L^{(0)} \qquad \omega_L^{(1)}$$

$$\omega_C - \alpha_1 \qquad \omega_C \quad \omega_A \qquad \omega_C + \alpha_1$$

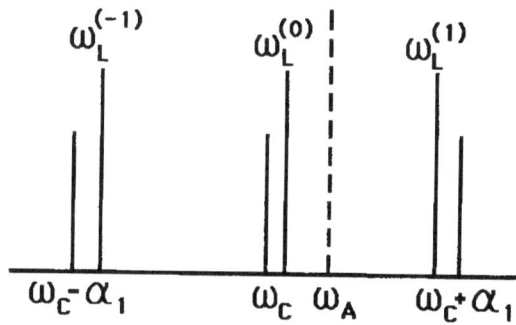

7.3   Schematic representation of the multiple stationary solutions. The lines labelled ω_C and $\omega_C \pm \alpha_1$ represent three adjacent cavity modes. The lines labelled $\omega_L^{(0)}$, $\omega_L^{(\pm 1)}$ indicate the positions of three coexisting steady state solutions. The dashed line at frequency ω_A indicates the position of the center of the atomic line.

An example of this variation is shown in Fig. 7.4. Here, the intermode spacing is large enough that the steady state solution j=0 is driven below threshold before the next possible steady state satisfies the threshold condition. This corresponds to the usual situation in a homogeneously broadened laser, which is often believed to be capable of supporting only one possible steady state at a time. Matters, in fact, are more complicated, as shown in Fig. 7.5 for a smaller intermode spacing. In this case, two different steady state configurations coexist, above threshold, for a certain range of values of the detuning parameter. We cannot determine the actual behavior of the system in the presence of competing steady states without the help of linear stability considerations but we can suggest possible scenarios.

One possibility correspond to the dominance of one steady state until its losses overcome the gain; at this point, the second steady state should take over. Another possibility is for both steady states to coexist, and to produce optical beats in a bichromatic configuration. Actually, both scenarios are possible depending on the parameters, as we will show.

We begin the linear stability analysis by setting

$$x_n = x_n^{(j)} \, \delta_{n,j} + \delta x_n \qquad\qquad (7.22a)$$

$$d_n = d_n^{(j)} \, \delta_{n,0} + \delta d_n \qquad\qquad (7.22b)$$

where x_n stands for f_n, f_n^*, p_n, or p_n^*, and we derive an infinite-dimensional linear system of equations for the fluctuation variables δx_n and δd_n. The remarkable aspect of this problem is that the infinite system breaks up into separate blocks of five equations each, where the fluctuation variables δf_{n+j}, δp_{n+j}, δf^*_{j-n}, δp^*_{j-n}, and δd_n, for fixed values of the steady state index j and modal index n, are coupled to one another, but not to any other variable. The result of a simple calculation is

$$\frac{d}{dt'} f_{n+j} = i \, \delta\Omega_j \, \delta f_{n+j} - \kappa \, (\delta f_{n+j} + 2C \, \delta p_{n+j}) \qquad\qquad (7.23a)$$

$$\frac{d}{dt'} f^*_{j-n} = - i \delta\Omega_j \, \delta f^*_{j-n} - \kappa \, (\delta f^*_{j-n} + 2C \, \delta p^*_{j-n}) \qquad\qquad (7.23b)$$

$$\frac{d}{dt'} p_{n+j} = - \gamma_\perp \left\{ f_j^{(j)} \, \delta d_n + \delta f_{n+j} \, d_0^{(j)} + [1 + i \, (\overline{\Delta}_j - \overline{\alpha}_n)] \, \delta p_{n+j} \right\} \qquad\qquad (7.23c)$$

$$\frac{d}{dt'} p^*_{j-n} = - \gamma_\perp \left\{ f_j^{(j)} {}^* \delta d_n + \delta f^*_{j-n} \, d_0^{(j)} + [1 - i \, (\overline{\Delta}_j + \overline{\alpha}_n)] \, \delta p^*_{j-n} \right\} \qquad\qquad (7.23d)$$

$$\frac{d}{dt'} d_n = i \, \alpha_n \, \delta d_n - \gamma_\parallel \left\{ - \frac{1}{2} (f_j^{(j)} {}^* \delta p_{n+j} + p_j^{(j)} \, \delta f^*_{j-n} + f_j^{(j)} \, \delta p^*_{j-n} + p_j^{(j)} {}^* \delta f_{n+j}) + \delta d_n \right\} \qquad (7.23e)$$

In order to solve Eqs. (7.23) we introduce the ansatz:

62

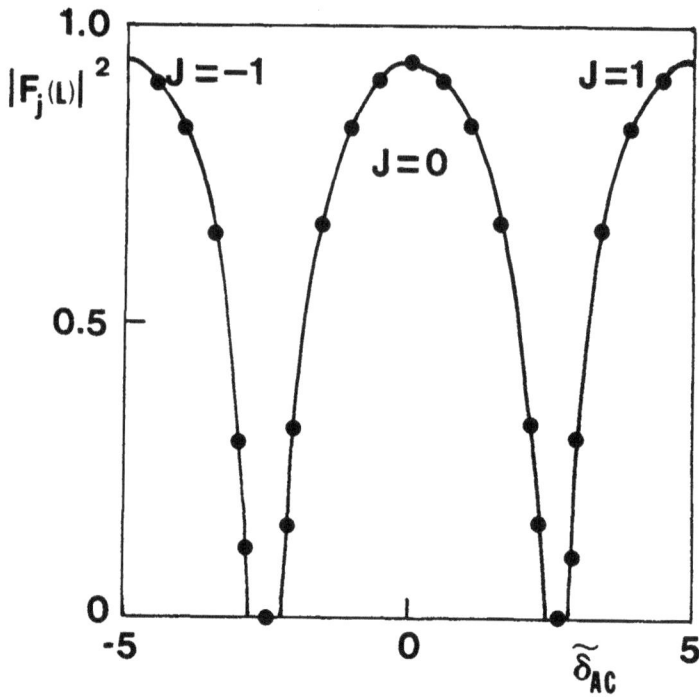

7.4   Steady state output intensity $|F_j(L)|^2$ for three stationary states labelled $j=0, \pm 1$ as a function of the detuning parameter $\overline{\delta}_{AC}$. The value $\overline{\delta}_{AC}= 0$ corresponds to resonance between the atomic transition frequency and the reference cavity mode. The dots correspond to steady state values obtained by direct integration of the Maxwell-Bloch equations. The parameters used in this calculation are $\alpha L = 2.0$, $R = 0.5$, $\overline{\gamma}= 2.0$, $\overline{\alpha}_1 = 5.0$. (The parameter $\overline{\gamma}= 2.0$ is irrelevant for the calculation of the steady state values, but is necessary for solving the Maxwell-Bloch equations).

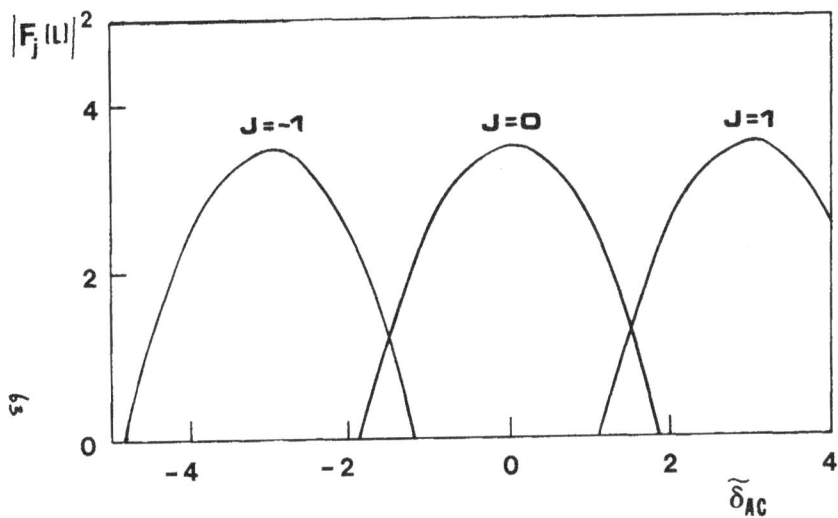

7.5   Steady state output intensity $|F_j(L)|^2$ for three stationary states labelled j=0, ±1, as a function of the detuning parameter $\bar{\delta}_{AC}$. The value $\bar{\delta}_{AC} = 0$ corresponds to resonance between the atomic transition frequency and the reference cavity mode. The parameters used in this calculation are $\alpha L = 2.0$, $R = 0.5$ and $\bar{\alpha}_1 = 3.0$.

$$
\begin{bmatrix}
\delta f_{n+j}(t') \\[4pt]
\delta f^{*}_{j-n}(t') \\[4pt]
\delta p_{n+j}(t') \\[4pt]
\delta p^{*}_{j-n}(t') \\[4pt]
\delta d_{n}(t')
\end{bmatrix}
= e^{\lambda t'}
\begin{bmatrix}
\delta f^{(0)}_{n+j} \\[4pt]
\delta f^{(0)\,*}_{j-n} \\[4pt]
\delta p^{(0)}_{n+j} \\[4pt]
\delta p^{(0)\,*}_{j-n} \\[4pt]
\delta d^{(0)}_{n}
\end{bmatrix}
\tag{7.24}
$$

and obtain a fifth-degree characteristic equation for λ of the form

$$
\sum_{j=0}^{5} A(j, \bar{\alpha}_n)\, \bar{\lambda}^j = 0
\tag{7.25}
$$

where $\bar{\alpha}_n = \alpha_n/\gamma_\perp$, $\bar{\lambda} = \lambda/\gamma_\perp$ and $A(j, \bar{\alpha}_n)$ are complicated, but explicit complex expressions that depend upon the stationary state parameters as well as the sideband frequency $\bar{\alpha}_n$. The steady state $j = \bar{j}$ which is being probed is stable if and only if the real parts of all five eigenvalues $\bar{\lambda}_{ni}$ are negative for all values of n. The appearance of a positive real part of an eigenvalue for a given value of n is an immediate indication that the selected steady state is unstable against a small perturbation. The instability manifests itself with the growth of sidebands at the frequencies $\pm\bar{\alpha}_n$. In this case, the field amplitude $F(z', t')$ begins to depart from its stationary configuration $\bar{F}_{st}$ and develops a space and time structure

Competitive effects between excited steady states become especially significant when the intermode spacing α_1 is of the order of a few atomic linewidths or smaller. In this case, unavoidably, the scaled cavity linewidth $\bar{\kappa} = \kappa/\gamma_\perp$ is considerably smaller than unity (this situation corresponds to the so-called good cavity configuration) because $\bar{\alpha}_1 = 2\pi\bar{\kappa}/T$ and $T \ll 1$. This is a point that should be kept in mind: uniform field multimode instabilities require the good cavity limit because $\bar{\alpha}_1$ is of order unity and T is very small.

A numerical study of the fifth-degree equation (7.25) shows that for every value of $\bar{\alpha}_n$, three of the five eigenvalues have a large negative real part. Thus, they can be ignored for the purpose of instability studies. The remaining two eigenvalues are responsible for possible unstable behaviors. One of these two eigenvalues has always a zero real part for $\bar{\alpha}_n = 0$; the real part of the other, instead, is different from zero. Because the phase of the steady state field is an irrelevant variable in an indifferent equilibrium state, we interpret the first type of eigenvalue as the phase eigenvalue. The second, for convenience, will be called the amplitude eigenvalue. This assignment, which may look fairly arbitrary in this context, could be proved rigorously by formulating the linear stability problem in terms of the field modulus and phase [for a general discussion of this problem, see Lugiato and Narducci (1986a,b)]. We will not do so here, but only suggest that a convincing test can be carried out in the resonance case ($\delta_{AC} = 0$) where the set of linearized equations for j=0 splits into two parts, one related to the amplitude variables (third degree equation), and the other related to the phases (second degree equation).

It is important to stress that in our previous discussion of the resonant laser, the phase eigenvalues have been entirely ignored by assuming real fluctuations for the variables of the system. This assumption will be made reasonable here by showing that in resonance, the largest phase eigenvalue has a negative real part, and thus, is irrelevant from the point of view of possible instabilities. Note that the term "large eigenvalue" is inaccurate because what we are really discussing are the real parts of the eigenvalues; for the sake of convenience we will continue to use this terminology when it is not likely to generate confusion.

To fix our attention on a familiar example, we begin our survey of the numerical analysis of Eq. (7.25) with a resonant case. Figures 7.6 and 7.7 display the two largest eigenvalues of Eq. (7.25) plotted as functions of the mode frequency $\bar{\alpha}_n$, viewed as a continuous variable, for $\bar{\delta}_{AC} = 0$ and two values of the unsaturated gain, αL. In the first case, (Fig. 7.6) the system is stable. In the case of Fig. 7.7, instead, the system develops an instability of the amplitude type which could have been predicted also on the basis of our resonant analysis. Two important points should be stressed:

   i)   Similar plots for the $j = \pm 1$ steady states (which are also above threshold for the chosen parameters) show that these are unstable for both figures. It follows that in the case of Fig. 7.7 no stable state is available; one may then conjecture that persistent self-pulsing will develop after a suitable transient.

   ii)   The instability predicted for the parameters of Fig. 7.7 is of the Risken-Nummedal type and emerges for high values of the gain when the sytem is well above threshold.

When $\bar{\delta}_{AC}$ is different from zero, the analysis of Sections 5 and 6 becomes inadequate, because the detuning is responsible for a new type of instability when the phase eigenvalue develops a positive real part. An important aspect of the phase instability is that it can emerge for much smaller gain values than in the case of the amplitude instability and, therefore, is much more interesting from an experimental point of view.

In both Fig. 7.8 and 7.9 we have selected a value of the gain which is too small to cause an instability of the amplitude type (in fact, the amplitude eigenvalues have both negative real parts for all values of $\bar{\alpha}_n$). The steady state j=0 is stable for the detuning value chosen in Fig. 7.8, but it becomes unstable for the larger value of $\bar{\delta}_{AC}$ of Fig. 7.9. Here the instability is associated with the growth of sidebands at $\bar{\alpha}_n = \bar{\alpha}_1$ and $\bar{\alpha}_{-n} = \bar{\alpha}_{-1}$. It is essential to stress that, unlike the resonant case, here the instability is brought about by the destabilization of the phase, i.e., by a very different physical mechanism from the one that is operative in connection with the Risken-Nummedal instability. This feature is also typical of all detuning scans that we have investigated [Narducci, Tredicce, Lugiato, Abraham and Bandy (1986)].

We are now in a position to consider the matter of competition among different steady states and the possible scenarios that one may expect from the direct numerical integration of the Maxwell-Bloch equations. If the intermode spacing is so large that only one steady at a time satisfies the threshold condition, the situation is not essentially different from that of a single-mode laser: above threshold for laser action the only stationary state is normally stable as long as $\bar{\kappa}$ is sufficiently small and the gain of the system is lower than required by the instability threshold.

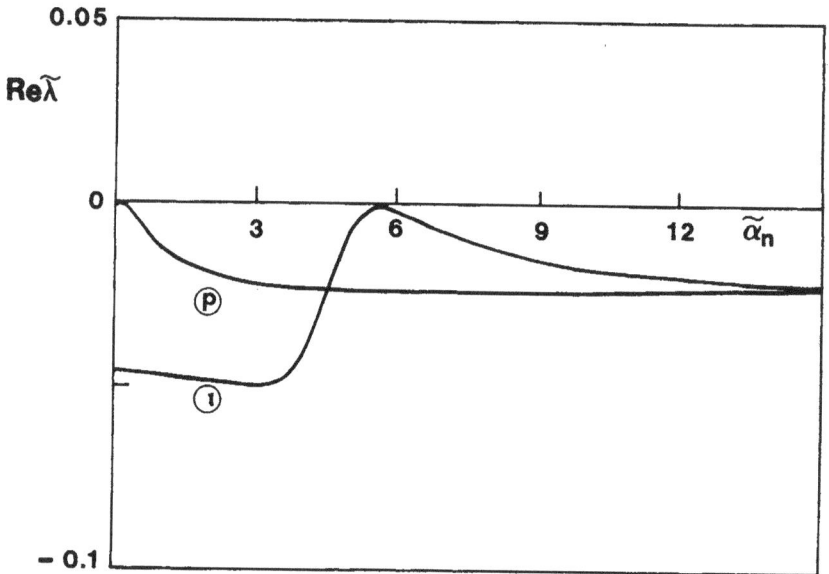

7.6   The two largest real parts of the eigenvalues of the linearized equations are plotted as functions of $\tilde{\alpha}_n$ viewed as a continuous variable for $\alpha L = 0.8$, $R = 0.95$, $\tilde{\alpha}_1 = 3.0$, $\bar{\gamma} = 1.5$ and $\bar{\delta}_{AC} = 0$. These eigenvalues describe the linear response of the system around the stationary state $j=0$ which is stable because $\text{Re } \bar{\lambda} < 0$ for all values of $\tilde{\alpha}_n$. The other two possible steady states $j=\pm1$, which are above threshold for the chosen parameters, are both unstable. Under the conditions of this simulation the $j=0$ steady state would be stable for long times. The line marked (a) denotes the amplitude eigenvalue; the line marked (p) denotes the phase eigenvalue.

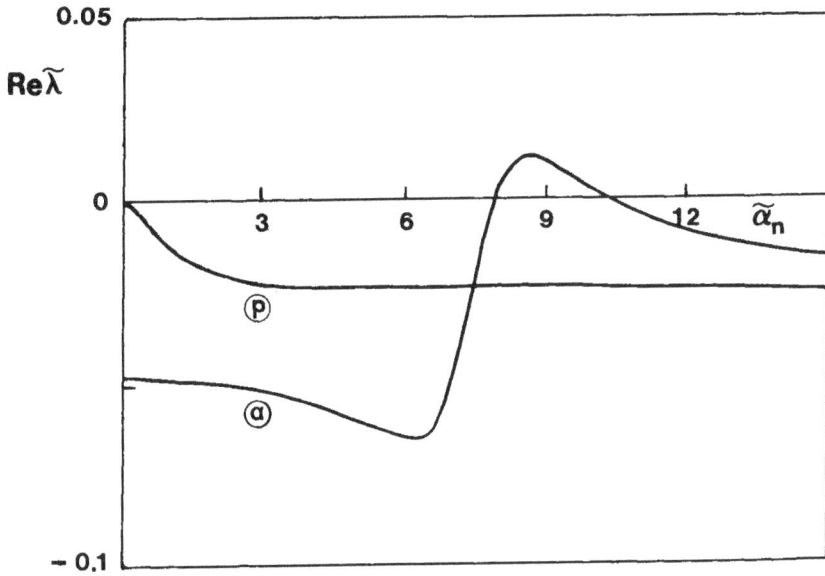

7.7 Same as Fig. 7.6 with $\alpha L = 2.0$. For sufficiently high values of the gain, the real part of the amplitude eigenvalue becomes positive and the $j=0$ steady state becomes unstable by developing sidebands at $\bar{\alpha}_{\pm3} = \pm3\bar{\alpha}_1$. Because the stationary states $j=\pm1$ are also unstable under these conditions, the system is expected to develop self-pulsing for sufficiently long times. The line marked (a) denotes the amplitude eigenvalue; the line marked (p) denotes the phase eigenvalue.

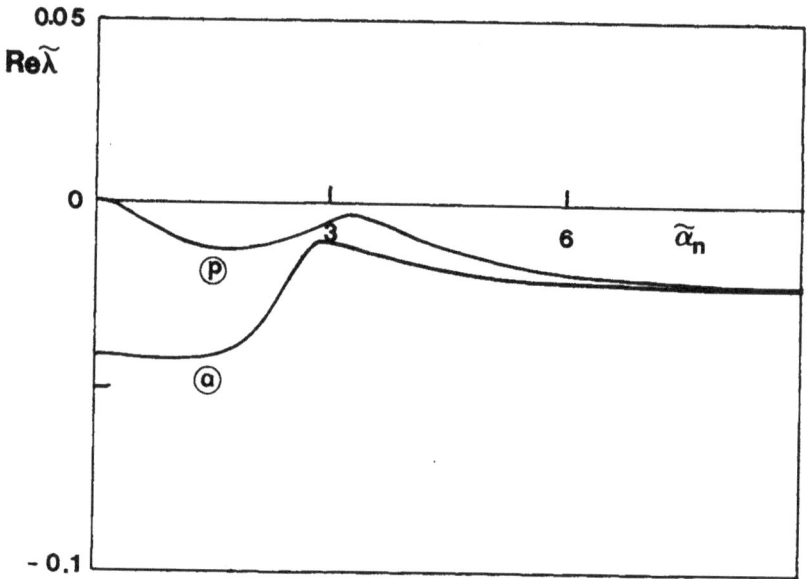

7.8   For increasing values of the detuning parameter $\bar{\delta}_{AC}$ the phase eigenvalue eventually develops a positive real part. In this case, the detuning is not sufficiently large to create an instability. The chosen parameters are $\alpha L = 0.5$, $R = 0.95$, $\bar{\alpha}_1 = 3.0$, $\bar{\gamma} = 0.8$ and $\bar{\delta}_{AC} = 0.7$.

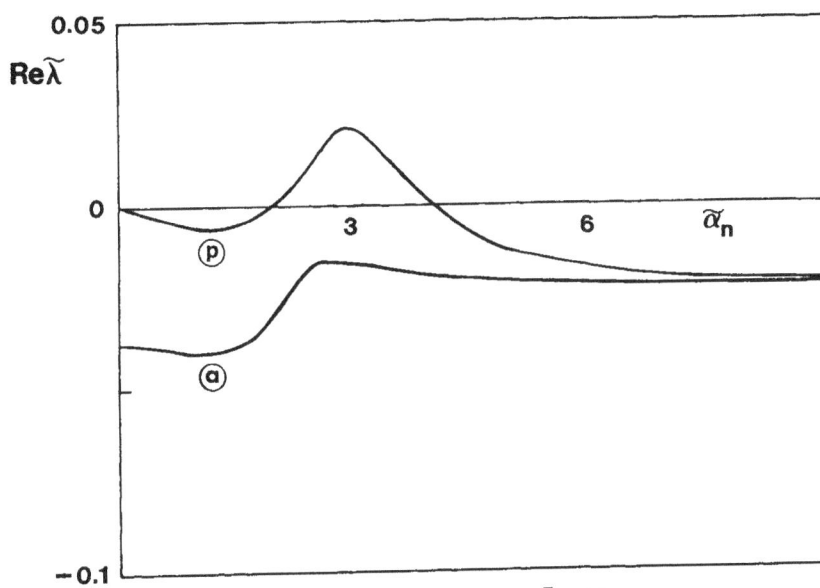

7.9   Same as Fig. 7.8 with $\bar{\delta}_{AC} = 1.2$.

For smaller values of $\bar{\alpha}_1$ (e.g., α_1 of the order of a few atomic linewidth, $\gamma_\perp$) it is easy to find gain values such that at most, two steady state configurations can coexist. This is the situation of greatest interest in a practical sense. In the following discussion, we imagine to carry out a detuning scan, with δ_{AC} being varied from 0 to a full free spectral range of the cavity (α_1). One could scan beyond the value $\delta_{AC} = \bar{\alpha}_1$, but of course, nothing essentially new would develop.

As indicated by the numerical study of Eq. (7.25) two different situations are possible which are schematically shown in Fig. 7.10a,b where we have displayed the stability range of the two steady states for different values of the detuning parameter δ_{AC}. In Fig. 7.10a we show schematically a case where the mode n=1 is stable for the steady state j=0 over the range $0 < \delta_{AC} \leq \delta_{thr}{}^{(0)}$, while the same mode is stable for the steady state j = 1 over the range $\delta_{thr}{}^{(1)} < \delta_{AC} < \bar{\alpha}_1$. (Note that the stability diagrams are symmetric relative to the middle of the free spectral range.) It is common to find only one cavity mode that can display a transition from a stable to unstable behavior (see, for example, Figs. 7.8 and 7.9) so that in Fig. 7.10a we focus only on the arbitrarily selected mode n = 1. The main point is that as δ_{AC} is varied from 0 to $\delta_{thr}{}^{(0)}$, the steady state j=0 is stable and retains control of laser operation. Beyond $\delta_{thr}{}^{(0)}$, the steady state j=0 becomes unstable, while j=1 is stable; at this detuning value one expects that the laser will switch from one configuration (j=0) to another (j=1). Because the two steady states are characterized by different output intensities and operating frequencies, the laser should undergo a discontinuous jump between steady state parameter values. Upon reversing the direction of the scan, a reverse transition occurs when δ_{AC} becomes equal to $\delta_{thr}{}^{(1)}$. Thus, the system exhibits hysteretic behavior.

Of course, a laser cannot not switch abruptly from one configuration to the other. It is reasonable to expect that, for example, at $\delta_{AC} = \delta_{thr}{}^{(0)}$ on the foward scan, the steady state j=1 will begin to grow, while the state j=0 begins to decay away. The transient evolution should be characterized by a beat pattern at the frequency difference between the mode-pulled frequencies of the two steady states. Eventually, the state j=0 must disappear and, with it, the beat pattern.

In Fig. 10b we have shown schematically the second possiblity. Here the domains of stability of the two steady states do not overlap so that both steady states are unstable in the range $\delta_{thr}{}^{(0)} < \delta_{AC} < \delta_{thr}{}^{(1)}$. A foward detuning scan should then show the emergence of steady oscillations in the common range of instability. Of course, unlike the previous case, these oscillations should persist instead of being just a transient phenomenon. Also, in this case, no hysteresis will be observed.

We must remark that, typically, case 10a is characteristic of systems with a small value of the ratio $\gamma_\parallel/\gamma_\perp$ while case 10b is favored by lasers whose value of $\gamma_\parallel/\gamma_\perp$ is of the order of unity. Because most known homogeneously broadened lasers tend to have small values of $\bar{\gamma}$, the prevailing situation should be the one displayed in Fig. 10a. Experimental exidence for this type of instability was given by Trodiooo, Narducci, Bandy, Lugiato and Abraham (1986) using a CO_2 laser.

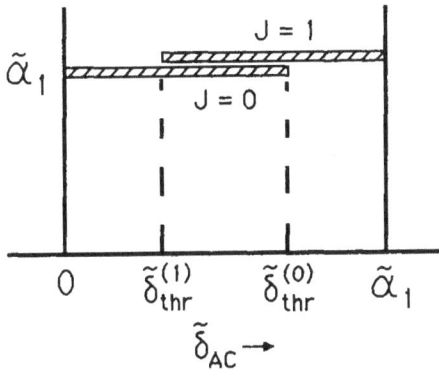

(a)

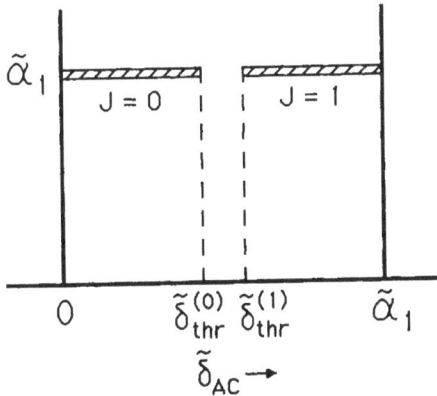

(b)

7.10   (a) Schematic representation of the stability domains of the j=0 and j=1 solutions as functions of the detuning parameter $\bar{\delta}_{AC}$. The solution j=0 is stable over the domain $0 < \bar{\delta}_{AC} < \bar{\delta}_{thr}^{(0)}$, while the solution j=1 is stable for $\bar{\delta}_{thr}^{(1)} < \bar{\delta}_{AC} < \bar{\alpha}_1$. A detuning scan in this case leads to discontinuous jumps between the steady states and hysteresis. (b) In this case the solutions j=0 and j=1 are both unstable over the range $\bar{\delta}_{thr}^{(0)} < \bar{\delta}_{AC} < \bar{\delta}_{thr}^{(1)}$; hence a detuning scan is expected to lead to self-pulsing over the common instability range of the two coexisting solutions.

8.   The Single Mode Laser: Deterministic Chaos

The resonant, single-mode model is the oldest theoretical description of the laser, dating back even further than the time of the laser itself (as we mentioned, the modern theory of the laser grew out of the earlier maser studies). The literature does not always stress the very restrictive assumptions that must be satisfied for this model to be consistent with the Maxwell-Bloch equations. On the other hand, the single-mode model has served us well as a theoretical testing ground for the study of laser instabilities and, according to recent evidence [Weiss (1986)], it may be also of some value in explaining specific experimental results.

The single-mode resonant model follows directly from Eqs. (7.14) in the uniform field limit [$\alpha L \to 0$, $T \to 0$ with $2C = \alpha L/T$ = arbitrary], after setting $\delta_{AC} = 0$, and upon ignoring all the modal amplitudes with indices different from zero [for a detailed review of this problem and references to numerous publications, see Abraham, Mandel and Narducci (1987)]. For simplicity, we set $f_0 = f$, $p_0 = p$, etc., and write the appropriate equations in the form

$$\frac{\partial f}{\partial t} = -\kappa (f + 2Cp) \tag{8.1a}$$

$$\frac{\partial f^*}{\partial t} = -\kappa (f^* + 2Cp^*) \tag{8.1b}$$

$$\frac{\partial p}{\partial t} = -\gamma_\perp (f\,d + p) \tag{8.1c}$$

$$\frac{\partial p^*}{\partial t} = -\gamma_\perp (f^* d + p^*) \tag{8.1d}$$

$$\frac{\partial d}{\partial t} = -\gamma_\parallel \left[-\frac{1}{2}(f^* p + f\,p^*) + d - 1 \right] \tag{8.1e}$$

Of course, the nature of the model is such that multiple solutions are ruled out by the required large value of the intermode spacing. Thus, only one steady state is possible; this is characterized by the atomic steady state values

$$p_{st} = -f_{st}\,\frac{1}{1 + |f_{st}|^2} \tag{8.2a}$$

$$d_{st} = \frac{1}{1 + |f_{st}|^2} \tag{8.2b}$$

and by the field intensity

$$|f_{st}|^2 = 2C - 1 \tag{8.3}$$

The phase of the field is arbitrary, as usual.

In order to analyze the stability of this steady state, we linearize the equations of motion (8.1) with the result

$$\frac{\partial}{\partial t} \delta f = -\kappa (\delta f + 2C \delta p) \tag{8.4a}$$

$$\frac{\partial}{\partial t} \delta p = -\gamma_\perp (f_{st} \delta d + d_{st} \delta f + \delta p) \tag{8.4b}$$

$$\frac{\partial}{\partial t} \delta d = -\gamma_\parallel \left[-\frac{1}{2} (\delta f^* p_{st} + f^* \delta p + c.c.) + \delta d \right] \tag{8.4c}$$

Similiar equations hold for δf^* and δp^*. Next, we seek solutions of the form

$$\begin{bmatrix} \delta f \\ \cdots \\ \cdots \\ \cdots \\ \delta d \end{bmatrix} = e^{\lambda t} \begin{bmatrix} a_1 \\ \cdots \\ \cdots \\ \cdots \\ a_5 \end{bmatrix} \tag{8.5}$$

and derive a system of five linear algebraic equations for the unknown quantities $a_1,...,$ a_5 which admits nontrivial solutions if and only if the determinant of the coefficients vanishes identically. The determinant conditions (characteristic equation) can be expressed as the fifth order polynomial equation

$$\bar{\lambda} (\bar{\lambda} + 1 + \bar{\kappa}) (\bar{\lambda}^3 + (\bar{\gamma} + 1 + \bar{\kappa}) \bar{\lambda}^2 + \bar{\gamma}(1 + \bar{\kappa} + |f_{st}|^2) \bar{\lambda}$$
$$+ 2\bar{\gamma} |f_{st}|^2 \bar{\kappa}) = 0 \tag{8.6}$$

which, as already mentioned, splits into the product of a cubic and a quadratic equation.

The two roots of the quadratic polynomial are given by

$$\bar{\lambda} = 0, \quad \bar{\lambda} = -(1 + \bar{\kappa}) \tag{8.7}$$

The first is associated with the absolute phase of the steady state field; the second with the relative phase between the field and the polarization. Both these eigenvalues cannot lead to dynamical instabilities. The amplitude eigenvalues are the three roots of the cubic polynomial whose coefficients are real. A convenient method to study the stability of this system is based on a famous test usually known as the Hurwitz criterion [see, for example, Haken (1983a)] which permits to establish the existence of eigenvalues with positive real part through a simple analysis of the coefficients of the polynomial equation. To be more precise, the Hurwitz criterion can be stated as follows: given the polynomial equation

$$\bar{\lambda}^n + c_1 \bar{\lambda}^{n-1} + \dots + c_n = 0 \tag{8.8}$$

with real coefficients. All zeros of the polynomial have negative real parts if and only if the following conditions are satisfied:

(i) $c_1 > 0, \ c_2 > 0, ..., \ c_n > 0$ (8.9a)

(ii)   The principal subdeterminants H_j of the quadratic scheme:

$$
\begin{vmatrix}
c_1 & 1 & 0 & 0 & \cdots & 0 & 0 \\
c_3 & c_2 & c_1 & 1 & \cdots & 0 & 0 \\
c_5 & c_4 & c_3 & c_2 & \cdots & 0 & 0 \\
- & - & - & - & - & - & - \\
0 & 0 & 0 & 0 & \cdots & c_{n-1} & c_{n-2} \\
0 & 0 & 0 & 0 & \cdots & 0 & c_n
\end{vmatrix}
$$

i.e., $H_1 = c_1,$
      $H_2 = c_1 c_2 - c_3, \ ...$
      $H_n = c_n \ H_{n-1},$

satisfy the inequalities

$$H_1 > 0 , \ H_2 > 0, ..., \ H_n > 0 \qquad\qquad (8.9b)$$

In our case, the quadratic form is

$$
\begin{vmatrix}
\bar{\gamma} + 1 + \bar{\kappa} & 1 & 0 \\
2\,\bar{\gamma}\,|f_{st}|^2\,\bar{\kappa} & \bar{\gamma}(1 + \bar{\kappa} + |f_{st}|^2) & \bar{\gamma} + 1 + \bar{\kappa} \\
0 & 0 & 2\,\bar{\gamma}\,|f_{st}|^2\,\bar{\kappa}
\end{vmatrix}
$$

and the subdeterminants are

$$H_1 = \bar{\gamma} + 1 + \bar{\kappa}$$

$$H_2 = \bar{\gamma}\,(\bar{\gamma} + 1 + \bar{\kappa})(1 + \bar{\kappa} + |f_{st}|^2) - 2\,\bar{\gamma}\,|f_{st}|^2\,\bar{\kappa}$$

$$H_3 = 2\,\bar{\gamma}\,|f_{st}|^2\,\bar{\kappa}\,H_2$$

so that the stability condition takes the form

$$(\bar{\gamma} + 1 + \bar{\kappa})(1 + \bar{\kappa}) + |f_{st}|^2(\bar{\gamma} + 1 - \bar{\kappa}) > 0 \qquad\qquad (8.10)$$

This condition is always satisfied if $\bar{\kappa} < 1 + \bar{\gamma}$. If, on the other hand, one is dealing with a "bad cavity" configuration, i.e., $\bar{\kappa} > 1 + \bar{\gamma}$, the system will remain stable only as long as

$$|f_{st}|^2 < \frac{(\bar{\gamma} + 1 + \bar{\kappa})(1 + \bar{\kappa})}{\bar{\kappa} - 1 - \bar{\gamma}}$$

The single mode equations, instead, will develop an amplitude instability if the bad cavity condition

$$\bar{\kappa} > 1 + \bar{\gamma} \qquad (8.11a)$$

is satisfied and if the gain 2C is large enough to fulfill the inequality

$$|f_{st}|^2 = 2C - 1 > \frac{(\bar{\gamma} + 1 + \bar{\kappa})(1 + \bar{\kappa})}{\bar{\kappa} - 1 - \bar{\gamma}} \qquad (8.11b)$$

The instability conditions (8.11a, b) are very stringent because they require simultaneously a low cavity Q and a high gain. It is instructive to plot the gain threshold value $2C_{thr}$ (the so-called second laser threshold) as a function of $\bar{\kappa}$ in order to identify the optimum conditions for the observation of the instability. The results are shown in Fig. 8.1. As with multimode instabilities of the amplitude type, also the single-mode instability is characterized by a high threshold; in addition, as we mentioned, it requires the bad cavity condition ($\bar{\kappa} > 1 + \bar{\gamma}$).

Equations (8-1) have provided a serious obstacle, since the very early days of the laser, against the interpretation of the output pulsations as the result of a dynamical instability of the single-mode type. Most laser systems cannot be made to operate in the bad cavity condition. Furthermore, unstable behavior in many solid state lasers is very common even at relatively low values of the gain. In fact, substantial efforts are required to reduce or eliminate the spiking action, or to favor ordinary relaxation oscillations. This is a problem because in ruby lasers, for example, the condition $\kappa <<$ $\gamma_\perp$ is satisfied, suggesting that single-mode operation should always lead to stable behavior.

Even at this time, a clear explanation of the common emergence of output pulsations at low gain in solid state lasers in lacking. Attractive arguments have been advanced by Casperson (1985), (1986) and by Zeghlache and Mandel (1985) beginning with the observation that instabilities can be produced if the system is perturbed strongly enough from the stationary state (for example by a strong pump or by a rapid transient), even when the linear stability analysis shows that the steady state is stable.

In spite of these open problems and because of the important role played by the single-mode model in the early development of laser physics, we continue our discussion with an analysis of its main dynamical predictions. Without loss of generality, we consider the reduced set of equations

$$\frac{\partial f}{\partial t} = -\kappa(f + 2Cp) \qquad (8.12a)$$

76

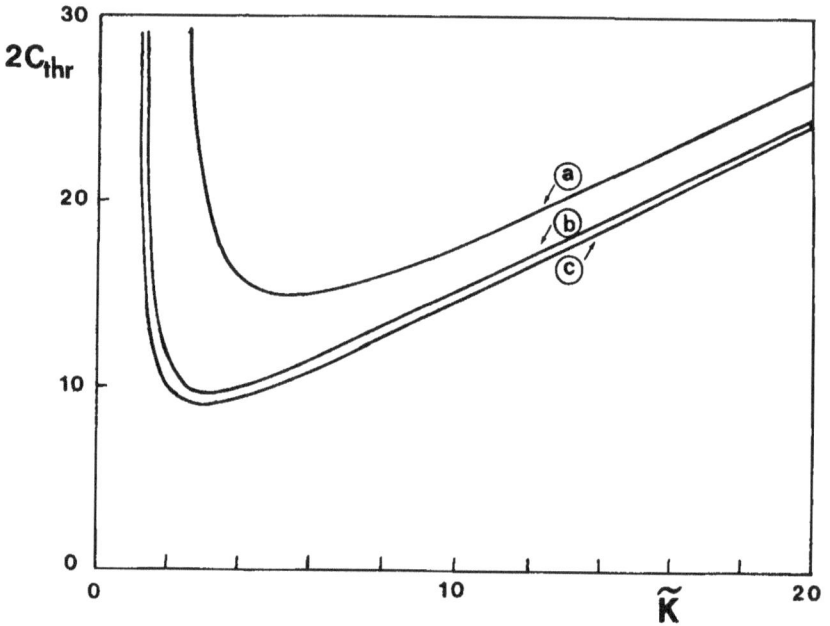

8.1 The instability boundary of the single-mode laser model is plotted in the plane of the parameters $\bar{\kappa}$ and 2C for (a) $\bar{\gamma} = 1.0$, (b) $\bar{\gamma} = 0.1$, (c) $\bar{\gamma} = 0.01$. The optimum (i.e., the most unstable) configuration corresponds to $\bar{\kappa} \cong 3$ and $2C_{thr} \cong 9$. Note that the threshold for ordinary laser action is $2C_{thr} = 1$.

$$\frac{\partial p}{\partial t} = -\gamma_{\perp}(f\,d + p)$$ (8.12b)

$$\frac{\partial d}{\partial t} = -\gamma_{\parallel}\left[-f\,d + d - 1\right]$$ (8.12c)

for the resonant evolution. These equations are isomorphic to the Lorenz model, first introduced to describe convective hydrodynamic instabilities [Lorenz (1963)]. In fact, remarkably, the first announcements of the unusual behavior of these equations in the context of laser and fluid dynamics studies came at approximately the same time, and entirely independently [we have already mentioned that the first numerical solutions of the single-mode laser equations were carried out by Grazyuk and Oraevskii (1964) and Buley and Cummings (1964)].

The threshold for laser action (first laser threshold) is $C_{thr} = 0.5$. For values of the gain just above C_{thr}, the laser approaches a stable steady state monotonically (Fig. 8.2). This behavior is not easy to justify directly from the cubic part of the polynomial equation (8.6), but it is immediate in the adiabatic limit $\gamma_{\perp} \gg \gamma_{\parallel}$, κ where the eigenvalues take the form

$$\lambda = -\frac{1}{2}\gamma_{\parallel}(1 + f_{st}^{2}) \pm \sqrt{\frac{1}{4}\gamma_{\parallel}^{2}(1 + f_{st}^{2})^{2} - 2\kappa\gamma_{\parallel}f_{st}^{2}}$$ (8.13)

and monotonic behavior is to be expected when f_{st}^{2} is very small. For larger values of the gain, with 2C still below the second laser threshold, the approach to the stable state is accompanied by relaxation oscillations whose origin can be traced to the appearance of an imaginary part in Eq. (8.13) [see Fig. 8.3]. A further increase in gain would lead to no additional effects according to Eq. (8.13). However, the adiabatic elimination breaks down because the relaxation oscillations begin to display higher and higher frequency, so that we have no option but to continue our discussion with the support of the exact cubic equation for the eigenvalues.

At the instability boundary, this equation predicts linearized eigenvalues whose real part vanishes, by definition, and whose imaginary parts are different from zero. This feature of the eigenvalues is a characteristic signature of a Hopf bifurcation. In fact, the mathematical literature [see, for example, Sattinger (1973)] distinguishes two types of Hopf bifurcations: a supercritical and subcritical type. A subcritical Hopf bifurcation generates oscillations above threshold whose amplitude grows gradually from zero; a supercritical Hopf bifurcation, instead, is responsible for the production of finite amplitude pulsations, even immediately above threshold. The bifurcation of the resonant single-mode model is of the supercritical type. In addition, the onset of an instability is accompanied by seemingly random pulsations.

The matter failed to attract widespread attention in the laser community until Haken (1975) pointed out the mathematical link between the laser and the Lorenz models. The latter had already stimulated a large number of mathematical investigations in an effort to understand the origin of the erratic behavior of the solutions [for an extensive review of the many aspects of this problem see, Sparrow (1982)]. Dynamical chaos, i.e. apparently random behavior even in the absence of external noise, is a very common feature of nonlinear differential equations and has

78

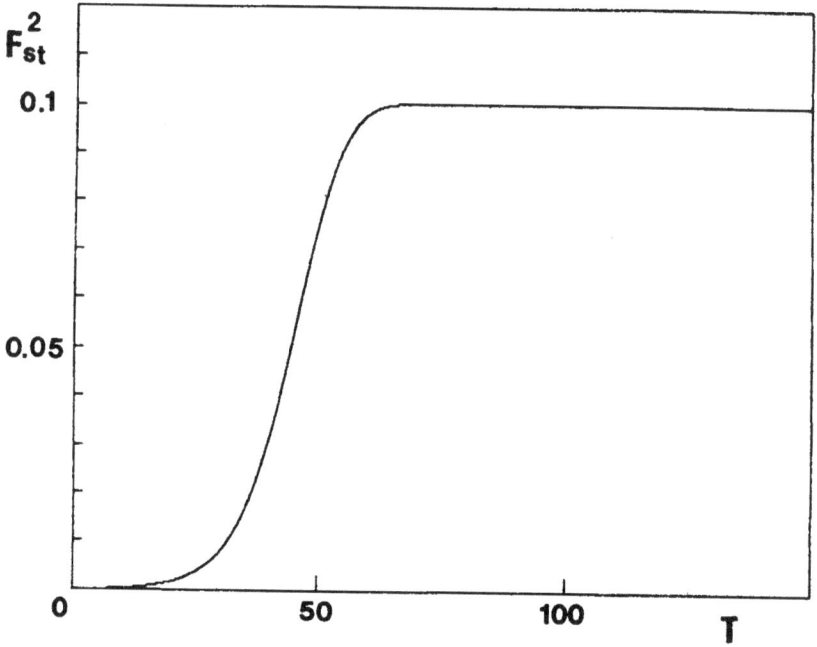

8.2   Time evolution of the output intensity according to Eqs. (8.12). When the gain is sufficiently small the approach to steady state is monotonic. The parameters chosen in this simulation are $2C = 1.1$, $\bar{\gamma} = 0.5$, $\bar{\kappa} = 4.0$.

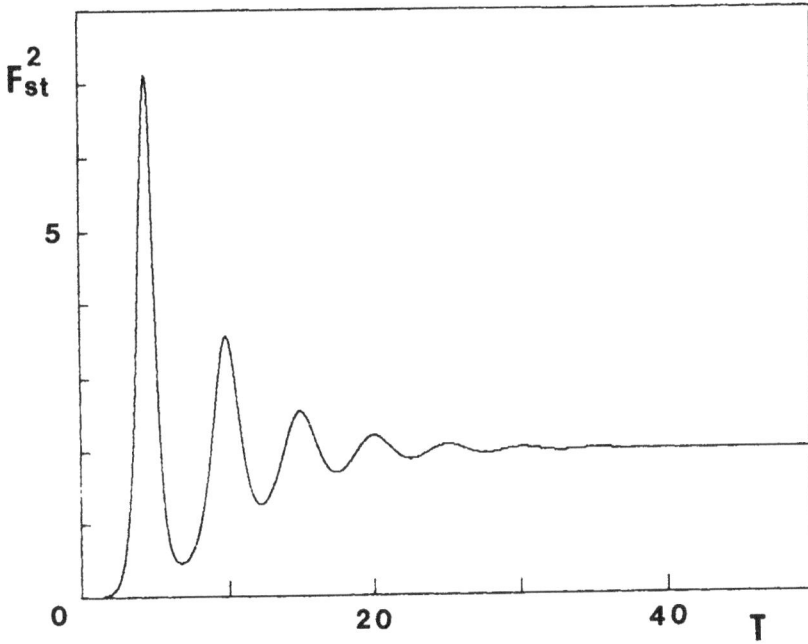

8.3   Time evolution of the output intensity according to Eqs. (8.12). The gain is
larger than in the simulation of Fig. 8.2 and the approach to steady state is oscillatory
(relaxation oscillations). The parameters chosen in this simulation are $2C = 3.0$, $\overline{\gamma} = 0.5$, $\widetilde{\kappa} = 4.0$.

become the focus of interest of a very active current branch of mathematics and theoretical physics. Here, we only observe that this type of mathematical behavior appears to be almost tailor-made to explain certain laser instabilities if one is willing to overlook the serious discrepancies between the calculated and the measured threshold values.

An example of the type of aperiodic oscillations that emerge immediately above the second laser threshold for most values of the system parameters is shown in Fig. 8.4. The phenomenology of the chaotic oscillations is very rich and complicated. For quite some time researchers believed that the only scenario predicted by the single-mode model was the discontinuous jump into a chaotic state of the type shown in Fig. 8.4. More recent studies [Narducci, Sadiky, Lugiato and Abraham (1985)] have shown instead that things can be considerably more complicated. In fact, the parameter $\bar{\gamma}$ plays an important role in selecting the kind of dynamics that one observes in numerical calculations. In particular, the discontinuous jump from a stationary state into a chaotic motion appears over a wide range of values of $\bar{\gamma}$ except for a domain approximately below $\bar{\gamma} = 0.2$.

Figure 8.5 illustrates this situation with the help of three important boundary lines. The solid line represents the usual boundary between stable (left) and unstable states (right) according to the linear stability analysis. Unlike the boundary line of Fig. 8.1, this is drawn for $\bar{\kappa} = 4$ and variable and 2C. As predicted by the linear stability analysis, a system characterized, for example, by $\bar{\gamma} = 0.5$ becomes unstable at 2C $\cong$ 12 and displays the kind of erratic oscillations shown in Fig. 8.4. If one uses the values acquired by the variables (f,p,d) at the end of a numerical run as initial conditions for the next computation corresponding to a smaller value of 2C (adiabatic scan), one finds that upon crossing the solid boundary from right to left the oscillations persist over a fairly wide range. This is a manifestation of the existence of what is known as a hard-mode excitation state that coexist with the stable solutions for the same parameter values. At the broken line, even the hard excitation state disappears, and the system returns to the stable configuration. Thus, the strip contained between the solid and the broken line is a domain of coexistence of solutions that are quite different in character. The stable state is only metastable in the sense that a sufficiently large perturbation can set up undamped random oscillations for most values of $\bar{\gamma}$.

In the range $0 < \bar{\gamma} < 0.2$, a new feature develops. Upon entering the instability domain, one finds periodic instead of chaotic solutions. These periodic solutions persists, under an adiabatic scan, even to the left of the solid boundary line and become chaotic only upon crossing the dotted line shown in Fig. 8.5. Over a very narrow range of values of $\bar{\gamma}$ the periodic, hard excitation solutions never become chaotic and simply disappear (apparently) at the broken boundary.

The behavior of the periodic solutions for increasing values of $\bar{\gamma}$ is extremely varied. Usually, periodic solutions develop in the form shown in Fig. 8.6a where we display the field amplitude instead of its modulus. We call these solutions symmetric for obvious reasons. The projection of the trajectory onto the (f, d) plane, produces a completely symmetric pattern as shown, for example, in Fig. 8.6b. For increasing values of $\bar{\gamma}$, the field amplitude undergoes a symmetric breaking transformation (Figs. 8.7a,b) with different positive and negative excursions. The intensity pattern would produce a regular alternation of higher and lower pulses, which would be very hard to distinguish from an ordinary period-doubled trajectory. This, of course, raises the issue of how it is possible to tell the difference between a period-doubling bifurcation and a

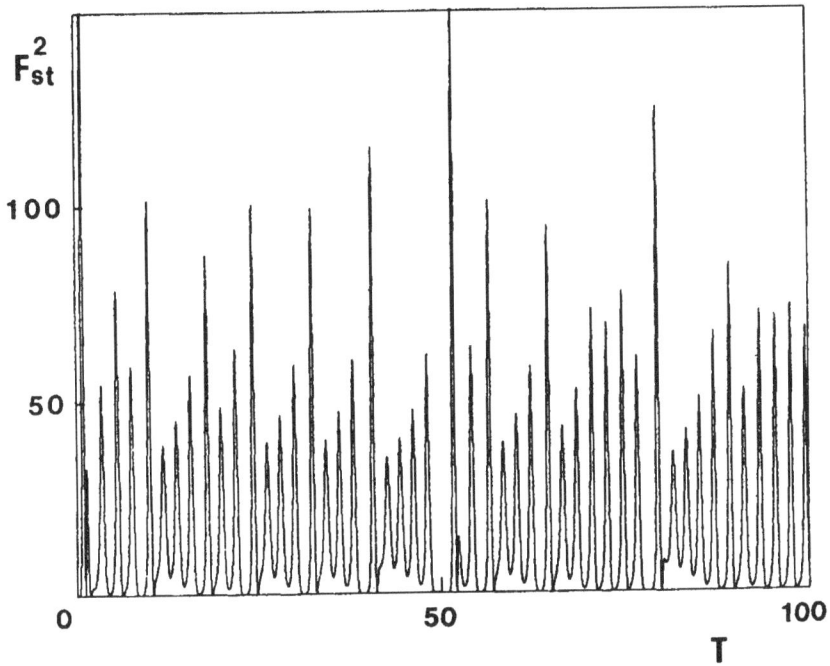

8.4   Time evolution of the output intensity according to Eqs. (8.12). When 2C
exceeds the second threshold value, the laser output often develops undamped
chaotic pulsations. The parameters chosen in this simulation are 2C = 15.0, $\bar{\gamma} = 0.5$, $\bar{\kappa}$
= 4.0.

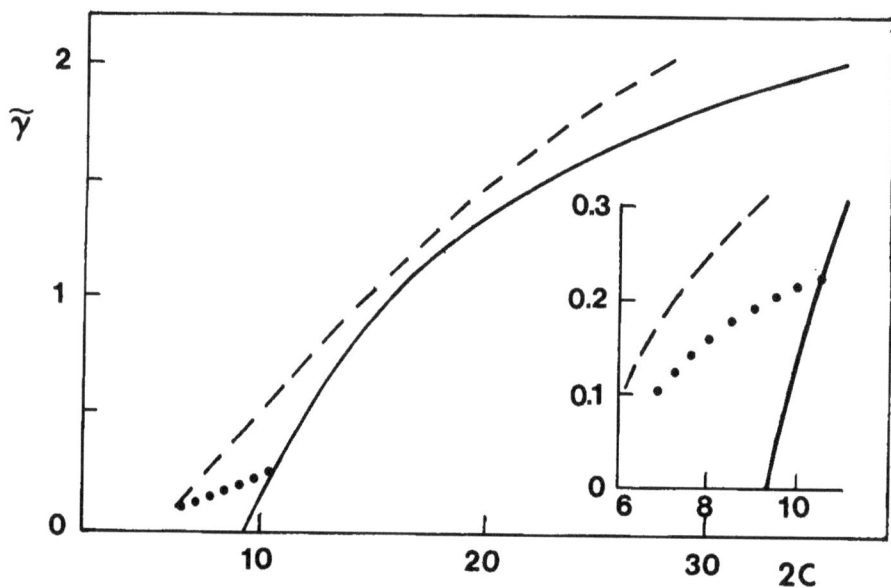

8.5   Instability threshold for the single mode laser in the $(2C, \tilde{\gamma})$ plane for $\bar{\kappa} = 4.0$. The solid line represents the locus of points where the time-independent stationary state loses stability against infinitesimal perturbations (the unstable domain lies to the right of this line). The broken line marks the end of the hard excitation domain. The dotted line denotes the approximate boundary between chaotic states (left of the dotted line) and periodic pulsing states. The inset shows an enlarged view of this domain for small values of $\tilde{\gamma}$.

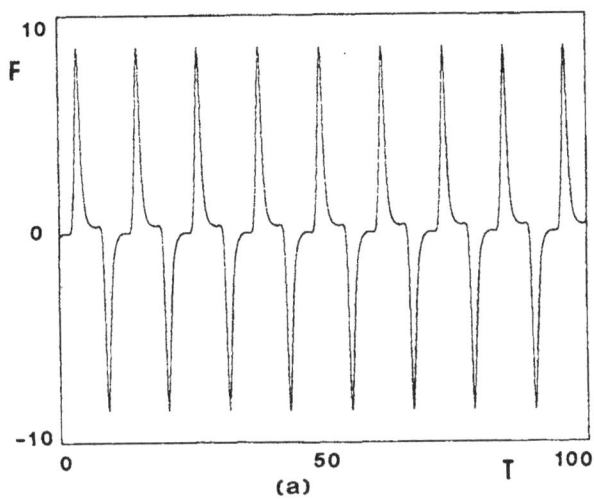

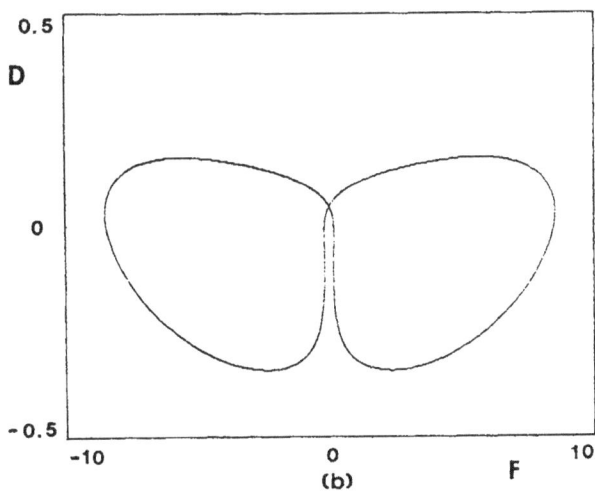

8.6   (a) Time dependence of the electric field amplitude f(t) for $\bar{\kappa} = 4$, $\bar{\gamma} = 0.14$ and 2C = 12. This is an example of a symmetric solution. (b) projection of the phase-space trajectory in the (f,d) plane.

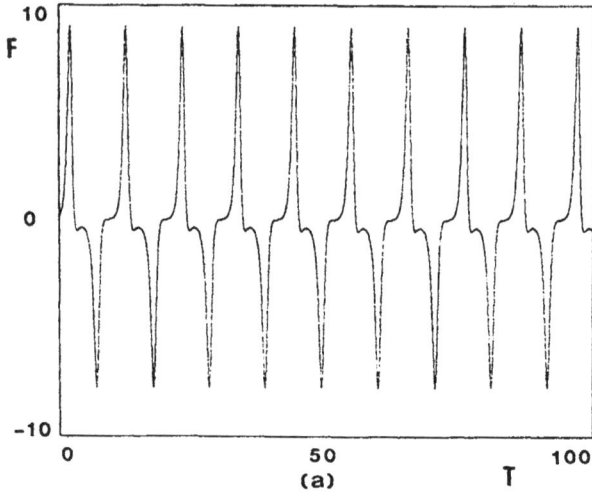

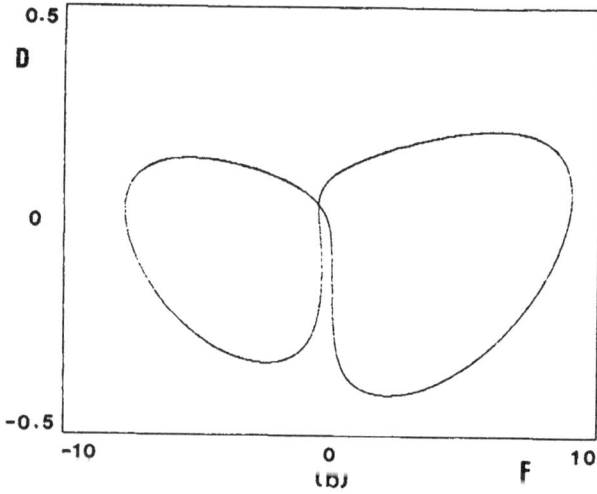

8.7   (a) Time-dependence of the electric field amplitude f(t) for $\bar{\kappa} = 4$, $\bar{\gamma} = 0.17$ and $2C = 12$. This is an example of asymmetric solution with period one. Note that the corresponding intensity pattern would appear, incorrectly to have undergone a period doubling bifurcation relative to Fig. 8.6a. (b) projection of the phase-space trajectory in the (f,d) plane.

symmetry breaking transformation from an experimental point of view. The answer is provided by the comparison between homodyne and heterodyne spectra, as we discuss in Section 10. True period-doubling bifurcations do emerge from asymmetric solutions. An example is shown in Figs. 8.8a,b. These are followed by progressively more complicated patterns which eventually merge into chaos.

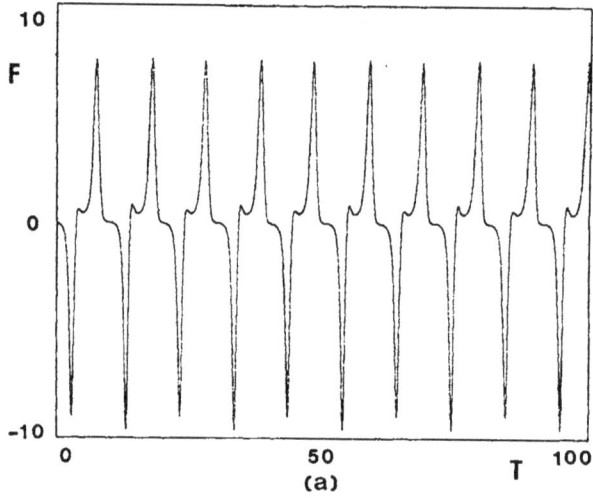

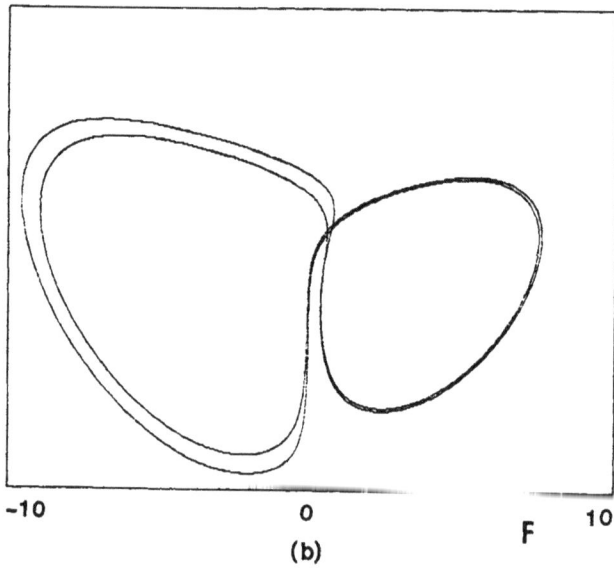

8.8   (a) Time-dependence of the electric field amplitude f(t) for $\bar{\kappa} = 4$, $\bar{\gamma} = 0.19$ and 2C = 12. This is an example of an asymmetric solution with period two. (b) projection of the phase-space trajectory in the (f,d) plane.

9. Dynamical Behavior of the Maxwell-Bloch Equations

The numerical simulation of the Maxwell-Bloch equations is a time-consuming effort without the help of a full-size computer. We have carried out most of our simulations using Eqs. (7.7) as the starting point. Of course, this version of the Maxwell-Bloch equations is equivalent to the original set (3.25). The transformed boundary conditions, however, are much easier to handle numerically.

When dealing with a large-scale computation of this type and a wide range of parameter values, it is essential to have a way of identifying potentially interesting combinations of operating conditions. In this connection, the linear stability analysis developed in earlier sections turns out to play a very useful role. A convenient procedure involves tracing, first, the outlines of the stability boundary and, then, scanning the interior of the unstable domain(s). Usually, after a certain amount of work "in the dark," patterns emerge which can be explored in greater depth by more refined scans of the parameters.

The most common tools in numerical exploration of this type include the numerical integration of the equations of motion, the calculation of power spectra (homodyne and heterodyne) of the output intensity, the construction of Poincare surfaces of section, and a variety of other tools that have been developed specifically in response to the growing interest in deterministic chaos.

We begin with a discussion of the numerical analysis of the Maxwell-Bloch equations. We have had considerable success with a numerical scheme adopted by Risken and Nummedal (1968a) for the study of a resonant model, and generalized by Narducci, Tredicce, Lugiato, Abraham and Bandy (1986) to include the dispersive effects of detuning.

As a starting point, we discretize the space and time derivatives at every grid point (m,n) in space-time using the Taylor expansion formulas

$$\frac{\partial}{\partial z'} X(m,n) = \frac{X(m,n) - X(m-1,n)}{\Delta z'} + \frac{1}{2} \Delta z' \frac{\partial^2}{\partial z'^2} X(m,n) \qquad (9.1a)$$

$$\frac{\partial}{\partial t'} X(m,n) = \frac{X(m,n+1) - X(m,n)}{\Delta t'} - \frac{1}{2} \Delta t' \frac{\partial^2}{\partial t'^2} X(m,n) \qquad (9.1b)$$

where X stands for F, P or D, and $\Delta z'$ and $\Delta t'$ are the elementary space-time steps. The second time derivatives follow directly from the equations of motion according to the obvious procedure

$$\frac{\partial}{\partial t'^2} X(m,n) = \frac{\partial}{\partial t'} \left(\frac{\partial}{\partial t'} X(m,n) \right) = \frac{\partial}{\partial t'} \text{ (right hand side of}$$

the X equation)

By selecting $\Delta z' = (cL/\Lambda) \Delta t'$ as the elementary space step, we can eliminate the mixed space-time derivatives; we select the time step in such a way that $\Delta t' = 2\pi/M\alpha_1$, where M is the number of spatial intervals in the active medium (M = L/$\Delta z'$) and α_1 is the empty cavity intermode spacing. A brief explanatory note and additional comments on these assignments can be found in the Appendix to the Risken-Nummedal paper of 1968a.

After considerable algebra, the discretized equations take the form:

$$F(m,n+1) = a_1 F(m,n) + a_2 F(m-1,n) + a_3 P(m,n) + a_4 P(m-1,n) +$$
$$a_5 D(m,n)\, F(m,n) \tag{9.2a}$$

$$P(m,n+1) = b_1 P(m,n) + b_2 D(m,n)\, F(m,n) + b_3 D(m,n)\, P(m,n) +$$
$$b_4 D(m,n)\, F(m-1,n) + b_5 P^*(m,n)\, F^2(m,n)\, E(m) +$$
$$b_6 P(m,n)\, F(m,n)\, F^*(m,n)\, E(m) + b_7 F(m,n) \tag{9,2b}$$

$$D(m,n+1) = c_1 D(m,n) + \mathrm{Re}\big(c_2 F^*(m,n)\, P(m,n)\, E(m)\big) +$$
$$c_3 P^*(m,n)\, P(m,n)\, E(m) + \mathrm{Re}\big(c_4 F(m-1,n)\, P^*(m,n)\, E(m)\big) +$$
$$c_5 D(m,n)\, F^*(m,n)\, F(m,n)\, E(m) + c_6 \tag{9.2c}$$

where

$$a_1 = \frac{1}{2}\Delta\tau^2\, \bar{\kappa}^2$$

$$a_2 = 1 - \bar{\kappa}\,\Delta\tau$$

$$a_3 = -\frac{1}{2}\, 2C\,\bar{\kappa}\,\Delta\tau\left(1 - \Delta\tau\,(1+i\,\bar{\delta}_{AC}) - \bar{\kappa}\Delta\tau\right)$$

$$a_4 = -\frac{1}{2}\, 2C\,\bar{\kappa}\,\Delta\tau$$

$$a_5 = -\frac{1}{2}\, 2C\,\bar{\kappa}\,\Delta\tau^2$$

$$b_1 = 1 - \Delta\tau\,(1+i\,\bar{\delta}_{AC}) + \frac{1}{2}(1+i\,\bar{\delta}_{AC})^2\Delta\tau^2$$

$$b_2 = \frac{1}{2}\Delta\tau\left(1 - \Delta\tau\,(1+i\,\bar{\delta}_{AC} + \bar{\gamma} + \bar{\kappa})\right)$$

$$b_3 = -\frac{1}{2}\Delta\tau^2\, 2C\,\bar{\kappa}$$

$$b_4 = \frac{1}{2}\Delta\tau$$

$$b_5 = -\frac{1}{4}\Delta\tau^2\,\bar{\gamma}$$

$$b_6 = b_5$$

$$b_7 = -\frac{1}{2}\Delta\tau^2\,\bar{\gamma}$$

$$c_1 = 1 - \bar{\gamma}\,\Delta\tau + \frac{1}{2}\Delta\tau^2\,\bar{\gamma}^2$$

$$c_2 = \frac{1}{2}\Delta\tau^2\,\bar{\gamma}(1+i\,\bar{\delta}_{AC}) + \frac{1}{2}\Delta\tau\,\bar{\gamma} + \frac{1}{2}\Delta\tau^2\,\bar{\gamma}\,\bar{\kappa} + \frac{1}{2}\Delta\tau^2\,\bar{\gamma}^2 - \bar{\gamma}\,\Delta\tau$$

$$c_3 = \frac{1}{2}\Delta\tau^2\,\bar{\gamma}2C\,\bar{\kappa}$$

$$c_4 = -\frac{1}{2}\Delta\tau\,\bar{\gamma}$$

$$c_5 = -\frac{1}{2}\Delta\tau^2\,\bar{\gamma}$$

$$c_6 = \frac{1}{2}\Delta\tau^2\,\bar{\gamma}^2 - \bar{\gamma}\,\Delta\tau$$

and where $\Delta\tau = \gamma_\perp\Delta t'$. The boundary conditions are imposed by setting

$$X(1,n) = X(m,n)$$

If necessary, the above discretized equations can be separated into real and imaginary parts.

A complete analysis of the interesting domain of parameter space is a monumental task. Matters are relatively simple when only one mode becomes unstable either through an amplitude or phase instability. In this case, the output intensity tends to undergo simple-looking oscillations with the frequency of the unstable sideband. This is precisely what Risken and Nummedal demonstrated in their important paper on multimode instabilities. An example of this behavior is shown in Figs. 9.1 and 9.2. If the unstable mode happens to be the nearest-neighbor of the resonant mode (as in the case in Figs. 9.1, 9.2) the transient pulsing pattern develops oscillations directly at the frequency $\bar{\alpha}_1$. If, instead, the unstable mode is futher removed from the resonant mode, the transient behavior can be more complicated. For example, in Figs. 9.3 and 9.4 we show the transient and the steady state oscillations of a laser whose unstable frequency is $\bar{\alpha}_2$. After a transient evolution where the dominant frequency is $\bar{\alpha}_1$, mode n=2 develops fully, and eventually dominates the picture.

When the intermode spacing is much smaller than the atomic linewidth, and many sidebands are unstable at the same time, the output intensity develops extremely complicated features. Figures 9.5 and 9.6a,b,c display the eigenvalues and the transient evolution for a selection of parameters such that three modes are simultaneouly unstable on each side of the resonant mode. At the end of the recorded

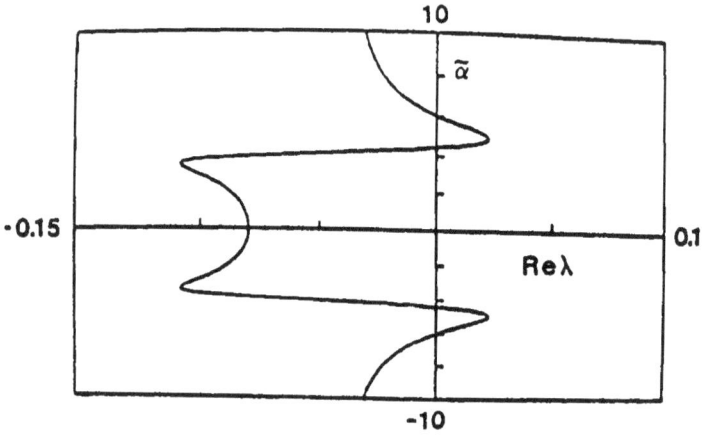

9.1   Real part of the unstable eigenvalue plotted as a function of $\bar{\alpha}_n$ viewed as a continuous variable. The parameters used in this study are $\alpha L=2.0$, $R=0.95$, $\bar{\gamma}=0.5$, $\bar{\delta}_{AC}=0$, and $\bar{\alpha}_1=5.0$. Note that $Re\bar{\lambda}>0$ for $\bar{\alpha}_n=5.0=\bar{\alpha}_1$.

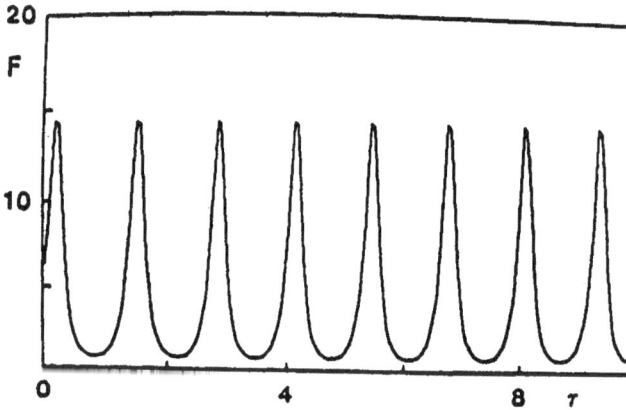

9.2   Time dependence of the output field modulus using the parameters of Fig. 9.1. The solution is displayed after all the transients have died off. The time is measured in units of $\gamma_\perp^{-1}$.

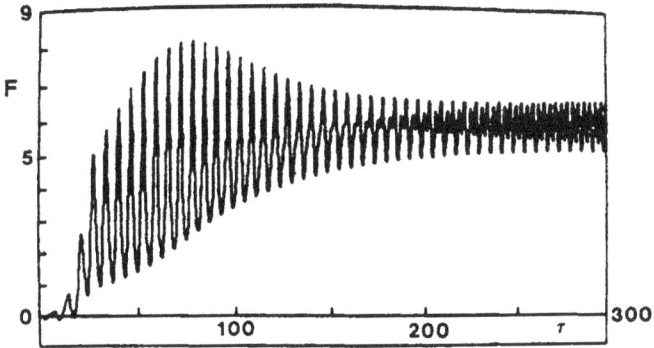

9.3   Transient evolution of the output field modulus for $\alpha L=2.0$, $R=0.95$, $\bar{\gamma}=0.08$, $\bar{\delta}_{AC}=0$ and $\bar{\delta}_1=1$. The initial state of the system corresponds to zero field, zero polarization and complete population inversion. The first unstable eigenvalue corresponds to $\bar{\alpha}_n=2.0=2\bar{\alpha}_1$. The initial pulsations have a radian frequency of the order of unity; after a while, one can see clear evidence of the onset of a higher frequency component.

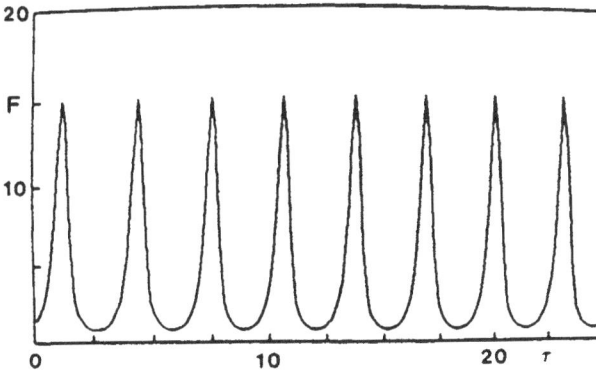

9.4   Long-time behavior of the output field modulus corresponding to the parameters of Fig. 9.3. The frequency of the pulsations is now very close to 2, in radian units (i.e., $2\bar{\alpha}_1$).

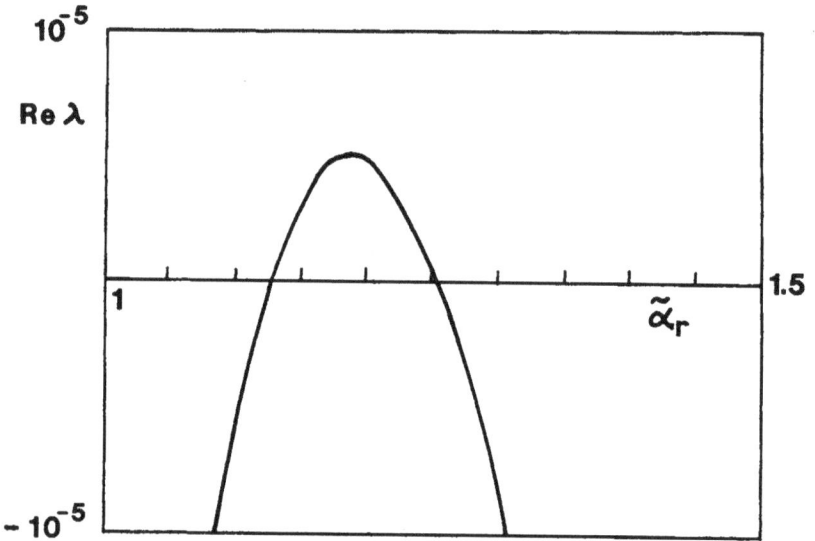

9.5   The real part of the unstable eigenvalue is plotted as a function of $\bar{\alpha}_n$ viewed as a continuous variable for $\alpha L = 0.54$, $R = 0.95$, $\bar{\gamma} = 0.1$, $\bar{\delta}_{AC} = 0$ and $\bar{\kappa} = 0.05$. The tic marks of the horizontal axis identify the position of the empty cavity modes. Three modes on either side of $\bar{\alpha}_n = 0$ are unstable according to the linearized stability analysis.

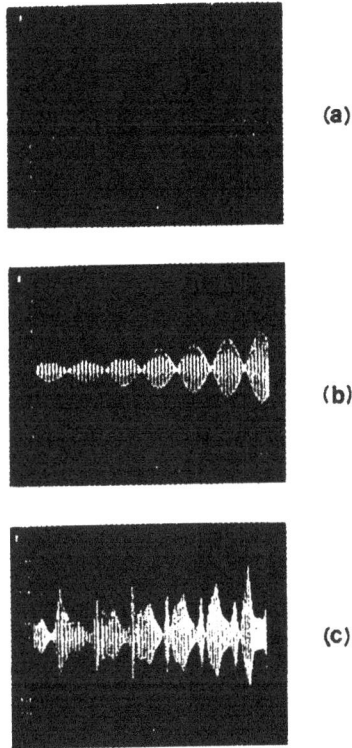

9.6   Time evolution of the output field modulus corresponding to the parameters of Fig. 9.5. The time is measured in units of $\gamma_{\perp}^{-1}$. Figures 9.6a,b,c represent snapshots of 1000 units of times, each, during the transient evolution and into steady state; the time frames are not consecutive.

evolution (Fig. 9.6c), the intensity pattern looks sufficiently complicated to suggest that many originally stable modes have actually been driven into oscillation through the mode-mode coupling mechanism that is inherent of the nonlinear equations. In fact, as we show in Section 10, approximately 20 distinct modal components can be identified in the homodyne spectrum of the output intensity of Fig. 9.6c.

Matters can become even more complicated at higher gain values because of the appearence of a chaotic attractor (Figs. 9.7, 9.8). Actually, the time-dependent output pattern does not look much more complicated than in Fig. 9.6c, but its power spectrum (See Section 10) will show the unmistakable signature of dynamical chaos.

The chaotic domain appears to be the neighbor (in some unclear way) of a very different type of attractor characterized by remarkably-looking square wave pulses. In Fig. 9.9 we show the initial behavior of a chaotic solution which, apparently under the influence of numerical noise, switches to an entirely different type of behavior (Fig. 9.10). Because in both Fig. 9.9 and 9.10 we have plotted the output field, instead of its modulus, it is clear that the average value of the field amplitude is very different in the two cases. Furthermore, the period of the square waves is consistent with the intermode spacing selected in the simulation ($\bar{\alpha}_1 = 0.05$) and is very different from the typical pulse-to-pulse separation that is typical of the chaotic transient. The significance of the square wave attractors in the context of this problem is still uncertain. We have observed, however, that the square wave solutions are strong attractors. To check this aspect of the problem we have perturbed the time-dependent solutions with the injection of numerical noise and observed the reappearance of the square wave pattern after the removal of the noisy component; in addition we have carried out an adiabatic scan for decreasing values of the gain and observed the persistence of this mode of oscillation practically as far down as the laser threshold.

Finally, we mention that sufficiently large amounts of numerical noise produced by a sequence of random numbers with a standard deviation of about 10% of the output intensity, can perturb the laser steady state, even under low gain conditions, and force the system to jump into a square wave attractor. These types of solutions are a useful reminder that the Maxwell-Bloch equations may still offer many additional surprises in future investigations.

We now turn our attention to a situation which we understand more closely: the competitive action of different modes in a homogeneously broadened laser with a low enough gain to prevent the occurence of amplitude instabilities of the Risken-Nummedal type. In this case, as we discussed in Section 7, new interesting effects arise when the detuning parameter is large enough to cause a destabilization of the phase eigenvalue. This requires that at least two steady states coexist under the gain curve of the active medium, and this, in turn, implies that the intermode spacing should be sufficiently small. In this case, according to the results of the linear stability analysis, self-pulsing or discontinuous jumps should develop.

We begin our survey with a set of parameters for which the overlapping steady states have a common instability range. Figure 9.11 shows the instability boundaries of the two only existing steady states, j=0 and j=1, in the plane (δ_{AC}, $\bar{\alpha}_n$). The selected intermode spacing is $\bar{\alpha}_n = 3$ (dashed line). The domain to the right of the solid line marked with j=0 is unstable for the steady state j=0, while the domain to the left of the second solid line is unstable for j=1. The only cavity mode that can be excited is $\bar{\alpha}_n=3$. If one imagines to carry out a detuning scan beginning with $\delta_{AC} = 0$, at first the laser

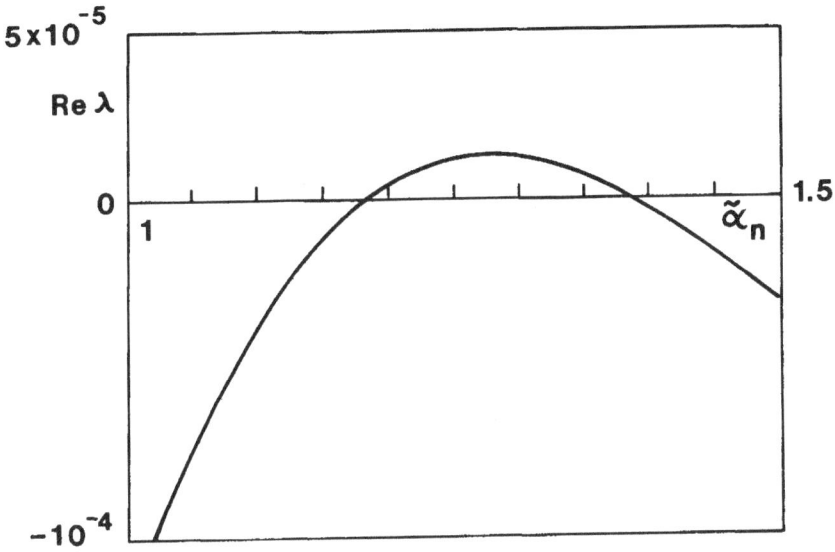

9.7   The real part of the unstable eigenvalue is plotted as a function of $\bar{\alpha}_n$ viewed as a continuous variable for $\alpha L=0.62$, $R=0.5$, $\tilde{\gamma}=0.1$, $\bar{\delta}_{AC}=0$ and $\bar{\alpha}_1=0.05$. The tic marks of the horizontal axis identify the position of the empty cavity modes. Four modes on either side of $\bar{\alpha}_n=0$ are unstable according to the linearized stability analysis.

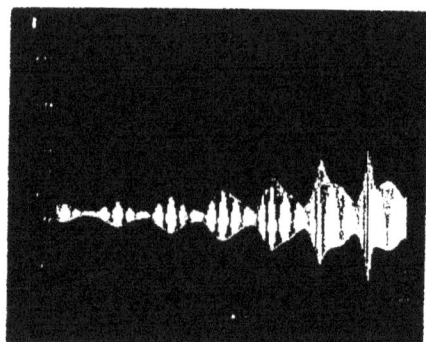

(a)

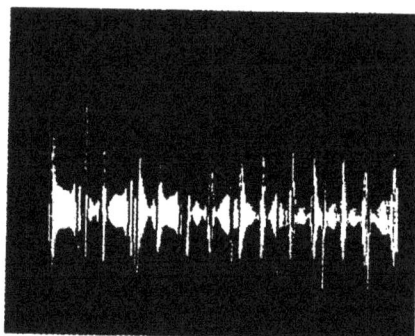

(b)

9.8   Time dependence of the modulus of the output field corresponding to the parameters of Fig 9.7. The solution in (b) is actually chaotic as confirmed by Fig. 10.13 (see accompanying discussion on power spectra in Section 10).

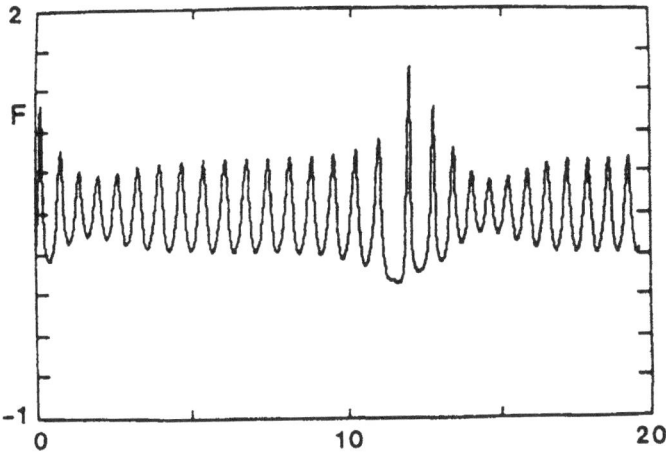

9.9 For sufficiently high values of the gain, the chaotic transient oscillations develop patterns that are reminiscent of intermittency phenomena. The parameters used in this simulation are $\alpha L=1.54$, $R=0.95$, $\bar{\gamma}=2$, $\bar{\delta}_{AC}=0$ and $\bar{\alpha}_1=0.05$. This figure displays the electric field, instead of its modulus. Note that $F(z',t')$ is real in resonance.

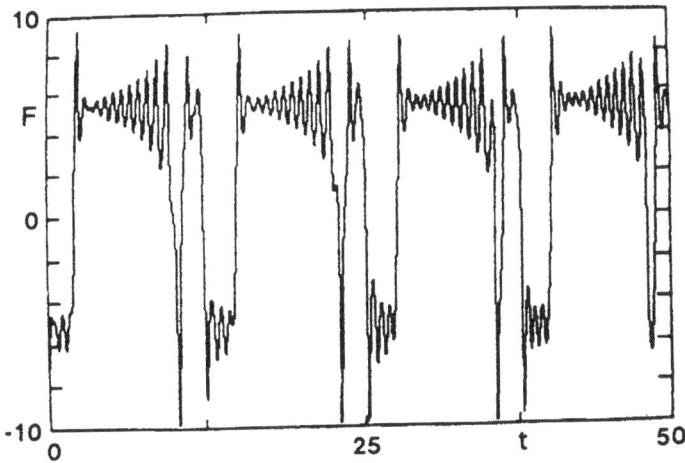

9.10 After a while, the oscillations shown in Fig. 9.9 switch over to a periodic wave form whose basic periodicity is linked to the round-trip time of light around the cavity (i.e., its radian frequency is very close to $\bar{\alpha}_1$).

98

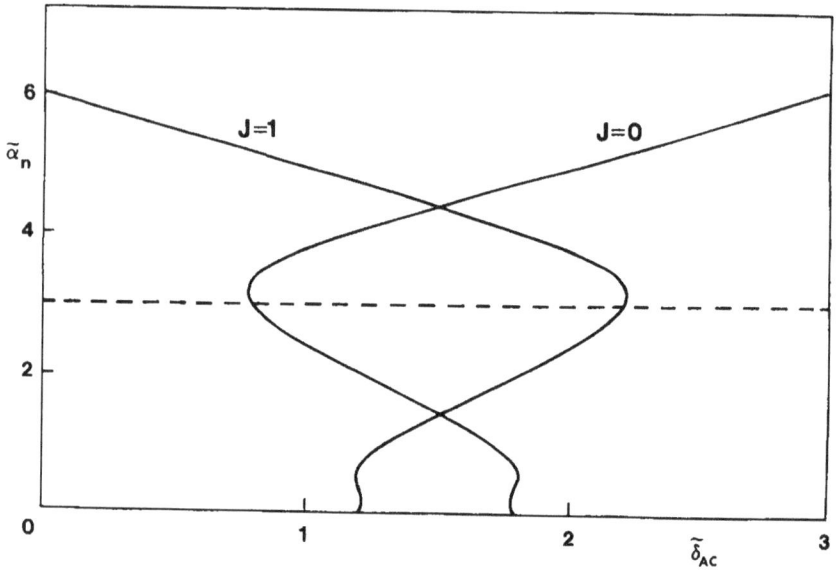

9.11 Instability boundaries for the steady states j=0 and j=1 as functions of the detuning parameter $\bar{\delta}_{AC}$. The steady state j=0 becomes unstable to the right of the solid line marked j=0. The steady state j=1 is unstable to the left of the corresponding solid line. The horizontal dashed line marks the position of the first off-resonant mode of the cavity. The parameters chosen in this simulation are αL=0.5, R=0.95, $\bar{\alpha}_1$=3.0, $\bar{\gamma}$=0.8.

operates stably in the resonant mode; at about $\bar{\delta}_{AC} = 0.8$, the steady state j=0 becomes phase-unstable. However, at this point, j=1 is also unstable so that self-pulsing is the only reasonable possibility for the long-time behavior of the system. In Fig. 9.12 we show the time evolution of the modulus of the output field beginning with an initial value of the detuning $\bar{\delta}_{AC} = 0.5$. This corresponds to a stable operating point of the j=0 steady state which is approached monotonically in this simulation.

If, on the other hand, the detuning parameter is increased to $\bar{\delta}_{AC} = 1.2$ (note that this is still less than half of the free spectral range) pulsations develop as shown in Fig. 9.13. We stress that this instability is caused by the simultaneous presence of unstable eigenvalues for the coexisting j=0 and j=1 steady states.

A different situation develops when a given steady state becomes unstable in the presence of a coexisting stable domain of another steady state. This is illustrated in Fig. 9.14 where the two domains of instability never overlap each other for the selected parameters. We stress that the vertical axis is labelled by $\bar{\alpha}_n$, viewed as a continuous variable. In fact, of course, only values of $\bar{\alpha}_n$ which are integer multiples of $\bar{\alpha}_1$ (the intermode spacing) have any physical significance. In this figure, the two horizontal dashed lines mark the position of the first two sidebands of the resonant mode, and, as in the case of Fig. 9.11, the j=0 unstable domain lies to the right of the j=0 solid line, while the j=1 unstable domain lies to the left of the corresponding boundary. It is clear that both the first and second off-resonant modes can become phase unstable for appropriate values of the detuning parameter.

In Fig. 9.15 we simulate a detuning scan by solving the Maxwell-Bloch equations for a given value of the detuning and by using the final values of the dynamical variables of a given run as the initial conditions for the next integration corresponding to a larger (or smaller) detuning value. As shown by the arrows, the output intensity decreases at first until it reaches a minimum value for $\bar{\delta}_{AC} = 1.2$. (Note that this minimum is well above zero, the operating threshold value.) At this point, an increase in detuning forces a discontinuous variation of the output intensity which then climbs monotonically, as the detuning scan continues. Upon reversing the direction of variation of $\bar{\delta}_{AC}$, the intensity undergoes a second jump at about $\bar{\delta}_{AC} = 0.8$, thus showing the presence of hysteresis and bistability. As indicated by the steady state and linear stability analysis, the jumps not only involve the intensity, but also the operating frequency of the laser. This effect is displayed in Fig. 9.16.

Of course, the steady state plots (figs. 9.15, 9.16), do not provide information on the behavior of the system during the transient from the unstable steady state to the new steady state attractor. This very interesting phase of the evolution is shown in Fig. 9.17 by plotting the real part of the field as a function of time. This solution was obtained using initial conditions which are typical of a stable configuration just before the jump, and a value of $\bar{\delta}_{AC}$ just to the right of the instability threshold. The recorded trace shows a low frequency envelope with a superposed fast modulation that grows in depth as time progresses. The low frequency component of the field corresponds to the laser oscillation in the j=0 state which is mode-pulled by the small amount $\bar{\kappa}\bar{\delta}_{AC}/(1+\bar{\kappa})$ from the resonant mode. The high frequency component is the beat note between j=0 and j=1 state. The beat note grows in strength because the system is leaving the j=0 state and moving in the direction of the j=1 state. The transient is characterized by a continuous weakening of the j=0 oscillations and a strengthening of the j=1 oscillations. Eventually, the slow frequency component in the pattern of the real part of the field disappears, and the solution becomes sinusoidal again at the higher

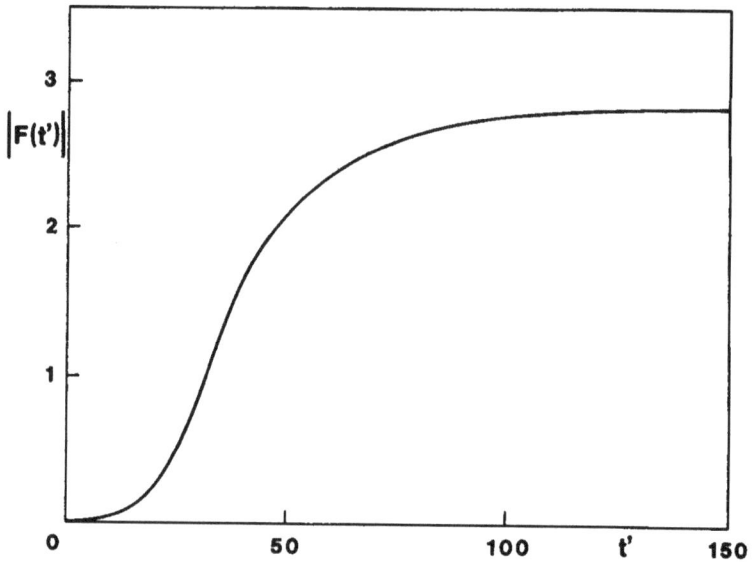

9.12   Time evolution of the modulus of the output field from a small initial value. The parameters used in this simulation are identical to those used in Fig. 9.11 with $\bar{\delta}_{AC}$=0.5. As expected, the system evolves into the stationary state j=0 and it remains in this state forever.

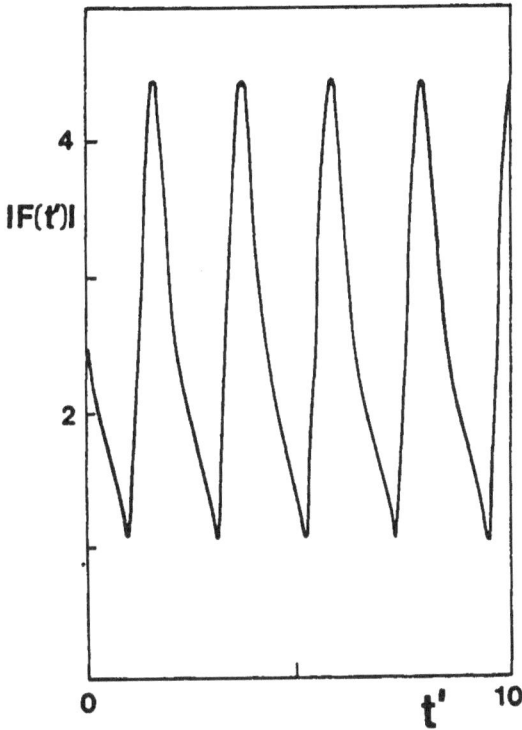

9.13 Time evolution of the modulus of the output field from a small initial value. The parameters used in this simulation are identical to those used in Fig. 9.11 with $\bar{\delta}_{AC}=1.2$. Because of the unstable nature of both phase eigenvalues for the j=0 and j=1 steady state, self-pulsing develops.

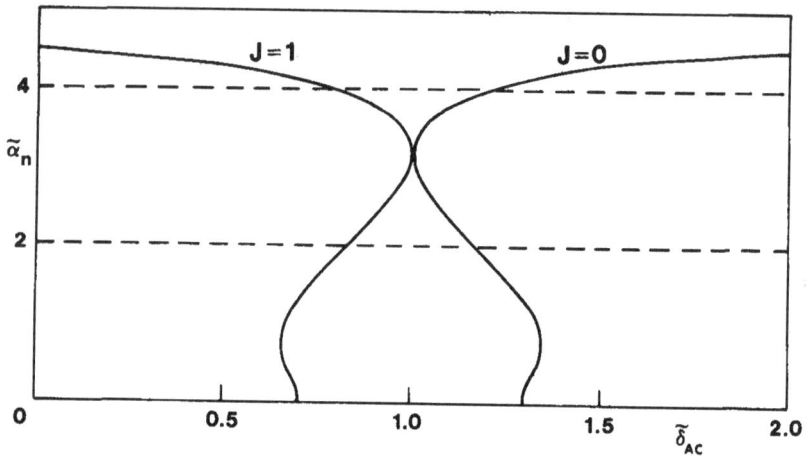

9.14   Instability boundaries for the steady states j=0 and j=1 as functions of the detuning parameter $\tilde{\delta}_{AC}$. The steady state j=0 becomes unstable to the right of the solid line marked j=0. The steady state j=1 is unstable to the left of the corresponding boundary. The horizontal dashed lines mark the first two off-resonant cavity modes. The parameters chosen for this simulation are $\alpha L=0.3$, $R=0.95$, $\tilde{\gamma}=2.0$, $\tilde{\alpha}_1=2.0$. Over the approximate range $0.85 < \tilde{\delta}_{AC} < 1.15$ both stationary states are stable, hence hysteresis and bistability are expected if one scans the detuning parameter.

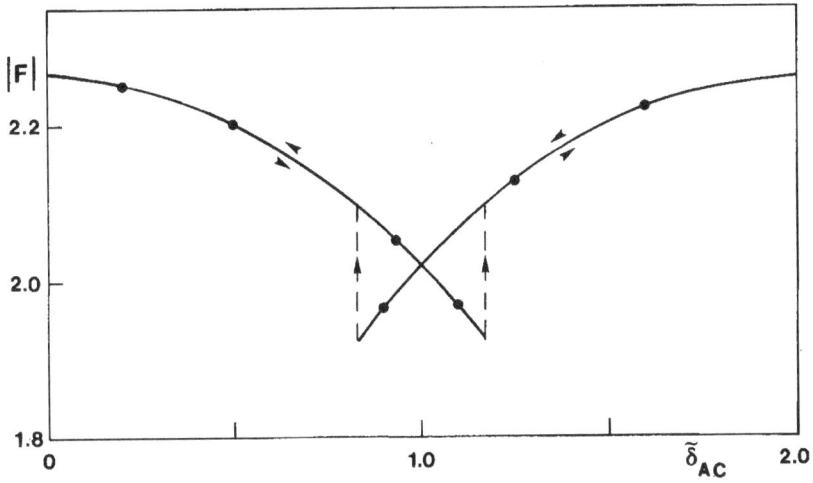

9.15 Simulated detuning scan corresponding to the parameter of Fig. 9.14. The dots represent the steady state values calculated numerically. The solid line is a plot of the modulus of the output field according to Eq. (7.18a) or of Eq. (4.10).

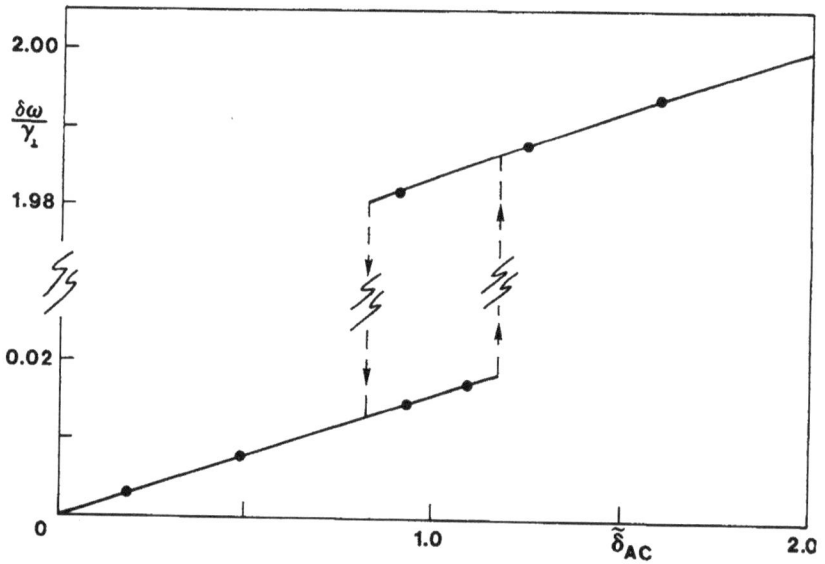

9.16   The variation of the frequency $\bar{\delta\Omega}$ [Eq. (7.20)] of the output field. The dots represents the operating frequency, measured from the selected reference frequency ω_c, and measured from the behavior of the real part of the output field. The solid line is a plot of Eq. (7.20).

frequency of the j=1 steady state.

This behavior has been tested experimentally [Tredicce, Narducci, Abraham, Bandy and Lugiato (1986)] with very good qualitative agreement with the above predictions using a CO_2 laser with more than one mode under the gain curve.

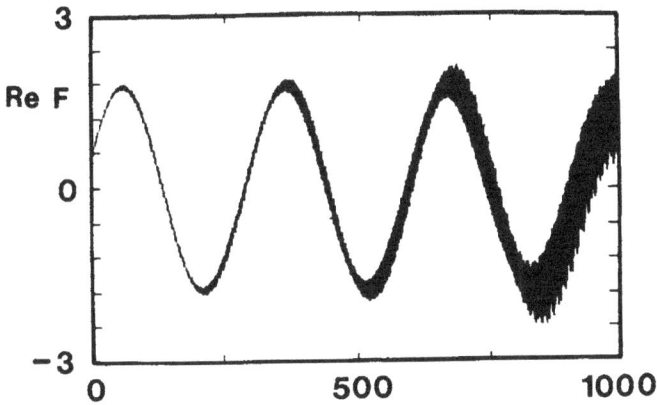

9.17 The real part of the electric field during the switching transient from one steady state to the other. The parameters of the simulation are $\alpha L=0.3$, $R=0.95$, $\bar{\gamma}=2.0$, $\bar{\alpha}_1=2.0$ and $\bar{\delta}_{AC}=1.3$. The initial conditions correspond to the stable steady state values for $\bar{\delta}_{AC}=1.1$.

10. The Spectral Analysis of the Output Field

A large amount of useful dynamical information which is not immediately available upon inspection of the time-dependent solutions can be extracted from their power spectra. In principle, of course, the time and frequency domains contain the same amount of information, in a strict mathematical sense; in practice, however, some features stand out more conspicuously in the time traces, while others are more obvious in the frequency domain. For this reason, power spectra have become a very common complementary tool to the time-dependent solutions in theoretical dynamical studies, not only in Quantum Optics but also in numerous other fields where nonlinear phenomena play a dominant role.

In fact, homodyne and heterodyne power spectral measurements have been common experimental tools for a long time. By homodyne spectrum we simply mean the power spectrum of the signal of interest (the output intensity, for example, if one is dealing with a laser). This is obtained as follows: the laser output field is detected by a square-law detector, and the resulting electrical signal, proportional to the field intensity, is fed into a spectrum analyzer which is just a tunable resonant circuit. This device responds by producing an output signal proportional to the power contained in a given narrow band of the incident signal. By varying the resonant frequency of the spectrum analyzer, one can probe the spectral content of the input signal over the required frequency range.

The numerical simulation of this measurement can be carried out with a Fast Fourier Transform routine that calculates the complex Fourier amplitudes of the time record and their square moduli. A good numerical code should be able to handle the Fourier Transform calculation accurately and efficiently; it must be fitted with a reliable "transmission function" (Fig. 10.1) whose task is to smooth out the discontinuities at the start and end of the time record; in addition it must be operated with carefully chosen temporal data sets and parameters to insure sufficient spectral resolution and the removal of aliasing problems.

Some general guidelines can be stated easily. By direct inspection of the time record, one can estimate the duration of the typical oscillations. Let τ_p denote the estimated characteristic time; a good rule of thumb is to sample the solution at such a rate as to collect between 6 and 10 points during a time τ_p, as sketched in Fig. 10.2. Often, the elementary step size adopted for the numerical solution of the differential equations is much smaller than τ_p/N_s, where N_s is the number of points collected in an interval τ_p. In this case, it is not good practice to retain all the calculated points in the time record, because this would lead to oversampling of the signal (i.e., accumulation of redundant information). Thus, for example, if the integration time is 10 times shorter than the interval τ_p/N_s, the recommended practice is to store only one point out of every 10 for the subsequent spectral processing.

At this point, one must select the appropriate length of the time series; the maximum size of the temporal record depends, of course, on the amount of computer memory that is available, on the amount of time that one is willing to wait for the completion of the calculation, and on the resolution that is required of the calculated spectrum. In general, it is a good idea to store as many points as possible in the time record to be transformed because the spectral resolution is directly related to the length of the record. The main limitation is that Fast Fourier Transforms usually

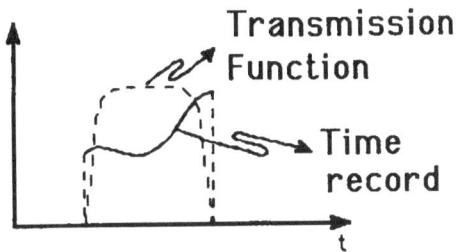

10.1   Schematic representation of a truncated time record and the smoothing transmission function.

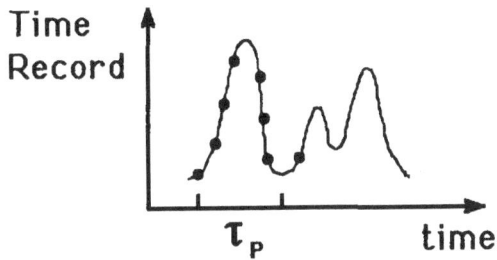

10.2   Illustration of the sampling procedure discussed in the text.

108

operate on time records containing a number of entries that is an exact power of 2 (e.g., 512, 1024, 2048, etc.), so that if N data points are barely adequate for purposes of resolution, it is not at all obvious that a computer will be able to process twice as many points. From our experience, records containing 2048 or 4096 data points are adequate for most studies of the Maxwell-Bloch problem, and only seldom have we had to use larger arrays.

If $(a_1, a_2,...., a_n)$ represents the string of selected values in the temporal record, the power spectrum consists of an equal number of spectral values $s_1, s_2,...., s_n$ However, only half of these are significant because the power spectrum produced by a Fast Fourier Transform routine consists of two symmetric replicas, with the line of symmetry placed in the middle of the free spectral range (Fig. 10.3). For this reason, it is important that all the relevant spectral feature under study be contained in the range (0, $\omega_{1/2}$). If this is not the case, the tails of the two spectral copies spill over the important portion of the frequency domain causing a great deal of confusion. The radian frequency is related to the index i of the spectral datum according to the formula:

$$\omega_i = 2\pi \frac{i}{N \Delta t} \tag{10.1}$$

where Δt is the time interval between two consecutive data points stored in the time record. As a result the important frequency $\omega_{1/2}$ is given by

$$\omega_{1/2} = \frac{2\pi (N/2)}{N \Delta t} = \frac{\pi}{\Delta t}$$

One must be sure that the maximum useful frequency in the spectral range of interest is sufficiently smaller than $\omega_{1/2}$. This is normally true if one employs typical records of 2048 or 4096 data points with a collection rates of about 6 to10 points for each major oscillation.

A very useful complement to the Fast Fourier Transform program is a routine that identifies automatically the central frequency of the individual spectral components. A convenient practical way to accomplish this task is based on the following procedure. After displaying the entire spectrum of interest on a monitor (or, less conveniently, on a XY-plotter) one assigns a discrimination line selected in such a way that all the peaks of interest are well separated from one another (Fig. 10.4). One can now scan the entire set of spectral values in the spectrum file and assign all the consecutive data point above the discrimination line to a given peak. In this way, the central frequency of each can be calculated as follows:

$$\omega_{peak} = \frac{\sum_{i_1}^{i_2} \omega_i s_i}{\sum_{i_1}^{i_2} s_i} \tag{10.2}$$

where i_1 and i_2 are the spectral indices corresponding to the beginning and to the end of the selected feature. Using Fig. 10.4 as a reference, it is clear that the lower selection of the discrimination line is not a good choice because it mixes together two different spectral lines. This problem does not arise, instead, if one selects the upper

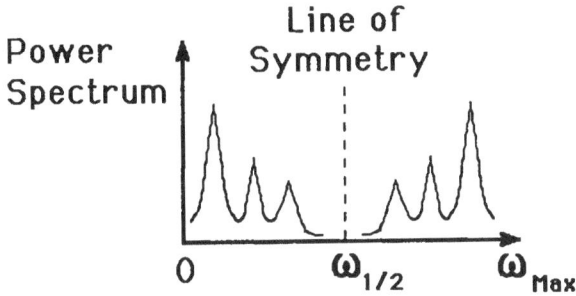

10.3   Schematic representation of a power spectrum. The data in the frequency domain form two identical replicas symmetrically placed on either side of one half of the free-spectral range.

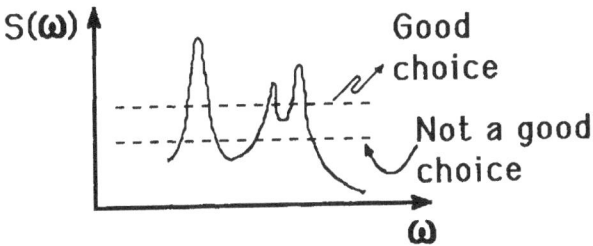

10.4   A proper choice of a discrimination level allows a correct processing of the spectral information. The top dashed line in the figure results in the automatic identification of three separate spectral components.

discrimination lines.

Homodyne power spectra are excellent diagnostic tools for the observation of subharmonic bifurcations, and the appearance of incommensurate frequencies. On the other hand, they are not immune from potential pitfalls. The most common is their inability to distinguish between symmetry breaking transitions and period doubling bifurcations.

A symmetry breaking transition occurs when the field envelope F changes from a symmetric oscillation pattern, of the type shown in Fig. 10.5a, to the asymmetric pattern shown in Fig. 10.5b. Usually this change is not accompanied by a significant variation of the basic period of oscillation, but only by a break of symmetry between the positive and negative excursions of the oscillating pattern. Because a square law detector is only sensitive to the field intensity and not to its envelope, the symmetry breaking transformation (a) → (b) can be easily confused with an ordinary period doubling bifurcation where the basic periodicity of an intensity pattern doubles an in Figs. 10c,d.

In principle, one might be able to tell the difference between the changes of pattern (a) → (b) and (c) → (d) because the time-dependent intensity in (a) and (b) vanishes at regular intervals while it does not in the latter case. The spectrum by itself, however, is not sensitive to the shape of the intensity profile, so that in both cases one observes the appearance of subharmonic frequencies, and may be led to the interpretation of both changes of patterns as period doubling bifurcations.

Obviously, what we need is an observation mechanism that is sensitive to the envelope of the variable of interest and not only to its square modulus. The heterodyne approach is an ideal technique, for this purpose. With reference to Figs. 10.5a and 10.5b, the main difference between the two envelopes is that the first has a zero average (i.e., no bias level), while the average of the second envelope is different from zero. It follows that a system capable of responding to the spectrum of the envelope will display a zero frequency component in case (b) but not in case (a). Thus, the emergence of a zero frequency component together with subharmonic spectral lines is an immediate indication that a symmetry breaking transformation has taken place. In cases 10.5c and 10.5d the intensity pattern is always different from zero; this implies that the field envelope will also have a non zero bias, so that the transformation (c) → (d) will display subharmonic spectral components in the presence of a zero frequency line in the heterodyne spectrum.

In principle, the heterodyne spectrum of an electromagnetic signal can be constructed by superimposing the field of interest with a reference signal having a fixed amplitude and carrier frequency. In practice, the design of a stable reference oscillator is not an easy matter from an instrumental point of view; or course, this is not a problem numerically. We denote by

$$\bar{E}_0(t) = E_0(t)\, e^{-i\omega_c t} + E_0^*(t)\, e^{i\omega_c t} \qquad (10.3)$$

and

$$\bar{E}_R(t) = A(t)\, e^{-i\omega_R t} + A^*(t)\, e^{i\omega_R t} \qquad (10.4)$$

the optical laser field and the reference signal, respectively. $E_0(t)$ is the slowly varying complex amplitude of the signal, ω_c its carrier frequency and ω_R is the frequency of the

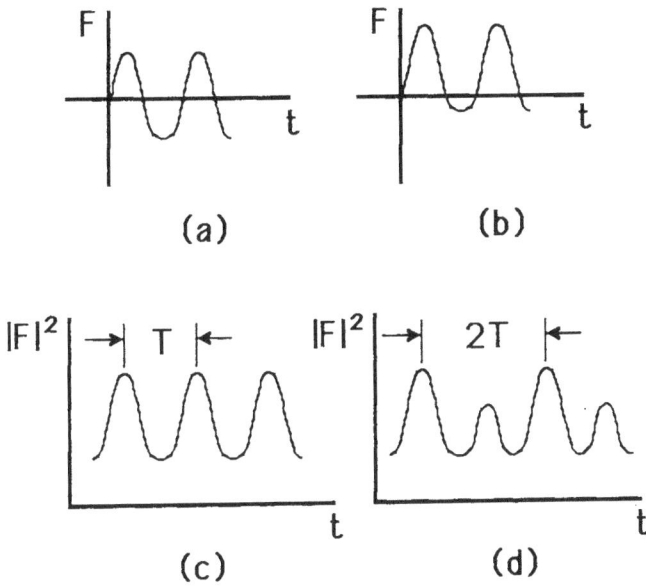

10.5   Schematic representation of the symmetry breaking process and of a real
period doubling bifurcation, as discussed in the text. Curves (a) and (b) illustrates a
symmetric and an asymmetric solution, respectively. Curves (c) and (d) illustrate
time-dependent intensity outputs which undergo a real period doubling bifurcation.

reference signal. The total intensity at the detector is given by

$$I(t) = 2\left\{|A|^2 + |E_0|^2 + A^*E_0(t)\, e^{-i(\omega_c - \omega_R)t} + AE_0^*(t)\, e^{i(\omega_c - \omega_R)t}\right\}$$ (10.5)

The power spectrum of the real time-dependent function I(t) consists of three distinct contributions: (i) a δ-function centered at zero frequency, (ii) the homodyne spectrum of the signal, and (iii) the spectrum that results from the beating of the signal and the reference wave (heterodyne component). Part (iii) of the spectrum is centered at the difference between the two carrier frequencies ω_c and ω_R, and its magnitude is proportional to the amplitude of the reference field. It is clear that if $(\omega_c - \omega_R)$ is sufficiently large and if the reference field is strong enough, the heterodyne component will be well removed from the heterodyne part of the spectrum and will be easily detectable.

In a practical numerical calculation it is preferable to subtract the quantity $2(|A|^2 + |E_0(t)|^2)$ from the total intensity to eliminate a possible overlap between the spectral components of interest and the homodyne contribution. A word of caution is needed with regard to the selection of the frequency difference $(\omega_c - \omega_R)$ which should not be too small if we want to avoid crowding of the spectrum near the origin of the frequency axis. At the same time this frequency difference must not be too large for the following reason: the time-dependent solution E(t) of the Maxwell-Bloch equations is represented by a string of complex numbers calculated at the discrete times $t_1, t_2,...,$ spaced by a constant interval δt. Following the rule of thumb advanced in our discussion of the homodyne spectrum, we anticipate that a reasonable sampling rate of the beat signal should involve about 6 to 10 points for every period $T = 2\pi/(\omega_c - \omega_R)$; this implies that the optimal sampling interval $(\delta t)_{opt}$ should be of the order of 0.1T. Thus, the frequency detuning should be selected with these two requirements in mind: (i) adequate separation of the spectral lines from the origin of the frequency axis, (ii) accurate sampling of the oscillating beat signal.

A good example of the type of problems that can arise with a careless use of the homodyne spectrum is given by the following analysis taken from the periodic domain of solutions of the single Lorenz model (Eqs. (8.12)). Figure 10.6a shows the time behavior of the output field amplitude for a periodic solution corresponding to as small value of $\bar{\gamma}$ and a gain value above the self-pulsing threshold. The corresponding homodyne spectrum of the intensity is shown in Fig. 10.6b and consists of a series of equispaced spectral lines with frequencies 1.09, 2.18, 3.27,... etc. The large number of harmonic components are expected from the sharp nature of the pulsation.

A larger value of $\bar{\gamma}$, for the same value of the gain, brings a qualitative change in the solution in the sense that the positive peaks are now smaller in size than the negative peaks (Fig. 10.7a). The square of this envelope differs from the square of the field shown in Fig. 10.6a because consecutive peaks alternate in size. The homodyne intensity spectrum of this solution (Fig. 10.7b) shows fundamental peaks at radian frequencies 1.194, 2.388,... etc. and subharmonic components at odd half-multiples of the fundamental (i.e., 0.596, 1.791, etc.). An uncritical look at the homodyne spectra of Fig. 10.6b and 10.7b without knowledge of the behavior of the time-dependent solution leads to the conclusion that the system has undergone a period doubling bifurcation. This is wrong, of course, because the field pattern has undergone, instead, a symmetry breaking transformation.

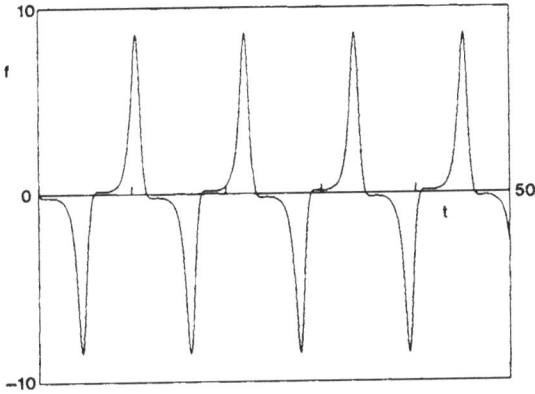

a

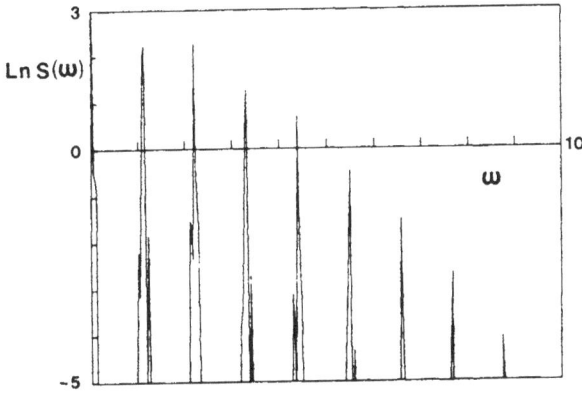

b

10.6   (a) Time evolution of the output field amplitude of the single mode laser model [Eqs. (8.12)] for C=6.0, $\bar{\kappa}$=4.0 and $\bar{\gamma}$=0.14. This is an example of a symmetric solution in the sense that the positive and negative peaks of the oscillating field are of equal height, and the time-averaged field is zero. (b) Intensity power spectrum corresponding to the time dependent solution shown in (a). The vertical axis displays the natural logarithm of the power spectrum. The horizontal axis is labelled by the radian frequency. A spectrum of this type is recognizable at once as typical of a simple periodic solution.

114

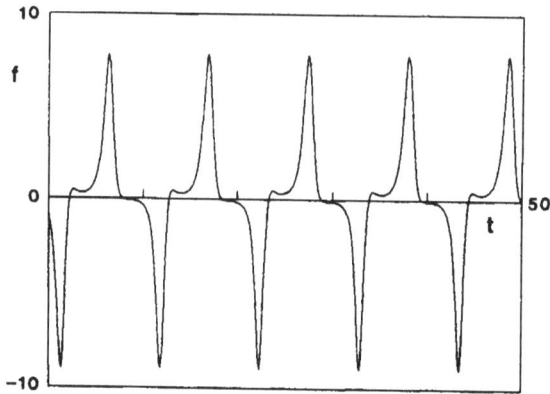

a

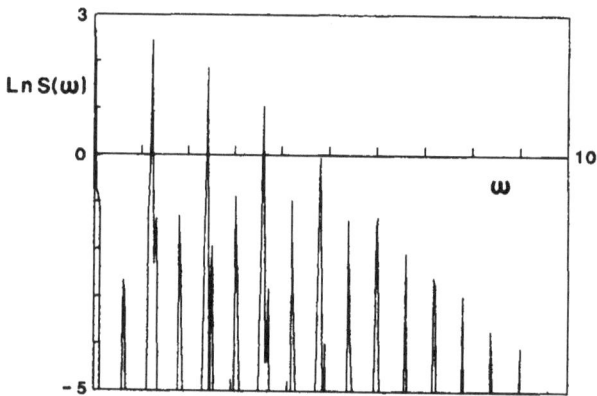

b

10.7 (a) Time evolution of the output field amplitude of the single mode laser model [Eqs. (8.12)] for $C=6.0$, $\bar{\kappa}=4.0$ and $\bar{\gamma}=0.17$. This is an example of an asymmetric solution in the sense that the positive and negative excursions of the oscillating field output are different in magnitude. In this case the time-averaged field is different from zero. (b) Intensity power spectrum corresponding to the time dependent solution shown in (a). The vertical axis displays the natural logarithm of the power spectrum. The horizontal axis is labelled by the radian frequency. The spectrum consists of regularly spaced fundamental and harmonic lines, alternating smaller peaks at the subharmonic component, and various combination tones. On the surface, this spectrum resembles the typical spectrum of a signal that has undergone a period doubling bifurcation.

The heterodyne spectra of the time dependent solutions in Figs. 10.6a and 10.7a show a clear clue of the nature of the change undergone by the temporal evolution. In Fig. 10.8 we show the heterodyne spectrum of the solution shown in Fig. 10.6a using a reference amplitude A = 100 and a frequency difference $(\omega_c - \omega_R) = 10\gamma_\perp$. The center of the spectrum is located at a radian frequency of 10, in units of $\gamma_\perp$, as expected, and shows no evidence of a component at this frequency. On the contrary, the heterodyne spectrum of Fig. 10.7a shown in Fig. 10.9 is also centered at a radian frequency of 10, but shows clear evidence of a component at this frequency. We stress that this component is a reflection of the presence of a nonzero bias value of the field envelope, and therefore a direct indication of the asymmetric nature of the solution.

Another clear example of the usefulness of the heterodyne spectrum for easy identification of the main frequency components that appear in a temporal record is taken from our study of the phase instability in multimode homogeneous lasers. With reference to Fig. 9.13 we recall that this oscillation pattern was induced by the simultaneous presence of two coexisting steady states. In this case, we have constructed the heterodyne spectrum by mixing the calculated field envelope with a local oscillator of amplitude A =100, whose carrier frequency ω_R is removed from the signal carrier frequency ω_c by $10\gamma_\perp$. The spectrum (Fig. 10.10) shows the presence of two fundamental oscillation frequencies $\omega^{(0)} = 9.986$ and $\omega^{(1)} = 7.005$, both measured in units of $\gamma_\perp$, which are easily traced to the frequencies associated with the unstable states j=0 and j=1, respectively. The remaining peaks are combination tones and harmonic components related to the nonsinusoidal nature of the time-dependent solution.

We conclude this section with three examples of homodyne spectra corresponding to the emergence of an amplitude instability of the Risken-Nummedal type. With reference to Fig. 9.4, we recall that the instability responsible for these pulsations was associated with the appearance of a positive real part in one of the eigenvalue of the mode n=2. Because the intermode spacing in this case was $\bar{\alpha}_1 = 1$, the oscillations should develop with a frequency of the order $\bar{\alpha}_2 = 2\bar{\alpha}_1 = 2$. In the absence of other neighboring modes that can be made unstable in the course of the nonlinear evolution, one expects the long term oscillations to display this fundamental frequency. This is demonstrated nicely in Fig. 10.11, where the fundamental radian frequency is very close to 2 (in units of $\gamma_\perp$). The remaining lines in this figure are higher harmonic of the nonsinusoidal oscillation.

When the mode spacing is small, more than one mode can become unstable for sufficiently large values of the gain. Under these conditions, it is also very easy for additional neighboring stable modes to be excited in the course of the nonlinear evolution. This is the main reason why a relatively simple-looking pattern such as shown in Fig. 9.6a can become as complicated as the trace shown in Fig. 9.6c. Not surprisingly, the homodyne spectrum is correspondingly complicated, while still retaining a highly ordered structure. Figure 10.12 shows the appearance of numerous evenly spaced lines (the spacing corresponds to $\bar{\alpha}_1$=0.05, the selected intermode spacing) that fall in a clean band. The second and third harmonics of the fundamental band are also present, in addition to a number of low frequency lines that represent combination and beat frequencies. At higher values of the gain, deterministic chaos makes its entrance. This is not readily visible from Fig. 9.8, although, in fact, the pattern looks more complicated than that of Fig. 9.6c. The homodyne spectrum shown in Fig. 10.13, on the other hand, is entirely different from that shown in Fig. 10.12. The nice regularity of the band structure has disappeared, to be replaced by the typical

116

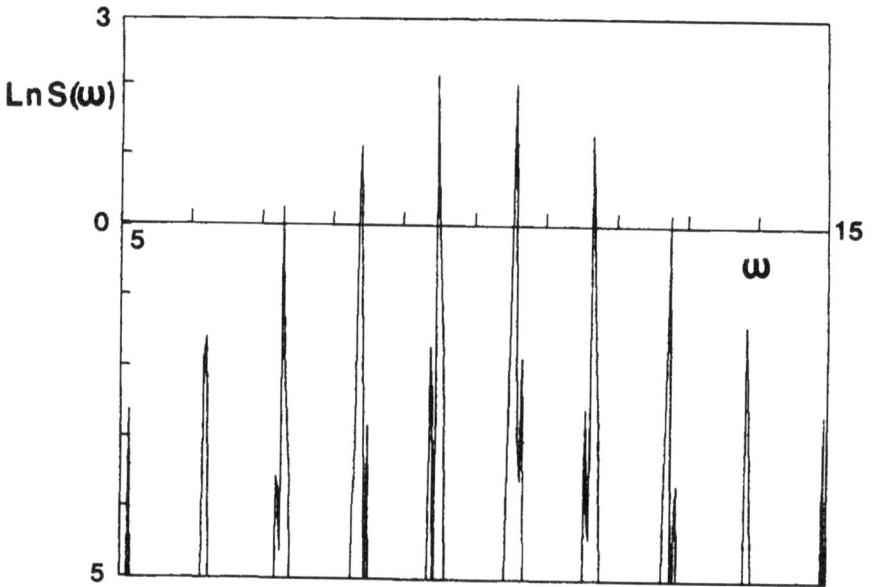

10.8   Heterodyne spectrum of the time-dependent solution shown in Fig. 10.6a. The reference amplitude is A=100 and the frequency difference, in units of $\gamma_\perp$, is 10. Note the absence of a spectral component at frequency 10.

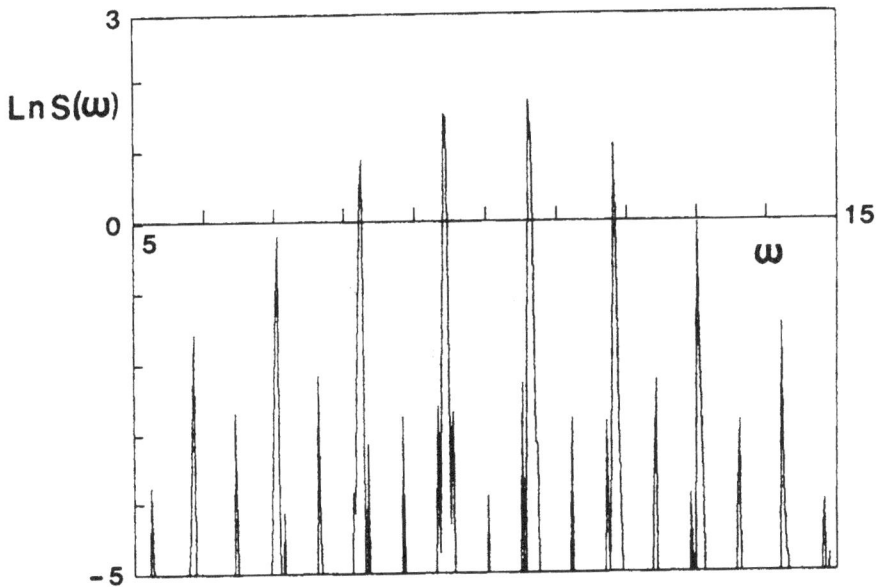

10.9   Heterodyne spectrum of the time-dependent solution shown in Fig. 10.7a. The reference amplitude is A=100 and the frequency difference, in units of $\gamma_\perp$, is 10. Note the appearance of a spectral component at frequency 10.

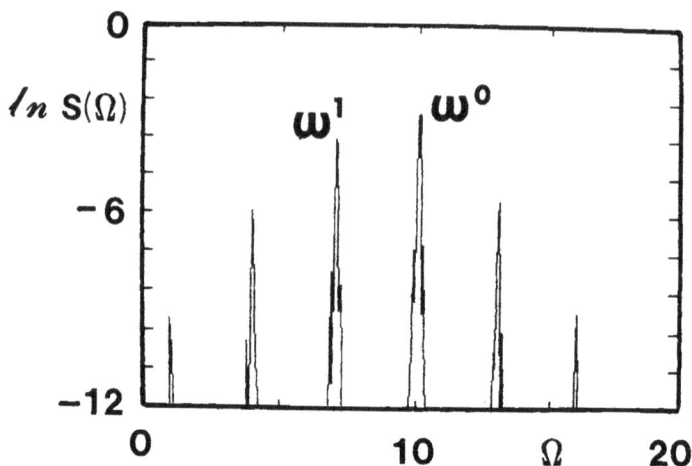

10.10 Heterodyne spectrum of the time-dependent solution Fig. 9.13 mixed with a local oscillator whose optical carrier is removed from ω_c by an amount $10\,\gamma_\perp$. The spectrum shows the presence of two fundamental oscillation frequencies at $\omega^{(0)}=$ 9.986 and $\omega^{(1)}=7.005$ which are easily traced to the natural oscillation frequencies associated with the unstable states $j=0$ and $j=1$.

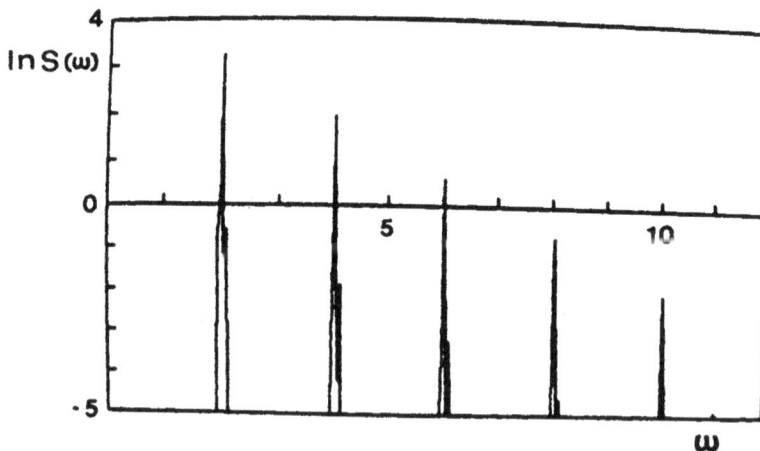

10.11 Homodyne power spectrum of the time dependent solution shown in Fig. 9.4. The fundamental oscillation frequency is $2\,\gamma_\perp$.

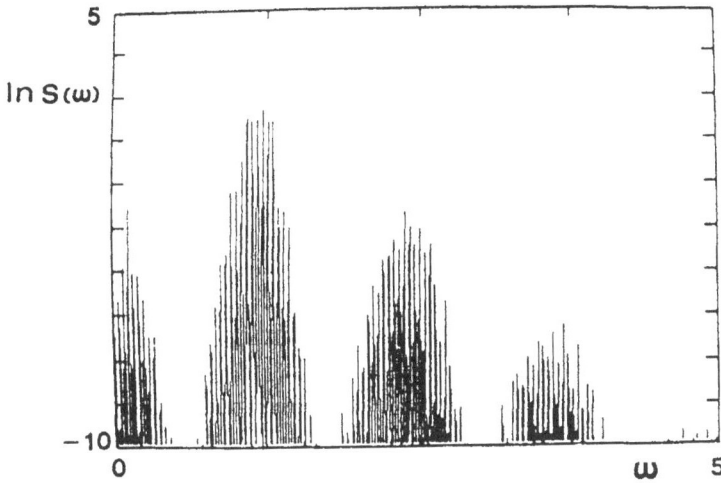

10.12 Homodyne power spectrum of the time dependent solution shown in Fig. 9.6c. The individual spectral component are spaced by a full free spectral range which in this simulation is $\bar{\alpha}_1 = 0.05$.

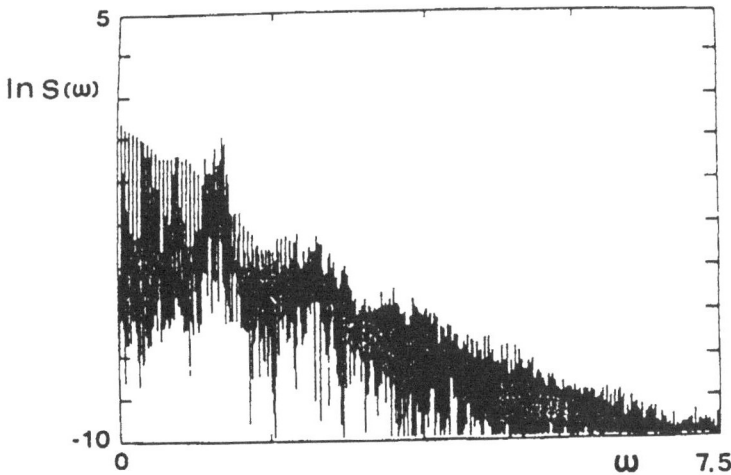

10.13 Homodyne power spectrum of the temporal solution shown in Fig. 9.8. Note the broadband structure of the spectrum which is characteristic of deterministic chaos. The low frequency evenly-spaced peaks are separated by a distance of $\bar{\alpha}_1 = 0.05$ from one another.

continuous structure of chaotic spectra. Note that sharp individual components, at a spacing of $\overline{\alpha}_1 = 0.05$, are still observable, but the entire pattern is definitely typical of the continous, broad featureless spectrum which one observes regularly in the presence of deterministic chaos.

11. Theory of a Laser with Inhomogeneous Broadening

The assumption that the active atoms are stationary inside the laser cavity and that their gain profile is homogeneously broadened is only a very rough approximation for many lasers. CO_2 and Far Infrared Lasers are somewhat exceptional [Siegman (1971), Svelto (1982)] because, even at moderate pressures, their gain lines are homogeneously broadened by collisional effects, and the theoretical description developed in the previous chapters applies to a reasonable approximation. In general, however, gain profiles are inhomogeneously broadened to the point that major corrections are needed in our theoretical description. In this section we discuss how we can modify our formalism to make it more compatible with the physics of a typical gas laser [Lamb (1964), Sargent, Scully and Lamb (1979), Stenholm (1971)].

First of all, we observe that moving atoms look spectroscopically different from one another to both an external observer and to the cavity field because of the Doppler effect. If the density of the gas is sufficiently low that collisions can be neglected, the frequency distribution of the emitted flourescence has a Gaussian profile with a center at the transition frequency of the stationary atom, and with a spread that is controlled by to the atomic or molecular mass and by the temperature of the medium. At higher gaseous concentrations, collisions begin to play a role and the low-density velocity distribution is usually altered in a significant way; a useful model of the spectral profile, in this case, is given by the convolution of the low-pressure Gaussian line and of a Lorentzian function simulating the effect of the collisions (the so-called Voight profile).

At higher pressures, the active atoms may suffer so many collisions and, therefore, so many velocity changes per unit time, that they become again approximately identical to one another, in a statistical sense. In this case the spectral profile of the gain line can be characterized again by a homogeneous distribution, as if the atoms were actually stationary. Thus, by controlling the pressure of the active atoms, or by the addition of various amounts of buffer gas, one may be able to transform a fully inhomogeneous (Doppler broadened) line into nearly fully homogeneous one. This is not always possible, or desirable, in a laser system because the broadening of a gain line is accompanied by a decrease of the small signal gain, so that gas lasers may not be able to operate at all under high pressure conditions; CO_2 and Xe lasers are examples of devices where the character of the gain profile can be controlled practically from one end of the spectrum to the other.

An important point is that, fundamentally, homogeneously and inhomogeneously broadened lineshapes are not really different from one another. In the former case the atoms are spectroscopically identical, at least in a statistical sense; in the latter, selected elements of the ensemble differ from each other because their resonant frequencies are different. Thus, in the latter case the active medium resembles a collection of slightly different homogeneously broadened radiators. At low pressure, if collision effects can be ignored, the situation is fairly simple because the spectral profile of the radiating ensemble mimics the well known Maxwell velocity distribution (the frequency shift of each resonant line is directly proportional to the velocity of the radiator). Thus, if $g(\delta)$ denotes the probability density that an atom emits (or absorbs) with a frequency shift δ relative to the center of the line, it is easy to verify that the functional form of $g(\delta)$ is given by

$$g(\delta) = \frac{1}{\sqrt{2\pi\sigma_D^2}} \, e^{-\frac{\delta^2}{2\sigma_D^2}}$$ (11.1)

where σ_D is the Doppler linewidth. This spectral profile is normalized to unity in the sense that

$$\int_{-\infty}^{+\infty} d\delta \, g(\delta) = 1$$

Note the δ measures the frequency shift from the center of the line so that its value, in principle, ranges from $-\infty$ to $+\infty$. The actual half-width at half-height, $\Delta\omega_D$, of the Doppler broadened spectrum is related to σ_D by

$$\Delta\omega_D = \sqrt{2 \ln 2} \; \sigma_D$$ (11.2)

Its dependence on the mass M of the radiator and the transition frequency ω_A is given by

$$\Delta\omega_D = \omega_A \sqrt{\frac{2kT}{Mc^2} \ln 2}$$ (11.3)

where k is Boltzmann's constant and T the absolute temperature of the gas. Two facts are immediately obvious; infrared lasers have a smaller absolute Doppler width than their visible or UV counterparts, and the temperature of the medium can effect (mildly) the width of the Doppler distribution. For the case of the 6328Å neon transition which is used in the He-Ne laser, and taking the temperature of the gas as 300 °K, the half width at half maximum of the Doppler distribution is, approximately, $\Delta\nu_D = \Delta\omega_D/2\pi$ =1.5x10⁹ Hz. The 3.39μm line of the same laser, instead, has a linewidth which is about five times narrower.

Now we turn our attention to the description of a ring laser with an inhomogeneously broadened line. The starting point is the usual set of Maxwell-Bloch equations for the slowly varying amplitudes of the field and of the atomic variables. Here, in contrast with the homogeneous case, the active atoms have different resonant frequencies because of their random motion. Thus, the total polarization is the sum of the individual polarization components over the entire line profile. A precise treatment should take its premises from the approach pioneered by Lamb in 1964 who labelled each atom with a random translational velocity to be averaged at the end of the calculation. This is, in fact, the traditional way to handle gas lasers in a precise way [In connection with instability studies, see for example the detailed calculations by Casperson (1978), (1985)]. Here, for simplicity, we model our system as a collection of stationary radiators with a distribution of homogeneous lines at different resonant frequencies. The situation can be illustrated schematically as shown in Fig. 11.1.

With reference to the cavity mode with frequency ω_c, the equations of motion are

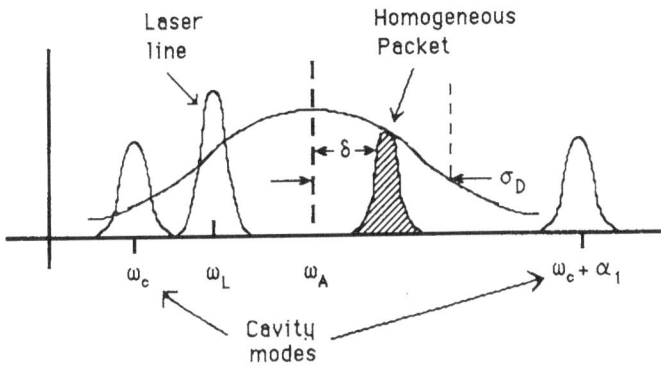

11.1   Schematic representation of the main components of an inhomogeneously broadened laser. The lines at frequencies ω_c and $\omega_c + \alpha_1$ represent two adjacent cavity resonances; the peak at frequency ω_L is the operating laser line; the broad line with a center frequency ω_A is the inhomogeneously broadened gain profile of the active medium. The dashed component of this line is a homogeneous packet.

$$\frac{\partial \overline{F}}{\partial z} + \frac{1}{c} \frac{\partial \overline{F}}{\partial t} = -\alpha \int_{-\infty}^{+\infty} d\overline{\delta}\, g(\overline{\delta})\, \overline{P}(\overline{\delta},z,t) \tag{11.4a}$$

$$\frac{\partial}{\partial t} \overline{P}(\overline{\delta},z,t) = -\gamma_\perp \left\{ \overline{F}\overline{D} + (1 + i\,(\overline{\delta}_{AC} + \overline{\delta}))\overline{P} \right\} \tag{11.4b}$$

$$\frac{\partial}{\partial t} \overline{D}(\overline{\delta},z,t) = -\gamma_\parallel \left\{ -\frac{1}{2} (\overline{F}^*\overline{P} + \overline{F}\,\overline{P}^*) + \overline{D} - 1 \right\} \tag{11.4c}$$

where $\overline{\delta}$ denotes the running variable $\delta/\gamma_\perp$.

These equations are not as simple to analyze as they may appear at first sight because they involve a set of atomic equations for $\overline{P}$, $\overline{P}^*$ and $\overline{D}$ for each of the selected velocity groups into which we have decomposed the Doppler line (in principle, of course, one is dealing with a continuum of velocity groups). The appropriate boundary conditions for the ring cavity are given by the usual equation

$$\overline{F}(0,t) = R\,\overline{F}(L, t - \frac{\Lambda - L}{c}) \tag{11.4d}$$

We consider first the steady state of this system. For this purpose we let

$$\overline{F}(z,t) = \overline{F}_{st}(z)\, e^{-i\delta\Omega t} \tag{11.5a}$$

$$\overline{P}(\overline{\delta},z,t) = \overline{P}_{st}(\overline{\delta},z)\, e^{-i\delta\Omega t} \tag{11.5b}$$

$$\overline{D}(\overline{\delta},z,t) = \overline{D}_{st}(\overline{\delta},z) \tag{11.5c}$$

where $\delta\Omega$ is the unknown frequency offset between the operating laser line and the reference cavity mode.

The stationary values of the atomic variables are

$$\overline{P}_{st}(\overline{\delta},z) = -\overline{F}_{st}(z)\, \frac{1 - i\,(\overline{\delta}_{AC} - \overline{\delta\Omega} + \overline{\delta})}{1 + (\overline{\delta}_{AC} - \overline{\delta\Omega} + \overline{\delta})^2 + |\overline{F}_{st}|^2} \tag{11.6a}$$

$$\overline{D}_{st}(\overline{\delta},z) = \frac{1 + (\overline{\delta}_{AC} - \overline{\delta\Omega} + \overline{\delta})^2}{1 + (\overline{\delta}_{AC} - \overline{\delta\Omega} + \overline{\delta})^2 + |\overline{F}_{st}|^2} \tag{11.6b}$$

while the field modulus and phase, according to the usual decomposition, $\overline{F}_{st}(z) = \rho e^{i\theta}$, are solutions of the stationary equations

$$\frac{d\rho}{dz} = \alpha \int_{-\infty}^{+\infty} d\overline{\delta}\, g(\overline{\delta})\, \frac{\rho}{1 + (\overline{\delta}_{AC} - \overline{\delta\Omega} + \overline{\delta})^2 + \rho^2} \tag{11.7a}$$

$$\frac{d\theta}{dz} = \frac{\delta\Omega}{c} - \alpha \int_{-\infty}^{+\infty} d\overline{\delta}\, g(\overline{\delta}) \, \frac{\overline{\delta}_{AC} - \overline{\delta\Omega} + \overline{\delta}}{1 + (\overline{\delta}_{AC} - \overline{\delta\Omega} + \overline{\delta})^2 + \rho^2} \qquad (11.7b)$$

with the boundary conditions

$$\rho(0)\, e^{i\,\theta(0)} = R\, \rho(L)\, e^{i\,\theta(L)}\, e^{i\,\delta\Omega\frac{\Lambda-L}{c}} \qquad (11.8)$$

Equation (11.8) can also be written in the form

$$\rho(0) = R\, \rho(L) \qquad (11.9a)$$

$$\theta(L) - \theta(0) = -\delta\Omega\,\frac{\Lambda-L}{c} + 2\pi j \qquad (11.9b)$$

where the additive term $2\pi j$ ($j=0, \pm1,...$) is an important part of the phase condition (11.9b). By ignoring it (i.e., by selecting $j=0$) one may lose a number of possible steady states.

The solution of Eqs. (11.7) with the boundary conditions (11.9) appears to be a significant undertaking, even numerically, and has not been carried out to the best of our knowledge; an exact numerical solution in resonance has been found recently by this lecturer using the algorithm described in Appendix D.

Here we focus on the uniform field limit, where Eqs. (11.7) take the form [this treatment parallels the studies by Abraham, Lugiato, Mandel, Narducci and Bandy (1985), and by Bandy, Narducci, Lugiato, Abraham (1985)]

$$\rho(L) - \rho(0) \cong \alpha L \int_{-\infty}^{+\infty} d\overline{\delta}\, g(\overline{\delta}) \, \frac{\rho(0)}{1 + (\overline{\delta}_{AC} - \overline{\delta\Omega} + \overline{\delta})^2 + \rho^2(0)} \qquad (11.10a)$$

$$\theta(L) - \theta(0) \cong \frac{\delta\Omega}{c} L - \alpha L \int_{-\infty}^{+\infty} d\overline{\delta}\, g(\overline{\delta}) \, \frac{\overline{\delta}_{AC} - \overline{\delta\Omega} + \overline{\delta}}{1 + (\overline{\delta}_{AC} - \overline{\delta\Omega} + \overline{\delta})^2 + \rho^2(0)} \qquad (11.10b)$$

These equations must be solved with the boundary conditions which, in the uniform field limit become

$$\rho(L) - \rho(0) \cong T\, \rho(0)$$

$$\theta(L) - \theta(0) \cong -\frac{\delta\Omega}{c}(\Lambda-L) + 2\pi j \qquad (j = 0, \pm1, \pm2,...) \qquad (11.10c)$$

After combining Eqs. (11.10a,b) with (11.10c), we obtain the coupled steady state equations for the output field modulus ρ_j and the frequency offset $\delta\Omega_j$ between the operating laser line and the reference cavity resonance

$$1 = 2C \int_{-\infty}^{+\infty} d\bar{\delta}\, g(\bar{\delta}) \; \frac{1}{1 + (\bar{\Delta}_j + \bar{\delta})^2 + \rho_j^2} \qquad (11.11a)$$

$$\frac{\bar{\Delta}_j - \bar{\delta}_{AC} + j\,\bar{\alpha}_1}{\bar{\kappa}} = -2C \int_{-\infty}^{+\infty} d\bar{\delta}\, g(\bar{\delta}) \; \frac{\bar{\Delta}_j + \bar{\delta}}{1 + (\bar{\Delta}_j + \bar{\delta})^2 + \rho_j^2} \qquad (11.11b)$$

The index j is a label for the possible coexisting steady states of the system. As in the case of the homogeneously broadened laser, the upper and lower bounds of j are set by the constraint $\rho_j > 0$, i.e., the threshold condition for the j-th steady state. The symbol $\bar{\Delta}_j$ denotes the difference δ_{AC}-$\delta\Omega_j$ in units of $\gamma_\perp$.

In the homogeneous limit ($\sigma_D \to 0$), the lineshape function reduces to a δ-function, and Eqs. (11.11) gives back the known results. In the general inhomogeneous case, there seems to be no way to calculate ρ_j and $\delta\Omega_j$ without some numerical labor. In our analysis of the steady state problem we have chosen to transform the steady state equations in the form

$$1 = \frac{\sqrt{2\pi}\,C}{\bar{\sigma}_D \xi_j} \, \text{Re}\, W\!\left(\frac{\bar{\Delta}_j + i\,\xi_j}{\sqrt{2}\,\bar{\sigma}_D}\right) \qquad (11.12a)$$

$$\frac{\bar{\Delta}_j - \bar{\delta}_{AC} - j\,\bar{\alpha}_1}{\bar{\kappa}} = \frac{\sqrt{2\pi}\,C}{\bar{\sigma}_D}\, \text{Im}\!\left(\frac{\bar{\Delta}_j + i\,\xi_j}{\sqrt{2}\,\bar{\sigma}_D}\right) \qquad (11.12b)$$

where W(z) is the error function of complex argument [Abramowitz and Stegun (1968)]

$$W(z) = e^{-z^2}\, \text{erfc}\,(-iz) \qquad (11.13)$$

and

$$\xi_j = \sqrt{1 + \rho_j^2}$$

Now we can solve the coupled transcendental equations (11.12) by a two-dimensional Newton-Raphson method. Some typical results are displayed in the following figures. Figure 11.2 shows the modulus of the output field as a function of the gain parameter C for a resonant situation. This curve is not qualitatively different from the corresponding ones obtained in the homogeneous limit, although the analytic expression leading to this result is much more complicated. In fact, not even the threshold condition can be expressed in closed analytic form. In resonance however, Eq. (11.12a) yields

$$2C_{thr} = \frac{2\bar{\sigma}_D}{\sqrt{2\pi}} \; \frac{\exp(-1/2\bar{\sigma}_D^2)}{1 - \text{erfc}\!\left(\dfrac{1}{\sqrt{2}\,\bar{\sigma}_D}\right)} \qquad (11.14)$$

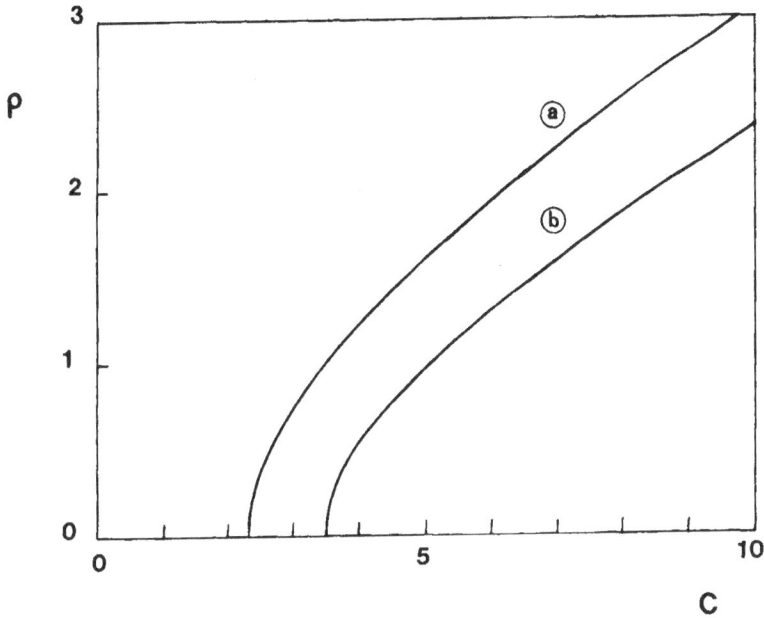

11.2 The modulus of the steady state output field is plotted as a function of the gain according to the uniform field equations (11.12) and the parameters $\bar{\sigma}_D = 5.0$, $\bar{\alpha}_1 = 5.0$, and $\bar{\delta}_{AC} = 0$. The curves displayed in this figure correspond to the steady state solutions (a) j=0 and (b) j=1. The output frequency of the j=0 state coincides with the cavity resonance ω_c. The frequency of the state j=1 varies very little as a function of C and is almost equal to $\omega_c + \alpha_1$ because the mode pulling is very small for the selected value of the cavity damping rate ($\bar{\kappa}=0.16$).

which is represented accurately for $\sigma_D > 0.5$ by the convenient empirical fit

$$2C_{thr} \cong 0.8\,\bar{\sigma}_D + 0.65 \tag{11.15}$$

In Fig. 11.3 we show an example of a detuned situation for four different steady state indices. It is clear that a large number of possible coexisting steady states are possible in an inhomogeneously broadened system. This fact was recognized from the early days of laser physics [See, for example, Bennett (1977)]. The argument advanced to justify this result was based on the low level of communication among different spectral segments of the gain profile. In effect, the multimode inhomogeneous laser is a device that can function as if different collections of atoms were responsible for the operation of different output lines. While, rigorously speaking, this is not quite true, it is not a bad qualitative approximation to say that the different modes operate in essentially independent ways if the intermode spacing is much larger than the homogeneous width of a single atomic packet. Note that, unlike the homogeneous laser, where the operating frequencies of the steady state are only functions of the detuning parameters and of the various decay rates, here the operating frequency is also a function of the output intensity. For the selected value of $\bar{\kappa}$ in Figs. 11.2, 11.3 the gain dependence of the output frequency is not very significant. It becomes much more pronounced, however, for large values of the cavity linewidth as shown in Fig. 11.4.

It is instructive to take a closer look at the steady state properties of the atomic variables. Equations (11.6a, b) describe the frequency and the intensity dependence of the polarization and population difference for a single homogeneous packet of atoms. The behavior of the atomic variables for the entire sample can be computed easily by multiplying $\bar{P}_{st}$ and $\bar{D}_{st}$ by the lineshape function $g(\bar{\delta})$. Thus, the macroscopic population difference of the entire sample $\bar{D}_M(\bar{\delta})$ in the uniform field limit is given by

$$\bar{D}_M(\bar{\delta}) = g(\bar{\delta})\,\bar{D}_{st}(\bar{\delta}) = \frac{\exp\left(-\dfrac{\bar{\delta}^2}{2\bar{\sigma}_D^2}\right)}{\sqrt{2\pi\bar{\sigma}_D^2}}\,\frac{1+(\bar{\delta}_{AC}-\bar{\delta}\Omega+\bar{\delta})^2}{1+(\bar{\delta}_{AC}-\bar{\delta}\Omega+\bar{\delta})^2+|\bar{F}_{st}(z)|^2} \tag{11.16}$$

When plotted as a function of frequency, this distribution has an interesting profile as shown qualitatively in Fig. 11.5. We see that the macroscopic population difference is distorted only slightly in the wings, but suffers large field-induced variations in the neighborhood of the operating laser frequency. A hole is "burned" in the unperturbed profile whose origin can be explained easily: initially, the population is inverted uniformly so that $\bar{D}_M(\bar{\delta})$, as a function of frequency has a Gaussian profile. When the laser is turned on, some of the original inversion is lost to electromagnetic energy. The pump mechanism attempts to reset the population inversion to its original value, but the stimulated emission keeps draining energy away mainly from the resonant and the neighboring homogeneous packets, until a dynamic equilibrium is established between the pump and the emission process. Of course, the atoms that are resonant with the carrier frequency of the emitted laser light are most readily affected than those that are further removed in frequency. The "hole burning" effect, one of the oldest phenomena to have been discovered in gas lasers [Bennett (1962)], owes its appearence to the saturation induced by the field inside the medium.

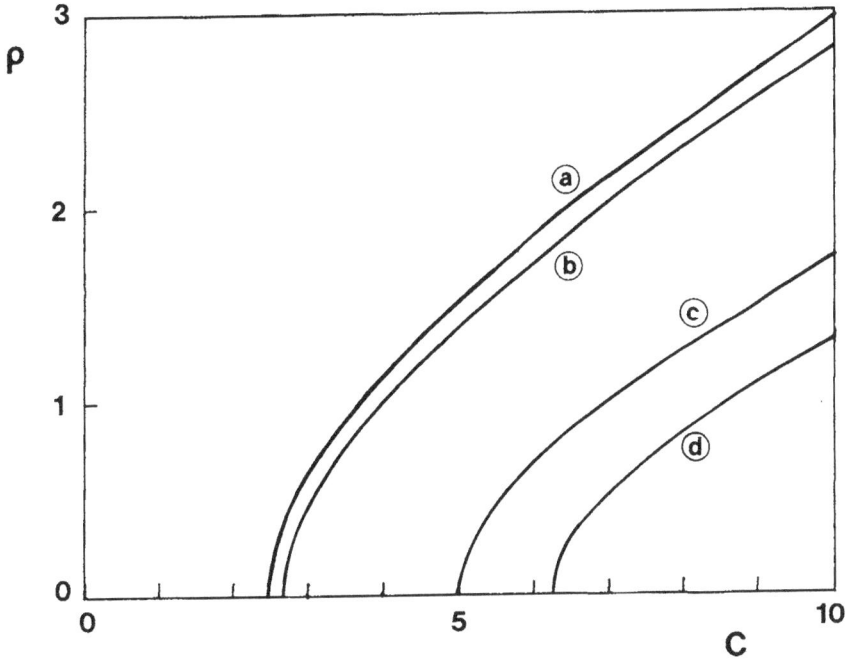

11.3   The modulus of the steady state output field is plotted as a function of the gain according to the uniform field equations (11.12) and the parameters $\bar{\sigma}_D = 5.0$, $\bar{\alpha}_1 = 5.0$, and $\bar{\delta}_{AC} = 2.0$. The curves displayed in this figure correspond to the steady state solutions (a) j=0, (b) j=1, (c) j=-1, and (d) j=2. The output frequencies of these solutions vary little as a function of C and are almost equal to $\omega_c + j\,\alpha_1$ because the mode pulling is very small for the selected value of the cavity damping rate ($\bar{\kappa}$=0.16).

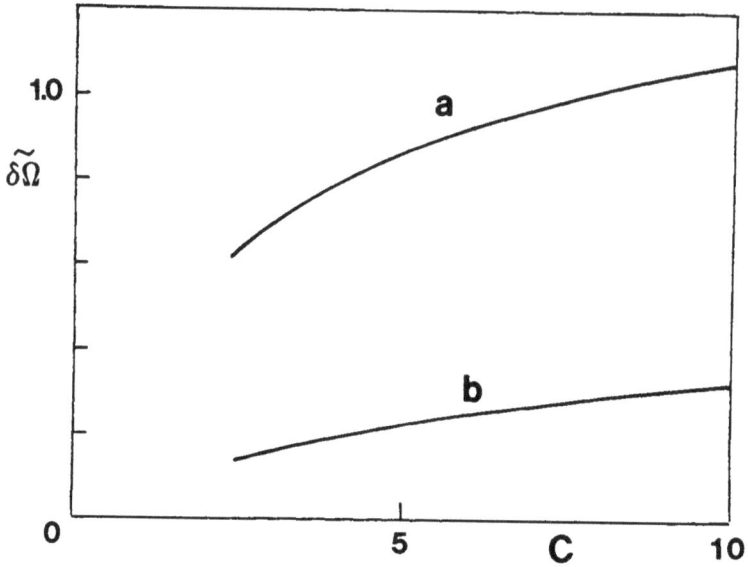

11.4   Behavior of $\bar{\delta}\Omega$ as a function of the gain parameter C for $\bar{\sigma}_D$ = 5.0, $\bar{\alpha}_1$ = 5.0, $\bar{\delta}_{AC}$ = 2.0 and j=0. Curve (a) corresponds to $\bar{\kappa}$ =3.0, while curve (b) refers to $\bar{\kappa}$ =0.5.

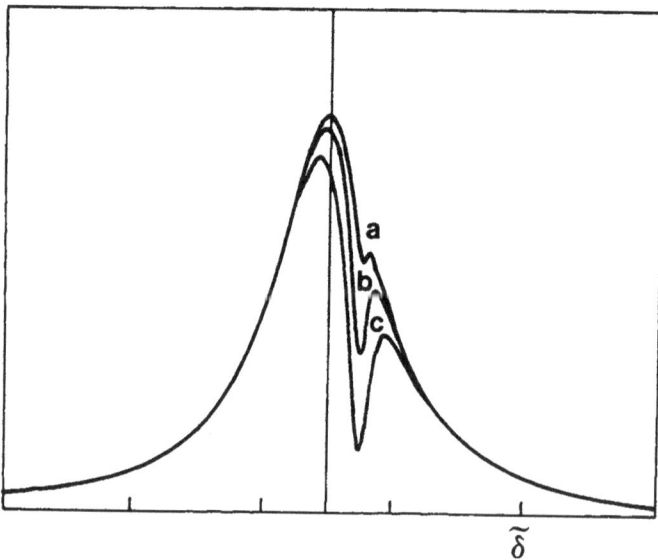

11.5   Behavior of the population difference (11.16) for different values of the output field intensity. Curves (a), (b), and (c) correspond to progressively larger values of the steady state field.

Figure 11.5 is also helpful in understanding why many of the results presented in this section do not differ qualitatively from those presented in our previous discussion of homogeneous systems. Clearly, the bulk of the laser interaction affects only a small region of the inhomogeneous line, almost as if the remainder of the active medium did not exist, at least as a rough approximation.

We now turn our attention to the linear stability analysis of this type of laser, following the same methods developed in earlier chapters for the study of the multimode homogeneous laser. For this purpose, we transform the space-time coordinates of the system according to

$$z' = z$$

$$t' = t + \frac{\Lambda - L}{c} \frac{z}{L} \tag{11.17}$$

and the dependent variables as follows

$$F(z',t') = \bar{F}(z',t') \, e^{-\frac{z'}{L} |\ln R|}$$

$$P(\bar{\delta},z',t') = \bar{P}(\bar{\delta},z',t') \, e^{-\frac{z'}{L} |\ln R|} \tag{11.18}$$

$$D(\bar{\delta},z',t') = \bar{D}(\bar{\delta},z',t')$$

The new Maxwell-Bloch equations in the uniform field limit become

$$\frac{\partial F}{\partial t'} + \frac{cL}{\Lambda} \frac{\partial F}{\partial z'} = -\kappa \left\{ F + 2C \int_{-\infty}^{+\infty} d\bar{\delta} \, g(\bar{\delta}) \, P \right\} \tag{11.19a}$$

$$\frac{\partial P}{\partial t'} = -\gamma_\perp \left\{ FD + (1 + i \, (\bar{\delta}_{AC} + \bar{\delta}))P \right\} \tag{11.19b}$$

$$\frac{\partial D}{\partial t'} = -\gamma_\parallel \left\{ -\frac{1}{2}(F^*P + FP^*) + D - 1 \right\} \tag{11.19c}$$

with the usual boundary conditions

$$F(0,t') = F(L,t') \tag{11.19d}$$

Now we introduce the modal decomposition

$$\begin{pmatrix} F(z',t') \\ P(\bar{\delta},z',t') \end{pmatrix} = e^{-i\,\delta\omega t'} \sum_{n=-\infty}^{+\infty} e^{ik_n z'} e^{-i\alpha_n t'} \begin{pmatrix} f_n(t') \\ p_n(\bar{\delta},t') \end{pmatrix} \tag{11.20a}$$

$$D(\bar{\delta},z',t') = \sum_n e^{ik_n z'} e^{-i\alpha_n t'} d_n(\bar{\delta},t') \tag{11.20b}$$

132

$$
\begin{pmatrix} F^*(z',t') \\ P^*(\bar{\delta},z',t') \end{pmatrix} = e^{i\delta\omega t'} \sum_{n=-\infty}^{+\infty} e^{-ik_n z'} e^{i\alpha_n t'} \begin{pmatrix} f_n^*(t') \\ p_n^*(\bar{\delta},t') \end{pmatrix}
\tag{11.20c}
$$

where $\delta\omega$ must be calculated from the steady state equation. The modal amplitudes satisfy the equations

$$
\frac{\partial f_n}{\partial t'} = i\,\delta\omega\, f_n - \kappa \left(f_n + 2C\int_{-\infty}^{+\infty} d\bar{\delta}\, g(\bar{\delta})\, p_n \right)
\tag{11.21a}
$$

$$
\frac{\partial f_n^*}{\partial t'} = -i\,\delta\omega\, f_n^* - \kappa \left(f_n^* + 2C\int_{-\infty}^{+\infty} d\bar{\delta}\, g(\bar{\delta})\, p_n^* \right)
\tag{11.21b}
$$

$$
\frac{\partial p_n}{\partial t'} = -\gamma_\perp \left\{ \sum_{n'} f_{n'} d_{n-n'} + (1 + i\,(\bar{\delta}_{AC} - \bar{\delta}\omega - \bar{\alpha}_n + \bar{\delta}))\, p_n \right\}
\tag{11.21c}
$$

$$
\frac{\partial p_n^*}{\partial t'} = -\gamma_\perp \left\{ \sum_{n'} f_{n'}^* d_{n-n'}^* + (1 - i\,(\bar{\delta}_{AC} - \bar{\delta}\omega - \bar{\alpha}_n + \bar{\delta}))\, p_n^* \right\}
\tag{11.21d}
$$

$$
\frac{\partial d_n}{\partial t'} = -i\,\alpha_n d_n - \gamma_\parallel \left\{ -\frac{1}{2} \sum_{n'} (f_{n'}^* p_{n+n'} + f_{n'} p_{n'-n}^*) + d_n - \delta_{n,0} \right\}
\tag{11.21e}
$$

where $\alpha_n = (2\pi c/\Lambda)n$ is the usual separation between the n-th empty cavity resonance and the selected reference mode. In steady state we seek solutions of the coupled equations (11.21) with all the time derivatives set equal to zero.

Consider the atomic equations, first. In principle, there is an infinite number of solutions of the type

$$
f_j^0 \neq 0, \quad f_n^0 = 0 \quad (n \neq j)
\tag{11.22a}
$$

$$
p_j^0 \neq 0, \quad p_n^0 = 0 \quad (n \neq j)
\tag{11.22b}
$$

$$
d_0^0 \neq 0, \quad d_n^0 = 0 \quad (n \neq 0)
\tag{11.22c}
$$

where the nonzero amplitudes are linked by the equations

$$f_j^0 \, d_0^0 + \left(1 + i \, (\overline{\delta}_{AC} - \overline{\delta\omega}_j - \overline{\alpha}_j + \overline{\delta})\right) p_j^0 = 0 \qquad (11.23a)$$

$$f_j^{0*} \, d_0^0 + \left(1 - i \, (\overline{\delta}_{AC} - \overline{\delta\omega}_j - \overline{\alpha}_j + \overline{\delta})\right) p_j^{0*} = 0 \qquad (11.23b)$$

$$\frac{1}{2} \left(f_j^{0*} \, p_j^0 + f_j^0 \, p_j^{0*} \right) - d_0^0 + 1 = 0 \qquad (11.23c)$$

whose solutions are

$$p_j^0 = -f_j^0 \, \frac{1 - i \, (\overline{\delta}_{AC} - \overline{\delta\omega} - \overline{\alpha}_n + \overline{\delta})}{1 + (\overline{\delta}_{AC} - \overline{\delta\omega} - \overline{\alpha}_n + \overline{\delta})^2 + |f_j^0|^2} \qquad (11.24a)$$

$$d_0^0 = \frac{1 + (\overline{\delta}_{AC} - \overline{\delta\omega} - \overline{\alpha}_n + \overline{\delta})^2}{1 + (\overline{\delta}_{AC} - \overline{\delta\omega} - \overline{\alpha}_n + \overline{\delta})^2 + |f_j^0|^2} \qquad (11.24b)$$

It is important to kep in mind that the upper index (j) plays the role of a multiplicity index that counts the possible steady states, while the lower index labels the nonzero modal components. The field modulus $|f_j^{(j)}|$ and the frequency offset $\overline{\delta\omega}_j$ must be calculated from the steady state equations

$$1 = 2C \int_{-\infty}^{+\infty} d\overline{\delta} \, g(\overline{\delta}) \, \frac{1}{1 + (\overline{\delta}_{AC} - \overline{\delta\omega} - \overline{\alpha}_n + \overline{\delta})^2 + |f_j^0|^2} \qquad (11.25a)$$

$$\frac{\overline{\delta\omega}_j}{\overline{\kappa}} = 2C \int_{-\infty}^{+\infty} d\overline{\delta} \, g(\overline{\delta}) \, \frac{(\overline{\delta}_{AC} - \overline{\delta\omega} - \overline{\alpha}_n + \overline{\delta})}{1 + (\overline{\delta}_{AC} - \delta\omega - \overline{\alpha}_n + \overline{\delta})^2 + |f_j^0|^2} \qquad (11.25b)$$

Note that, in the homogeneous limit $[g(\overline{\delta}) \rightarrow \delta(\overline{\delta})]$, one has the familiar result

$$|f_j^0|^2 = 2C - \left(1 + (\overline{\delta}_{AC} - \overline{\delta\omega} - \overline{\alpha}_n + \overline{\delta})^2\right) \qquad (11.26a)$$

$$\overline{\delta\omega}_j = \frac{\overline{\kappa}}{1 + \overline{\kappa}} (\overline{\delta}_{AC} - \overline{\alpha}_j) \qquad (11.26b)$$

Hence, the oscillation frequency of the j-th steady state solution is given by

$$\omega_j = \omega_c + \delta\omega + \alpha_j = \frac{(\omega_c + \alpha_j) \, \gamma_\perp + \kappa \omega_A}{\gamma_\perp + \kappa} \qquad (11.27)$$

This is the standard mode pulling formula. The frequency offset $\delta\omega_j$ measures the separation between the frequency of the j-th stationary state and that of the j-th empty cavity mode, i.e., $\delta\omega_j = (\omega_L^{(j)} - \alpha_j)/\gamma_\perp$.

In the inhomogeneously broadened case, we must solve Eqs. (11.25) which are essentially identical in structure to the steady state equations already discussed in this section. Now, with reference to the modal equations (11.21) we let

$$X_n = X_n^0 \delta_{n,j} + \delta X_n \tag{11.28}$$

where X_n denotes f_n, f_n^*, p_n, or p_n^* and

$$d_n = d_n^0 \delta_{n,0} + \delta d_n \tag{11.29}$$

and obtain the linearized equations

$$\frac{\partial}{\partial t'} \delta f_n = i\, \delta\omega_j \delta f_n - \kappa \left(\delta f_n + 2C \int_{-\infty}^{+\infty} d\bar{\delta}\, g(\bar{\delta})\, \delta p_n \right) \tag{11.30a}$$

$$\frac{\partial}{\partial t'} \delta f_n^* = -i\, \delta\omega_j \delta f_n^* - \kappa \left(\delta f_n^* + 2C \int_{-\infty}^{+\infty} d\bar{\delta}\, g(\bar{\delta})\, \delta p_n^* \right) \tag{11.30b}$$

$$\frac{\partial}{\partial t'} \delta p_n = - \gamma_\perp \left\{ f_j^0\, \delta d_{n-j} + d_0^0\, \delta f_n + \left(1 + i\, (\bar{\delta}_{AC} - \bar{\delta}\omega - \bar{\alpha}_n + \bar{\delta}) \right) \delta p_n \right\} \tag{11.30c}$$

$$\frac{\partial}{\partial t'} \delta p_n^* = - \gamma_\perp \left\{ f_j^{0*}\, \delta d_{n-j}^* + d_0^0\, \delta f_n^* + \left(1 - i\, (\bar{\delta}_{AC} - \bar{\delta}\omega - \bar{\alpha}_n + \bar{\delta}) \right) \delta p_n^* \right\} \tag{11.30d}$$

$$\frac{\partial}{\partial t'} \delta d_n = i\, \alpha_n\, \delta d_n - \gamma_\parallel \left\{ -\frac{1}{2} (f_j^{0*} \delta p_{n+j} + p_j^0\, \delta f_{j-n} + f_j^0\, \delta p_{j-n}^* + \right.$$

$$\left. p_j^{0*} \delta f_{j+n}) + \delta d_n \right\} \tag{11.30e}$$

In order to avoid confusion, we must keep in mind the following facts and conventions. The system of linearized equations (11.30) is comprised of an infinite number of equations for the fluctuation variables δf_n, δf_n^*, etc. (this is not unexpected because the original Maxwell-Bloch equations are partial differential equations). These fluctuations, in general, will be different from zero, even if the associated steady state modal amplitudes vanish. There is one system of equations of the type (11.30) for each steady state configuration. The different systems are recognizable because the steady state variables are different from one another (thus, $f_n^{(1)}$, $n=0, \pm 1, \pm 2,...$ is quite different from $f_n^{(2)}$, $n=0, \pm 1, \pm 2,...$). After selecting the given steady state configuration of interest, for example j=0, we can explore its stability by solving for the eigenvalues of the linearized problem (11.30). These are also infinite in number, and can be thought of as being grouped into sets of five (as many as the linearized amplitudes for each mode). If any of these infinite eigenvalues develops a positive real part, the j-th steady state of interest is unstable. From a physical point of view, if Re $\lambda_{n,k}^{(j)}$ is positive for some values of n,k and j (k runs from 1 to 5), we can say that the j-th state is unstable because a fluctuation of the n-th modal amplitude grows exponentially for short times. Thus, once again, as for a homogeneously broadened laser, the instabilities of inhomogeneous system can be attributed to the growth of sidebands in correspondence with selected cavity resonances.

Fortunately, we do not have to solve an infinitely dimensional problem every time we face Eqs. (11.30) because, upon close inspection, we see that the fluctuation variables are not all coupled to one another. In fact, this set of linear equations can be written explicitly in such a way as to show that only five fluctuation variables at a time are coupled to one another. In fact, after some minor manipulation of indices, we can write Eqs. (11.30) in the form

$$\frac{\partial}{\partial t'} \delta f_{n+j} = i\, \delta\omega_j \delta f_{n+j} - \kappa\left(\delta f_{n+j} + 2C \int_{-\infty}^{+\infty} d\bar{\delta}\, g(\bar{\delta})\, \delta p_{n+j}\right) \tag{11.31a}$$

$$\frac{\partial}{\partial t'} \delta f_{j-n}^* = -i\, \delta\omega_j \delta f_{j-n}^* - \kappa\left(\delta f_{j-n}^* + 2C \int_{-\infty}^{+\infty} d\bar{\delta}\, g(\bar{\delta})\, \delta p_{j-n}^*\right) \tag{11.31b}$$

$$\frac{\partial}{\partial t'} \delta p_{n+j} = -\gamma_\perp \left\{ f_j^0 \delta d_n + d_0^0 \delta f_{n+j} + \left(1 + i\,(\bar{\delta}_{AC} - \bar{\delta\omega} - \bar{\alpha}_n + \bar{\delta})\right) \delta p_{n+j} \right\} \tag{11.31c}$$

$$\frac{\partial}{\partial t'} \delta p_{j-n}^* = -\gamma_\perp \left\{ f_j^{0*} \delta d_n^* + d_0^0 \delta f_{j-n}^* + \left(1 - i\,(\bar{\delta}_{AC} - \bar{\delta\omega} - \bar{\alpha}_n + \bar{\delta})\right) \delta p_{j-n}^* \right\} \tag{11.31d}$$

$$\frac{\partial}{\partial t'} \delta d_n = i\, \alpha_n \delta d_n - \gamma_\parallel \left\{ -\frac{1}{2} \left(f_j^{0*} \delta p_{n+j} + p_j^0 \delta f_{j-n}^* + f_j^0 \delta p_{j-n}^* + \right. \right.$$

$$\left. \left. p_j^{0*} \delta f_{n+j} \right) + \delta d_n \right\} \tag{11.31e}$$

which is a closed set of five linear equations for the fluctuation variables δf_{n+j}, δf_{j-n}^*, δp_{j+n}, etc. If we now let

$$\begin{bmatrix} \delta f_{n+j} \\ \delta f_{j-n}^* \\ \delta p_{n+j} \\ \delta p_{j-n}^* \\ \delta d_n \end{bmatrix} = e^{\bar{\lambda}\tau} \begin{bmatrix} A \\ B \\ P \\ Q \\ D \end{bmatrix} \tag{11.32}$$

where $\tau = \gamma_\perp t'$ and introduce the symbol $\bar{\Delta}_j \equiv \bar{\delta}_{AC} - \bar{\delta\omega}_j - \bar{\alpha}_j$ we obtain

$$\bar{\lambda}A = -\bar{\kappa}\left\{\left(1 + i\,\frac{\overline{\Delta}_j - \overline{\delta}_{AC} + \overline{\alpha}_j}{\overline{\kappa}}\right) A + 2C\int_{-\infty}^{+\infty} d\overline{\delta}\, g(\overline{\delta})\, P\right\}$$ (11.33a)

$$\bar{\lambda}B = -\bar{\kappa}\left\{\left(1 - i\,\frac{\overline{\Delta}_j - \overline{\delta}_{AC} + \overline{\alpha}_j}{\overline{\kappa}}\right) B + 2C\int_{-\infty}^{+\infty} d\overline{\delta}\, g(\overline{\delta})\, Q\right\}$$ (11.33b)

$$\bar{\lambda}P = -f_j^0 D - d_0^0 A - \left(1 + i\,(\overline{\Delta}_j - \overline{\alpha}_n + \overline{\delta})\right) P$$ (11.33c)

$$\bar{\lambda}Q = -f_j^{0*} D - d_0^0 B - \left(1 + i\,(\overline{\Delta}_j - \overline{\alpha}_n + \overline{\delta})\right) Q$$ (11.33d)

$$\bar{\lambda}D = i\,\overline{\alpha}_n D + \frac{1}{2}\,\overline{\gamma}\left(f_j^{0*}P + p_j^0 B + f_j^0 Q + p_j^{0*}A\right) - \overline{\gamma}D$$ (11.33e)

The atomic equations can be solved immediately with the result

$$P = T_1(\overline{\alpha}_n)\, A + T_2(\overline{\alpha}_n)\, B$$ (11.34a)

$$Q = T_2^*(-\overline{\alpha}_n)\, A + T_1^*(-\overline{\alpha}_n)\, B$$ (11.34b)

where

$$T_1(\overline{\alpha}_n) = \frac{c_1^*(-\overline{\alpha}_n)\, c_3(\overline{\alpha}_n) - c_2(\overline{\alpha}_n)\, c_4^*(-\overline{\alpha}_n)}{c_1(\overline{\alpha}_n)\, c_1^*(-\overline{\alpha}_n) - c_2(\overline{\alpha}_n)\, c_2^*(-\overline{\alpha}_n)}$$ (11.35a)

$$T_2(\overline{\alpha}_n) = \frac{c_1^*(-\overline{\alpha}_n)\, c_4(\overline{\alpha}_n) - c_2(\overline{\alpha}_n)\, c_3^*(-\overline{\alpha}_n)}{c_1(\overline{\alpha}_n)\, c_1^*(-\overline{\alpha}_n) - c_2(\overline{\alpha}_n)\, c_2^*(-\overline{\alpha}_n)}$$ (11.35b)

and

$$c_1(\overline{\alpha}_n) = \overline{\lambda} + 1 + i\,(\overline{\Delta}_j - \overline{\alpha}_n + \overline{\delta}) + \frac{1}{2}\,\overline{\gamma}\,\frac{|f_j^0|^2}{\overline{\lambda} + \overline{\gamma} - i\,\overline{\alpha}_n}$$ (11.36a)

$$c_2(\alpha_n) = \frac{1}{2}\,\gamma\,\frac{f_j^{02}}{\overline{\lambda} + \overline{\gamma} - i\,\overline{\alpha}_n}$$ (11.36b)

$$c_3(\bar\alpha_n) = d_0^0 - \frac{1}{2}\,\bar\gamma\,\frac{f_j^0\,p_j^{0*}}{\bar\lambda + \bar\gamma - i\,\bar\alpha_n} \tag{11.36c}$$

$$c_4(\bar\alpha_n) = -\frac{1}{2}\,\bar\gamma\,\frac{f_j^0\,p_j^0}{\bar\lambda + \bar\gamma - i\,\bar\alpha_n} \tag{11.36d}$$

Note that the operation of complex conjugation <u>does not</u> involve λ i.e., in calculating, for example, $c_1{}^*(-\alpha_n)$ we handle λ <u>as if</u> it were a real variable. The reason for this peculiar rule is easily traced through the calculation.

If we now substitute Eqs. (11.34) into (11.33a, b) we arrive at the characteristic equation

$$\left(\bar\lambda + \bar\kappa(1 + i\,\theta_j) + 2C\bar\kappa \int_{-\infty}^{+\infty} d\bar\delta\, g(\bar\delta)\, T_1(\bar\alpha_n)\right)\left(\bar\lambda + \bar\kappa(1 - i\,\theta_j) + 2C\bar\kappa \int_{-\infty}^{+\infty} d\bar\delta\, g(\bar\delta)\, T_1^*(-\bar\alpha_n)\right) -$$

$$- (2C\bar\kappa)^2 \left(\int_{-\infty}^{+\infty} d\bar\delta\, g(\bar\delta)\, T_2(\bar\alpha_n)\right)\left(\int_{-\infty}^{+\infty} d\bar\delta\, g(\bar\delta)\, T_2^*(-\bar\alpha_n)\right) = 0 \tag{11.37}$$

where we have introduced the symbol

$$\theta = \frac{\bar\Delta_j - \bar\delta_{AC} + \bar\alpha_j}{\bar\kappa} \tag{11.38}$$

The solution of the characteristic equation (11.37) is cumbersome even by numerical methods. Here we consider two special cases of interest.

i)   Single mode laser.

In this case, the intermode spacing is large enough that the only possible steady state above threshold is the j=0 state, and the only cavity mode that can become unstable is the resonant mode n=0. Under these conditions, the characteristic equation (11.37) takes the form

$$\left(\bar\lambda + \bar\kappa(1 + i\theta_0) + 2C\bar\kappa \int_{-\infty}^{+\infty} d\bar\delta\, g(\bar\delta)\, T_1(0)\right)\left(\bar\lambda + \bar\kappa(1 - i\theta_0) + 2C\bar\kappa \int_{-\infty}^{+\infty} d\bar\delta\, g(\bar\delta)\, T_1^*(0)\right) -$$

$$- (2C\bar\kappa)^2 \left(\int_{-\infty}^{+\infty} d\bar\delta\, g(\bar\delta)\, T_2(0)\right)\left(\int_{-\infty}^{+\infty} d\bar\delta\, g(\bar\delta)\, T_2^*(0)\right) = 0 \tag{11.39}$$

This equation (with different notation) has been analysed by Abraham, Lugiato, Mandel, Narducci and Bandy (1985). The main conclusion of this study is that even a small amount of inhomogeneous broadening has the effect of lowering the threshold for unstable behavior by large amounts. This result was already known from earlier studies [See, for example, Idiatulin and Uspenskii (1973), Casperson (1978), (1980), (1981)].

The situation is summarized in Fig. 11.6 and 11.7 with several graphs displaying the instability boundaries for different values of the gain linewidth, and the behavior of the relevant eigenvalues for a homogeneously and an inhomogeneously broadened laser as a function of the gain parameters. These predictions are in good qualitative agreement with many observations carried out with Doppler broadened lasers since about 1976. For a review of this problem see Casperson (1983), Abraham et a. (1983).

ii)   Multimode lasers.

In the multimode case, and for the same reason advanced in our discussion of homogeneous lasers, ther parameter $\bar{\kappa}$ is necessarily much smaller than unity. This suggests that we seek solutions of Eqs (11.37) in the form

$$\bar{\lambda} \cong \bar{\lambda}^{(0)} + \bar{\kappa}\,\lambda^{(1)} \tag{11.40}$$

where $\bar{\lambda}^{(0)}$ is one of the solutions of the characteristic equation with $\bar{\kappa}=0$ and $\lambda^{(1)}$ is the first order correction. With this ansatz, Eq. (11.37) becomes

$$\left(\bar{\lambda}^{(0)} + \bar{\kappa}\,\lambda^{(1)} + \bar{\kappa}\,\rho(\bar{\alpha}_n, \bar{\lambda}^{(0)}+\bar{\kappa}\lambda^{(1)})\right)\left(\bar{\lambda}^{(0)} + \bar{\kappa}\lambda^{(1)} + \bar{\kappa}\,\rho^*(-\,\bar{\alpha}_n, \bar{\lambda}^{(0)}+\bar{\kappa}\lambda^{(1)})\right) -$$

$$\bar{\kappa}^2\,\tau(\bar{\alpha}_n, \bar{\lambda}^{(0)}+\bar{\kappa}\lambda^{(1)})\,\tau^*(-\,\bar{\alpha}_n, \bar{\lambda}^{(0)}+\bar{\kappa}\lambda^{(1)}) = 0 \tag{11.41}$$

where we have introduced the symbols

$$\rho(\bar{\alpha}_n, \bar{\lambda}) = 1 + i\,\theta_j + 2C\int_{-\infty}^{+\infty} d\bar{\delta}\,g(\bar{\delta})\,T_1(\bar{\alpha}_n, \bar{\lambda}) \tag{11.42a}$$

$$\tau(\bar{\alpha}_n, \bar{\lambda}) = 2C\int_{-\infty}^{+\infty} d\bar{\delta}\,g(\bar{\delta})\,T_2(\bar{\alpha}_n, \bar{\lambda}) \tag{11.42b}$$

To order $\bar{\kappa}^0$ we obtain the double root $\bar{\lambda}^{(0)} = 0$, while to order $\bar{\kappa}^2$ we find

$$\left(\lambda^{(1)} + \rho(\bar{\alpha}_n, 0)\right)\left(\lambda^{(1)} + \rho^*(-\,\bar{\alpha}_n, 0)\right) - \tau(\bar{\alpha}_n, 0)\,\tau^*(-\,\bar{\alpha}_n, 0) = 0 \tag{11.43}$$

The solutions of this equation are

$$\lambda^{(1)}_{\pm} = \frac{1}{2}\left\{-\left(\rho(\bar{\alpha}_n,0) + \rho^*(-\,\bar{\alpha}_n,0)\right) \pm \left(\,(\rho(\bar{\alpha}_n,0) - \rho^*(-\,\bar{\alpha}_n,0))^2 +\right.\right.$$

$$\left.\left. 4\,\tau(\bar{\alpha}_n,0)\,\tau^*(-\,\bar{\alpha}_n,0)\right)^{1/2}\right\} \tag{11.44}$$

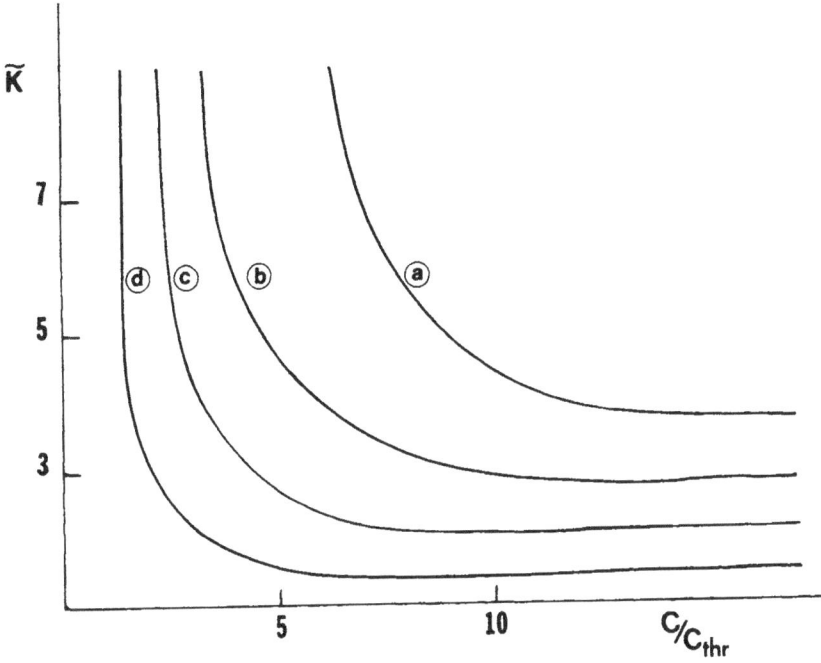

11.6   Instability boundaries for a resonant inhomogeneously broadened laser in the space of the control parameters $\tilde{\kappa}$ and C/C_{thr}, where C_{thr} is the threshold gain for ordinary laser action. Curves (a) through (d) correspond to $\bar{\sigma}_D = 1, 2, 5$, and ∞, respectively.

140

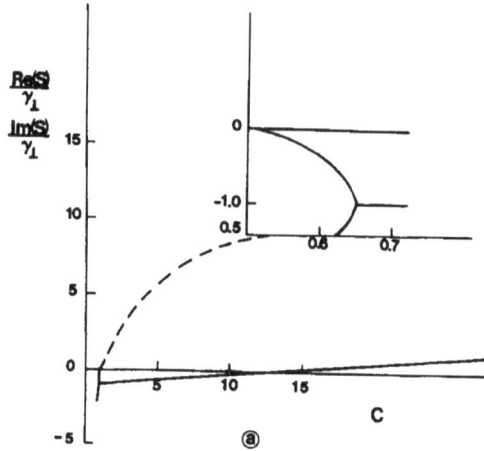

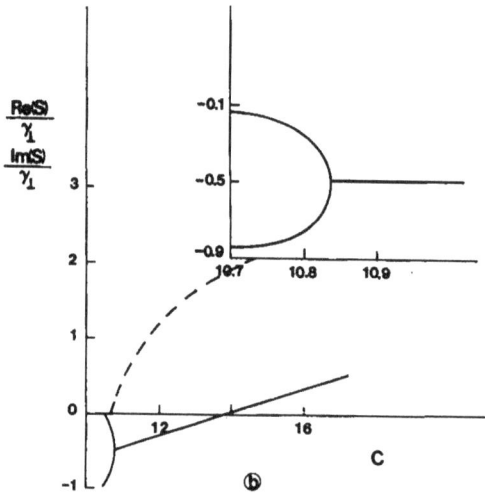

11.7   The real and imaginary parts of the unstable eigenvalue corresponding to j=0 and n=0 (single-mode approximation) are plotted as function of the gain parameter for a homogeneously (a) and an inhomogeneously (b) broadened laser. In both cases, the real parts of the unstable eigenvalues become positive with nonzero imaginary parts (Hopf bifurcation). In the homogeneously broadened case the ratio C/C_{thr} at the instability threshold is considerably larger.

The quantities ρ and τ can be calculated for different values of $\bar{\alpha}_n$ by numerical techniques. If for any $\bar{\alpha}_n = n\,\bar{\alpha}_1$, Re $\lambda_\pm^{(1)} > 0$ the system is unstable. A numerical study of Eq. (11.44) shows that the multimode instability threshold is much lower for an inhomogeneously broadened than for a homogeneously broadened laser. In addition, just as we have seen in our earlier analyses, the observed instabilities can be of the amplitude or phase type. However, unlike the homogeneous limit, phase and amplitude instabilities usually coexist for an inhomogeneous laser.

Appendix A: The Laser Rate Equations

An improved, but conceptually identical version of the laser rate equations introduced in Section 1 follows from the Maxwell-Bloch equations (3.25) if we assume that the relaxation rate of the atomic polarization is much larger than the rates of decay of the population difference and of the cavity field. Under this condition, we can set the time derivative of the polarization equal to zero in Eq. (3.25b), as discussed in Section 6; the result is

$$\bar{P} = -\frac{\bar{F}\,\bar{D}}{1 + i\,\bar{\delta}_{AC}}, \qquad \bar{\delta}_{AC} = \frac{\delta_{AC}}{\gamma_{\perp}} \tag{A.1}$$

If we now substitute Eq. (A.1) into Eqs. (3.25a, c) we obtain

$$c\frac{\partial \bar{F}}{\partial z} + \frac{\partial \bar{F}}{\partial t} = -\alpha c\,\frac{\bar{F}\,\bar{D}}{1 + i\,\bar{\delta}_{AC}} \tag{A.2a}$$

$$\frac{\partial \bar{D}}{\partial t} = -\gamma_{\parallel}\,\frac{|\bar{F}|^2\,\bar{D}}{1 + \bar{\delta}^2_{AC}} - \gamma_{\parallel}(\bar{D} - 1) \tag{A.2b}$$

We can obtain a closed set of equations by introducing the new variable $I = |\bar{F}|^2$ with the result

$$c\frac{\partial I}{\partial z} + \frac{\partial I}{\partial t} = -\frac{2\alpha c}{1 + \bar{\delta}^2_{AC}}\,I\,\bar{D} \tag{A.3a}$$

$$\frac{\partial \bar{D}}{\partial t} = -\gamma_{\parallel}\left(\frac{I\,\bar{D}}{1 + \bar{\delta}^2_{AC}} + \bar{D} - 1\right) \tag{A.3b}$$

Apart from details, these equations have the same form as Eqs. (1.14) that we derived in Section 1 on the basis of phenomenological arguments. The field loss term is not present in Eq. (A.3a) because, more accurately, the losses emerge as a result of the boundary conditions

$$I(0,t) = R^2\,I\,(L, t - \frac{\Lambda - L}{c}) \tag{A.4}$$

Appendix B - An Improved Single-Mode Model

One of the main drawbacks of the single-mode approximation discussed in Section 8 is not so much the possible lack of longitudinal uniformity of the electrical field envelope, but the longitudinal dependence of the atomic variables. In fact, in the new reference system (7.4) and in terms of the field amplitude $F(z',t')$, the cavity field remain fairly uniform even reasonably far away from the uniform field limit, but the atomic variables usually do not. The problem can be immediately recognized by inspecting the steady state equations (4.4) after transformation to the new variables $\bar{P}_{st}(z')$ and $\bar{D}_{st}(z')$. It is clear that both $\bar{P}_{st}$ and $\bar{D}_{st}$ develop a non-negligible longitudinal dependence when the reflectivity becomes significantly different from unity. Thus, at this point, we face the following situation: the near field-uniformity allows the single-mode approximation for the field amplitude to a good approximation (by single mode approximation we mean the neglect of all the Fourier components in Eq. (7.9a) except for n=0); however, the lack of uniformity for the atomic variables suggests that a similar step, as applied to $\bar{P}(z',t')$ and $\bar{D}(z', t')$ will not be as accurate.

We have developed a scheme to overcome this problem [Lugiato, Narducci, Bandy and Tredicce (1986)]. Here we summarize our considerations and encourage the reader to consult the above source for more detail. The starting point of our analysis is a set of coupled equations [See Eqs. (7.7)]

$$\frac{\partial F}{\partial t'} + \frac{cL}{\Lambda} \frac{\partial F}{\partial z'} = - \kappa \, (F + 2CP) \tag{B.1a}$$

$$\frac{\partial P}{\partial t'} = -\gamma_\perp (FD + (1+i \, \bar{\delta}_{AC})P) \tag{B.1b}$$

$$\frac{\partial D}{\partial t'} = -\gamma_\parallel \left\{ - \frac{1}{2} (F^*P + FP^*) \, e^{\frac{2z'}{L}|\ln R|} + D - 1 \right\} \tag{B.1c}$$

We remove the oscillatory part of the field, that originates from the laser operating at a frequency different from ω_c, with the transformation

$$F(z',t') = e^{-i\delta\Omega t'} f(z',t') \tag{B.2a}$$

$$P(z',t') = e^{-i\delta\Omega t'} p(z',t') \tag{B.2b}$$

$$D(z',t') = d(z',t') \tag{B.2c}$$

where

$$\delta\Omega = \frac{\bar{\kappa}}{1+\bar{\kappa}} \, \delta_{AC} \tag{B.3}$$

and obtain the equations of motion

$$\frac{\partial f}{\partial t'} + \frac{cL}{\Lambda}\frac{\partial f}{\partial z'} - i\,\delta\Omega\,f = -\,\kappa\,(f + 2Cp) \tag{B.4a}$$

$$\frac{\partial p}{\partial t'} = -\gamma_{\perp}\left\{fd + \left(1 + i\,(\bar{\delta}_{AC} - \delta\Omega)\right)p\right\} \tag{B.4b}$$

$$\frac{\partial d}{\partial t'} = -\gamma_{\parallel}\left\{-\frac{1}{2}\,(f^{*}p + fp^{*})\,e^{\frac{2z'}{L}|hR|} + d - 1\right\} \tag{B.4c}$$

Now we introduce the modal expansion

$$f(z',t') = \sum_{n} e^{ik_{n}z'}\,e^{-i\alpha_{n}t'}\,f_{n}(t') \tag{B.5}$$

for the field, but not for the atomic variables, so that the equations of motion (B.4) become

$$\frac{\partial f_{m}}{\partial t'} - i\,\delta\Omega\,f_{m} = -\,\kappa\left\{f_{m} + 2C\,e^{\alpha_{m}t'}\frac{1}{L}\int_{0}^{L}dz'\,p(z',t')\right\} \tag{B.6a}$$

$$\frac{\partial p}{\partial t'} = -\gamma_{\perp}\left\{d\sum_{n}e^{ik_{n}z'}\,e^{-i\alpha_{n}t'}\,f_{n} + \left(1 + i\,(\delta_{AC} - \delta\Omega)\right)p\right\} \tag{B.6b}$$

$$\frac{\partial d}{\partial t'} = -\gamma_{\parallel}\left\{-\frac{1}{2}\,e^{\frac{2z'}{L}|hR|}\left(p\sum_{n}e^{-ik_{n}z'}\,e^{i\alpha_{n}t'}\,f^{*} + \text{c.c.}\right) + d - 1\right\} \tag{B.6c}$$

At this point, we make the single mode approximation for the field and obtain

$$\frac{\partial f_{0}}{\partial t'} = i\,\delta\Omega\,f_{0} - \kappa\left\{f_{0} + 2C\int_{0}^{L}dz'\,p(z',t')\right\} \tag{B.7a}$$

$$\frac{\partial p}{\partial t'} = -\gamma_{\perp}\left\{df_{0} + \left(1 + i\,(\bar{\delta}_{AC} - \delta\Omega)\right)p\right\} \tag{B.7b}$$

$$\frac{\partial d}{\partial t'} = -\gamma_{\parallel}\left\{-\frac{1}{2}\,e^{\frac{2z'}{L}|hR|}\left(p\sum_{n}e^{-ik_{n}z'}\,e^{i\alpha_{n}t'}\,f^{*} + \text{c.c.}\right) + d - 1\right\} \tag{B.7c}$$

A few comments are in order:

i)   the remaining field amplitude f_{0} is only a function of time as a result of the single-mode approximation;

ii) the atomic variables appear with their full space-time dependence;

iii)   however, the space dependence is parametric, in the sense that z' enters the equations of motion as a "label" and not as a running variable (such as t', for example);

iv)   from a formal point of view, this model is reminiscent of the inhomogeneously broadened extension of the Maxwell-Bloch equations (See Section 11).

The steady state configuration of this model can be derived easily by setting, first, the time derivatives of p and d equal to zero, with the result

$$p_{st} = -f_0 \frac{1 - i\bar{\Delta}}{1 + \bar{\Delta}^2 + |f_0|^2 \exp(\frac{2z'}{L}|\ln R|)} \tag{B.8a}$$

$$d_{st} = \frac{1 + \bar{\Delta}^2}{1 + \bar{\Delta}^2 + |f_0|^2 \exp(\frac{2z'}{L}|\ln R|)} \tag{B.8b}$$

where $\bar{\Delta} = \bar{\delta}_{AC}/(1+\bar{\kappa})$, and then substituting Eq. (B.8a) into the field steady state equation. The real and imaginary parts of the field equation take the form

$$1 = 2C \frac{1}{L} \int_0^L dz' \frac{1}{1 + \bar{\Delta}^2 + |f_0|^2 \exp(\frac{2z'}{L}|\ln R|)} \tag{B.9a}$$

$$\delta\Omega = \frac{\bar{\kappa}}{1+\bar{\kappa}} \delta_{AC} \tag{B.9b}$$

Equation (B.9a) can be integrated easily to yield

$$|f_0|^2 = (1+\bar{\Delta}^2) \frac{1 - R^2 \exp(\frac{|\ln R|}{c}(1+\bar{\Delta}^2))}{\exp(\frac{|\ln R|}{c}(1+\bar{\Delta}^2)) - 1} \tag{B.10}$$

As expected, in the limit R →1, we recover the uniform field results. Figure (B.1) shows an example of the variations of the output field modulus with respect to the gain. The line marked 1 has been calculated according to the uniform field limit, the line marked 2 represents the improved single-mode solution (B.10), and, finally, the line marked 3 represents the results of the exact solution, Eq. (4.10).

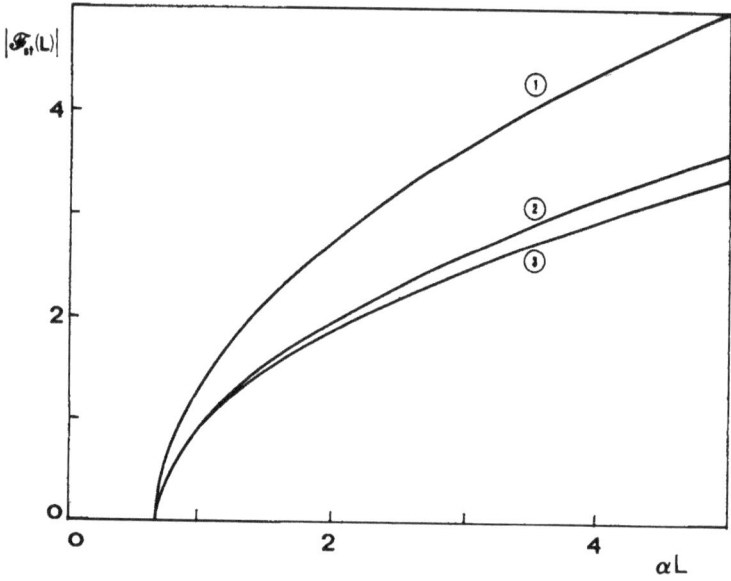

B.1   Dependence of the modulus of the output field on the laser gain αL for $R = 0.5$, $\bar{\alpha}_1 = 20.0$ and $\bar{\delta}_{AC} = 0$. Curve (1) has been obtained in the uniform field limit, curve (2) is the result of a calculation based on Eq. (B.10), while curve (3) is the exact solution of Eq. (4.10).

Appendix C - Bidirectional Propagation

As explained in our introductory comments, the selection of a unidirectional model for our analysis of the laser and its instabilities is motivated by reasons of simplicity; in addition, the unidirectional ring analyzed extensively in our main text is an experimentally accessible and interesting system. Yet, of course, most lasers operate in a bidirectional mode, either because the cavity is designed as a Fabry-Perot resonator or because, in a ring configuration, no special precaution is adopted to prevent the growth of both counterpropagating waves.

The theory of a bidirectional laser is more complicated than that of a unidirectional system because the interference between counterpropagating waves creates a spatial modulation in the medium that acts as an efficient backscatterer, at least under certain conditions. Thus the forward and backward waves are coupled to one another through a more complicated mechanism than might be expected at first sight.

Theoretical descriptions of Fabry-Perot lasers are as old as the laser itself and numerous versions are now available in advanced texbooks. A rigorous treatment of this problems including several improvements was provided recently by Lugiato and Narducci (Zeit. Phys., to be published). In this Appendix we discuss a few introductory aspects of this problem with the help of a simplified and approximate description.

The starting point is the usual wave equation

$$\frac{\partial^2 E}{\partial z^2} - \frac{1}{c^2}\frac{\partial^2 E}{\partial t^2} = \mu_0 \frac{\partial^2 P}{\partial t^2} \tag{C.1}$$

with the following, more realistic representation for the field and for the atomic polarization

$$E(z,t) = \frac{1}{2}\left(E_F(z,t)\,e^{i(k_c z - \omega_c t)} + E_F^*(z,t)\,e^{-i(k_c z - \omega_c t)}\right) + \frac{1}{2}\left(E_B(z,t)\,e^{-i(k_c z + \omega_c t)} +\right.$$

$$\left. E_B^*(z,t)\,e^{i(k_c z + \omega_c t)}\right) \tag{C.2a}$$

$$P(z,t) = \frac{1}{2}N\mu i\left(P_F(z,t)\,e^{i(k_c z - \omega_c t)} - P_F^*(z,t)\,e^{-i(k_c z - \omega_c t)}\right) + \frac{1}{2}N\mu i\left(P_B(z,t)\,e^{-i(k_c z + \omega_c t)} -\right.$$

$$\left. P_B^*(z,t)\,e^{i(k_c z + \omega_c t)}\right) \tag{C.2b}$$

where the indices F and B label the foward and backward components, respectively. The representation of the polarization is a truncation of an infinite Fourier expansion containing wave vectors $k_{cn} = nk_c$ ($n=\pm1, \pm2,...$) whose origin can be easily understood by inspection of the Bloch equations: the mixing of the lowest order terms of E and P creates population terms with a spatial dependence of the type $e^{\pm2ik_c}$; these are responsible for higher order polarization components, which in turn, produce additional corrections for the population difference, and so on.

We consider the lowest nontrivial approach to this problem, in which Eqs. (C.2a) and (C.2b) are assumed to be accurate enough, and where the population difference is represented by

$$D(z,t) = N \left\{ D_0(z,t) + \frac{1}{2} \left(D_2(z,t) \, e^{2ik_c z} + D_2^*(z,t) \, e^{-2ik_c z} \right) \right\} \qquad \text{(C.2c)}$$

Thus, in essence, the only main difference with respect to the unidirectional model is the appearence of the second harmonic terms D_2 and D_2^* in the population variable.

These terms are descriptive of a very important new physical effect which suggests that, even at the lowest order of approximation, our procedure is likely to be useful and to yield new results; the second harmonic terms D_2 and D_2^* are responsible for the creation of a spatial grating-like structure on top of the background contribution D_0. This is a consequence of the interference between the foward and backward waves. The importance of the second harmonic term in the population difference is that it acts as a source of backscattered radiation, as the equations of motion demonstrate, and provides an active dynamical coupling between the two cavity waves. It is now possible to imagine situations where the cavity becomes unstable against the growth of one of the travelling waves and develops pulsations and beating phenomena which are ruled out by the unidirectional model.

In the slowly varying envelope approximation, the wave equation takes the form

$$\frac{\partial E_F}{\partial z} + \frac{1}{c} \frac{\partial E_F}{\partial t} = - \frac{N\mu\omega_c}{2\varepsilon_0 c} P_F \qquad \text{(C.3a)}$$

$$\frac{\partial E_B}{\partial z} + \frac{1}{c} \frac{\partial E_B}{\partial t} = - \frac{N\mu\omega_c}{2\varepsilon_0 c} P_B \qquad \text{(C.3b)}$$

as we may expect on physical grounds. The new version of the Bloch equations can be derived from the usual amplitude equations

$$\frac{da_1}{dt} = - i \frac{E_1}{\hbar} a_1 + i \frac{\mu}{\hbar} E a_2 \qquad \text{(C.4a)}$$

$$\frac{da_2}{dt} = - i \frac{E_2}{\hbar} a_2 + i \frac{\mu}{\hbar} E a_1 \qquad \text{(C.4b)}$$

We define

$$D = N(a_2^* a_2 - a_1^* a_1) \qquad \text{(C.5)}$$

and equate $D(z, t)$ with the truncated expansion (C.2c) with the result

$$\frac{\partial D_0}{\partial t} = \frac{1}{2} \frac{\mu}{\hbar} \left(P_F E_F^* + P_F^* E_F + P_B E_B^* + P_B^* E_B \right) \qquad \text{(C.6a)}$$

$$\frac{\partial D_2}{\partial t} = \frac{\mu}{\hbar} \left(P_F E_B^* + P_B^* E_F \right) \qquad \text{(C.6b)}$$

In arriving at Eqs. (C.6) we have introduced the identification

$$a_1^* a_2 = \frac{1}{2}\left(P_F\, e^{i(k_c z - \omega_c t)} + P_B\, e^{-i(k_c z - \omega_c t)}\right)$$
(C.7)

and we have neglected systematically all the higher harmonic terms. In a similiar way, we can derive the polarization equations

$$\frac{\partial P_F}{\partial t} = -i\,\delta_{AC} P_F + \frac{\mu}{\hbar}\left(D_0 E_F + \frac{1}{2} D_2 E_B\right)$$
(C.8a)

$$\frac{\partial P_B}{\partial t} = -i\,\delta_{AC} P_B + \frac{\mu}{\hbar}\left(D_0 E_B + \frac{1}{2} D_2^* E_F\right)$$
(C.8b)

Now we let

$$F_{\binom{F}{B}} = \frac{\mu}{\hbar\sqrt{\gamma_\perp \gamma_\parallel}}\, E_{\binom{F}{B}}$$
(C.9a)

$$\bar{P}_{\binom{F}{B}} = \sqrt{\frac{\gamma_\perp}{\gamma_\parallel}}\, P_{\binom{F}{B}}$$
(C.9b)

$$\alpha = \frac{N\mu\omega_c}{2\hbar\epsilon_0 c\gamma_\perp}$$
(C.9c)

and finally we obtain the required approximation to the bidirectional equations of motion:

$$\frac{\partial F_F}{\partial z} + \frac{1}{c}\frac{\partial F_F}{\partial t} = -\alpha\bar{P}_F$$
(C.10a)

$$-\frac{\partial F_B}{\partial z} + \frac{1}{c}\frac{\partial F_B}{\partial t} = -\alpha\bar{P}_B$$
(C.10b)

$$\frac{\partial \bar{P}_F}{\partial t} = -\gamma_\perp\left(D_0 F_F + \frac{1}{2} D_2 F_B + (1+i\,\bar{\delta}_{AC})\bar{P}_F\right)$$
(C.10c)

$$\frac{\partial \bar{P}_B}{\partial t} = -\gamma_\perp\left(D_0 F_B + \frac{1}{2} D_2 F_F + (1+i\,\bar{\delta}_{AC})\bar{P}_B\right)$$
(C.10d)

$$\frac{\partial D_0}{\partial t} = -\gamma_\parallel\left\{-\frac{1}{2}\left(\bar{P}_F F_F^* + \bar{P}_F^* F_F + \bar{P}_B F_B^* + \bar{P}_B^* F_B\right) + D_0 - 1\right\}$$
(C.10e)

$$\frac{\partial D_2}{\partial t} = -\gamma_\parallel\left\{-\bar{P}_F F_B^* - \bar{P}_B^* F_F + D_2\right\}$$
(C.10f)

Note that the new Maxwell-Bloch equations show evidence of effects that are absent from the unidirectional model. For example, Eq. (C.10c) shows that the foward polarization wave is driven by two different source terms: one, $D_0 F_F$, is produced by the coupling of the foward wave with the background population difference, while the second $D_2 F_F$, is due to the backscattering of the backward wave against the spatial grating produced by D_2. A similar comment can be made with regard to the source terms of D_0 and D_2.

The equations of motion (C.10) must be supplemented by appropriate boundary conditions. In the case of ring cavity (See Fig. C.1a), we require that the following conditions be satisfied

$$F_F(0,t) = R\, F_F(L, t - \frac{\Lambda - L}{c}) \tag{C.11a}$$

$$F_B(0,t) = R\, F_B(L, t - \frac{\Lambda - L}{c}) \tag{C.11b}$$

In case of a Fabry-Perot (See Fig C.1b) we have

$$F_F(0,t) = \sqrt{R}\; F_B(0,t) \tag{C.12a}$$

$$F_B(L,t) = \sqrt{R}\; F_F(L,t) \tag{C.12b}$$

The analysis of the equations of motion (C.10) is a fairly complicated task and will not be considered in these lectures.

(a)

(b)

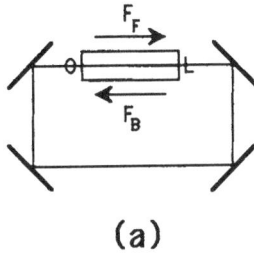

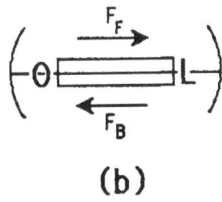

C.1   (a) Schematic representation of a bidirectional ring cavity; (b) schematic representation of a Fabry-Perot resonator.

Appendix D - Exact Steady State Equations:
Resonant linhomogeneously Broadened Laser

The problem to be discussed in this appendix is the following: given the steady state equation

$$\frac{d\rho}{dz} = \alpha \int_{-\infty}^{+\infty} d\bar{\delta}\, g(\bar{\delta}) \frac{1}{1 + \bar{\delta}^2 + \rho^2}\, \rho \tag{D.1}$$

for the field modulus $\rho(z)$, we want to calculate the output field $\rho(L)$ in accordance with the boundary conditions

$$\rho(0) = R\, \rho(L) \tag{D.2}$$

The numerical procedure adopted for the solution of this problem can be summarized as follows. We first introduce the scaled variable $\eta = z/L$ and let

$$x = \frac{\bar{\delta}}{\sqrt{2}\, \bar{\sigma}_D} \tag{D.3}$$

Equation (D.1) takes the form

$$\frac{d\rho}{d\eta} = \frac{\alpha L}{\sqrt{\eta}} \int_{-\infty}^{+\infty} dx\, e^{-x^2} \frac{\rho}{1 + 2\bar{\sigma}_D^2 x^2 + \rho^2} \tag{D.4}$$

with

$$\rho(0) = R\, \rho(1) \tag{D.5}$$

For a given value of $\rho(0)$ (usually an appropriate guess), we can solve the space-dependent equation for ρ by a standard Runge-Kutta method, and evaluate the integral using a Gauss-type integration routine [See, for example, Abramowitz and Stegun (1968) Eq. 25.5.46]. The selected value of $\rho(0)$, will produce, in general, a value of $\rho(1)$ which is not compatible with eq. (D.5), that is, such that the difference

$$\delta = \rho(0) - R\, \rho(1) \tag{D.6}$$

is different from zero. Now we can select a larger value of $\rho(0)$, generate another value for $\rho(1)$, and repeat this procedure until δ changes sign. This indicates that, for some $\rho(0)$ contained between the last two values used in the iteration, the difference δ was equal to zero, as required. By stepping back to the old value of $\rho(0)$ and going forward with a smaller step size, one can arrive by linear interpolation to a very accurate value of the required output field modulus $\rho(L)$ for the chosen values of the operating parameters αL, $\bar{\sigma}_D$ and R. As a test of this procedure we have adopted this algorithm to the case $\bar{\sigma}_D = 0$, for which an exact analytic solution exist, and obtained excellent results.

REFERENCES

Abraham, N.B., 1983, Laser Focus, 19, 73.
Abraham, N.B., et al., in Third New Zealand Symposium on Laser Physics, vol. 182 of Springer Lecture Notes in Physics, 1983, (Springer Verlag, Berlin) p. 107.
Abraham, N.B., J.P. Gollub and H.L. Swinney, 1984, Meeting Report-Testing Nonlinear Dynamics, Physica D11, 252.
Abraham, N.B., L.A. Lugiato and L.M. Narducci, 1985, J. Opt. Soc. Am. B2, 7.
Abraham, N.B., L.A. Lugiato, P. Mandel, L.M. Narducci and D.K. Bandy, 1985, J. Opt. Soc. Am. B2, 35.
Abraham, N.B., P. Mandel, L.M. Narducci, in Progress in Optics, edited by E. Wolf (North Holland, Amsterdam), to be published.
Abramowitz, M., and I. Stegun, Handbook of Mathematical Functions, 1968, (Dover Publications).
Ackerhalt, J.R., P.W. Milonni, and M.L. Shih, 1985, Phys. Rep. 128, 205.
Bandy, D.K., L.M. Narducci, L.A. Lugiato, N.B. Abraham, 1985, J. Opt. Soc. Am. B2, 56.
Beck, R., W. Englisch, K. Gurs, 1980, Tables of Laser Lines in Gases and Vapors, (Springer Verlag, Berlin).
Bennett, W.R., Jr., 1962, Phys. Rev. 126, 580.
Bennett, W.R., Jr., 1977, The Physics of Gas Lasers, (Gordon and Breach, New York).
Bonifacio, R., L.A. Lugiato, 1978, Lett. Nuovo Cimento, 21, 505.
Boyd, R.W., M.G. Raymer, and L.M. Narducci, Editors, 1986, Optical Instabilities, (Cambridge University Press).
Buley, E.R. and F.W. Cummings, 1964, Phys. Rev. 134, A1454.
Carmichael, H.J., 1983, Phys. Rev. A28, 480.
Casperson, L.W., 1978, IEEE J. Quantum Electron. QE-14, 756.
Casperson, L.W., 1980, Phys. Rev. A21, 911.
Casperson, L.W., 1981, Phys. Rev. A23, 248.
Casperson, L.W., in Third New Zealand Symposium on Laser Physics, vol. 182 of Springer Lecture Notes in Physics, 1983, (Springer Verlag, Berlin) p. 88.
Casperson, L.W., 1985, J. Opt. Soc. Am. B2, 62.
Casperson, L.W., 1985, J. Opt. Soc. Am. B2, 72.
Casperson, L.W., 1985, J. Opt. Soc. Am. B2, 993.
Casperson, L.W., 1986, in Optical Instabilities, edited by R.W. Boyd, M.G. Raymer, L.M. Narducci (Cambridge University), p. 58.
Chrostowski, J. and N.B. Abraham, Editors, 1986, Optical Chaos, SPIE Vol. 667, (SPIE Bellingham).
Cohen-Tannoudji, C., B. Diu, and F. Laloe, Quantum Mechanics, Vol I and II (Wiley Interscience, New York).
Einstein, A., 1917, Phys. Z., 18, 121.
Englund, J.C., R.R. Snapp, W.C. Schieve, 1984, in Progress in Optics, Vol XXI, Edited by E. Wolf., (North Holland, Amsterdam).
Fain, V.M., 1958, Sov. Phys. JETP 6, 726.
Frehland, E., Editor, 1984, Synergetics-From Microscopic to Macroscopic Order (Springer Verlag, Berlin).
Graham, R., and H. Haken, 1968, Z. Phys. 213, 420.
Grazyuk, A.Z., and A.N. Oraevskii, 1964a, Radio. Eng. Electron. Phys. 9, 424.
Grazyuk, A.Z., and A.N. Oraevskii, 1964b, in Quantum Electronics and Coherent Light, Edited by P.A. Miles, (Academic Press, New York), p. 192.
Haken, H., and H. Sauermann, 1963a, Z. Phys. 173, 261.

Haken, H., and H. Sauermann, 1963b, Z. Phys. 176, 47.
Haken, H., and H. Sauermann, 1964, in Quantum Electronics and Coherent Light, edited by P.A. Miles, (Academic Press, New York), p. 111.
Haken, H., 1966, Z. Phys. 190, 327.
Haken, H., 1970, Handbuch der Physics, edited by L. Genzel, Vol. XXV/2c, (Springer Verlag, Berlin).
Haken, H., 1975, Phys. Lett. 53A, 77.
Haken, H., 1982, Evolution of Order and Chaos in Physics, Chemistry, Biology, (Springer Verlag, Berlin).
Haken, H., 1983a, Synergetics-An Introduction, Third Edition, (Springer Verlag, Berlin).
Haken, H., 1983b, Advanced Synergetics, (Springer Verlag, Berlin).
Haken, H., 1986, in Optical Instabilities, edited by R.W. Boyd, M.G. Raymer, L.M. Narducci (Cambridge University), p. 1.
Harrison, R.G., and D.J. Biswas, 1985, Prog. Quantum Electron. 11, 127.
Hillman, L.W., J. Krasinski, R.W. Boyd and C.R. Stroud, Jr., 1984, Phys. Rev. Lett. 52, 1605.
Hofelich-Abate, E., and F. Hofelich, 1968, Z. Phys. 209, 13.
Journal of the Optical Society of America, 1985, Special Issue on Instabilities in Active Optical Media, Volume B2.
Klische, W., C.O. Weiss, W. Al-Soufi, G. Huttmann, 1986, in Optical Instabilities, edited by R.W. Boyd, M.G. Raymer, L.M. Narducci (Cambridge University), p. 237.
Lamb, W.E., Jr., 1964, Phys. Rev. 134, 1429.
Lax, M., 1968, in Statistical Physics, Phase Transitions and Superfluidity, Vol. II, Brandeis University Summer Institute in Theoretical Physics, Edited by M. Chretien, E.P. Gross, and S. Deser, (Gordon and Breach, New York), p. 269.
Lorenz, E., 1963, J. Atm. Sci. 20, 130.
Lugiato, L.A., 1980a, Opt. Comm. 33, 108.
Lugiato, L.A., 1980b, Lett. Nuovo Cimento 29, 375.
Lugiato, L.A., 1984, in Progress in Optics, Vol. XXI, edited by E. Wolf (North Holland, Amsterdam) p. 71.
Lugiato, L.A., P. Mandel, and L.M. Narducci, 1984, Phys. Rev. A29, 1438.
Lugiato, L.A., L.M. Narducci, E.V. Eschenazi, D.K. Bandy and N.B. Abraham, 1985, Phys. Rev. A32, 1563.
Lugiato, L.A., L.M. Narducci, and M.F. Squicciarini, 1986, Phys. Rev., A34, 3101.
Lugiato, L.A., L.M. Narducci, D.K. Bandy, and J.R. Tredicce, 1986, Phys. Rev. A33, 1109.
Lugiato, L.A., L.M. Narducci, 1986, in Optical Instabilities, edited by R.W. Boyd, M.G. Raymer, L.M. Narducci (Cambridge University), p. 34.
Lugiato, L.A., M.L. Asquini, and L.M. Narducci, 1986, in Optical Chaos, edited by J. Chrostowski and N.B. Abraham, SPIE Vol. 667, (SPIE Bellingham), p. 132.
Narducci, L.M., H. Sadiky, L.A. Lugiato and N.B. Abraham, 1985, Opt. Comm. 55, 370.
Narducci, L.M., J.R. Tredicce, L.A. Lugiato, N.B. Abraham, and D.K. Bandy, 1986, Phys. Rev. A33, 1842.
Nicolis, G. and I. Prigogine, 1977, Self-Organization in Nonequilibrium Systems-From Dissipative Structures to Order Through Fluctuations, (Wiley, New York).
Oraevskii, A.N., 1959, Radio. i. Elek. Tek. Kor. 4, 718.
Oraevskii, A.N., 1964, Molecular Oscillators, (Nauka, Moskow).
Oraevskii, A.N., and A.V. Uspenskii, 1968, in Quantum Electronics in Lasers and Masers, edited by D.V. Skobel'tsyn, Vol 31 (Consultants Bureau, New York), p.87.
Oraevskii, A.N., 1981, Kvan. Elekt. 8, 130.
Risken, H., C. Schmidt, and W. Weidlich, 1966, Z. Phys. 194, 337.

Risken, H., and K. Nummedal, 1968a, J. Appl. Phys. 39, 4662.
Risken, H., and K. Nummedal, 1968b, Phys. Lett. 26A, 275.
Risken, H., 1986, in Optical Instabilities, edited by R.W. Boyd, M.G. Raymer, L.M.
    Narducci (Cambridge University), p. 20.
Sargent, M., III, M.O. Scully and W.E. Lamb, Jr., 1979, Laser Physics, (Addison Wesley,
    Reading).
Sattinger, D., 1973, Topics in Stability and Bifurcation Theory, (Springer Verlag,
    Berlin)
Scully, M.O., and W.E. Lamb, Jr., 1967, Phys. Rev. 159, 208.
Scully, M.O., and W.E. Lamb, Jr., 1968a, Phys. Rev. 166, 246.
Scully, M.O., and W.E. Lamb, Jr., 1968b, Phys. Rev. 166, 368.
Siegman, A.E., 1971, An Introduction to Lasers and Masers, (McGraw Hill, New York).
Sparrow, C.T., 1982, The Lorenz Equations: Bifurcations. Chaos and Strange
    Attractors, (Springer Verlag, Berlin).
Stenholm, S., 1971, in Progr. in Quantum Electron., Vol. 1, edited by J.H. Sanders and
    K.W.H. Stevens (Pergamon Press) p. 187.
Stroud, C. R., Jr., K. Koch, S. Chakmakjian, L.W. Hillman, 1986, in Optical Chaos,
    edited by J. Chrostowski and N.B. Abraham, SPIE Vol. 667, (SPIE Bellingham), p.
    47.
Stroud, C. R., Jr., K. Koch, S. Chakmakjian, 1986, in Optical Instabilities, edited by
    R.W. Boyd, M.G. Raymer, L.M. Narducci (Cambridge University), p. 274.
Svelto, O., 1982, Principles of Lasers, Second Edition, (Plenum Press, New York).
Swinney, H.L. and J.L. Gollub, Editors, 1981, Hydrodynamic Instabilities and Transition
    to Turbulence, (Springer Verlag, Berlin).
Tang, C.L., 1963, J. Appl. Phys. 34, 2935.
Tang, C.L., H. Statz, and G. DeMars, 1963, Appl. Phys. Lett. 2, 222.
Uspenskii, A.V., 1963, Radio. Eng. Electron. Phys. (USSR), 8, 1145.
Uspenskii, A.V., 1964, Radio. Eng. Electron. Phys. (USSR), 9, 605.
Weiss, C.O., 1986, in Optical Chaos, edited by J. Chrostowski and N.B. Abraham, SPIE
    Vol. 667, (SPIE Bellingham), p. 26.
Yariv, A., 1985, Optical Electronics, Third Edition, (Holt, Rinehart and Winston, New
    York).
Zeghlache, H. and P. Mandel, 1985, J. Opt. Soc. Am. B2, 18.

SELECTED ILLUSTRATIONS AND INTERPRETATIONS
OF LASER PHYSICS AND LASER INSTABILITIES
IN EXPERIMENTAL SYSTEMS

by

Neal B. Abraham

Department of Physics
Bryn Mawr College
Bryn Mawr, PA 19010 USA

158

I. Foreword

The history of the pedagogy of laser physics is punctuated by successive generations of textbooks designed to explain the basic principles of field-atom interactions, to introduce appropriate formalisms for dealing with resonant cavities, and to illustrate various points with results from specific laser systems. When successful, a good approach teaches formal techniques and provides the researcher with a well-developed (and trained) intuition which can be used to interpret both theoretical and experimental results.

In the approach taken at this school we hope to provide the interested researcher with the most modern and formal semiclassical approach to laser theory which will deal as carefully as possible with such topics as single mode and multimode operation and with transitions from constant intensity output to pulsating intensity. In counterpoint to the formal approach taken by Professor Narducci, my lectures are designed to provide the complementary details of many experiments and a more intuitive, "experimentalist's" picture of both theoretical and experimental results. We do not intend to urge two views of the physical world, rather we hope that by our showing some contrasting styles each reader can more effectively develop a personal synthesis.

In our collaborations over the last three years (which have also included Professor Lugiato of the University of Milan, Dr. Bandy of Drexel University and Bryn Mawr College, Dr. Tarroja of Bryn Mawr College, and Professor Tredicce of The National Institute of Optics in Florence, Italy and Drexel University) Professor Narducci and I have found a rewarding and creative interplay in our mixing of experimental results, the rigorous derivations and results of analytical and numerical theoretical studies, and the interpretations and intuitions of experimentalists and theoreticians.

It is our hope that some of that interplay will be set forth in our lectures and discussions.

I wish to express my thanks to my collaborators in our experiments, notably Dr. Tarroja, Professor Tredicce, and Professor Gioggia at Widener University and to students Mari Maeda, Lois Hoffer, Nancy Halas, Emily Fisch, Su-Nin Liu and Joel Wesson whose thesis projects were major contributions. I also wish to thank fellow experimentalists who have kindly shared their latest results with us. I am particularly indebted to Ann Daudert and LeAnn Davis for their patient typing and artistic contributions for this manuscript.

II. Background Notes for the Discussion of Laser Systems

A. Rate Equations for Multilevel Systems

It now appears that a laser can be made from almost any substance. (Solids, liquids, gases and even jello have been used as media.) The important characteristic is that energy can be stored in the medium in a suitable nonequilibrium fashion. Whenever the population of an excited energy level exceeds the population of a lower energy level, then net energy may be extracted by interaction with a resonantly tuned radiation field which stimulates emission of radiation. Such emissions increase the population of the lower level, and, unless the population of this level rapidly decays to even lower levels, then the process of net energy extraction by stimulated processes ceases. In contrast, flourescence processes (spontaneous emissions) can continue to occur so long as there is energy stored in the excited states. Laser materials must, therefore, be excited from equilibrium and may be used in a pulsed or continous manner depending on the available energy level schemes and their decay rates. In general, one finds a good lasing transition when the upper level lifetime is much longer than the lower level lifetime, that is, when the decay rates $\gamma_a = 1/T_a$ and $\gamma_b = 1/T_b$ (as indicated in Figure 1) are arranged so that $\gamma_b \gg \gamma_a$. Note that γ_a is not limited to representing only spontaneous transitions from level a to level b, as transitions to other lower levels are included in γ_a. We distinguish the direct a → b spontaneous transition rate by γ_{ab}, as indicated in Figure 2.

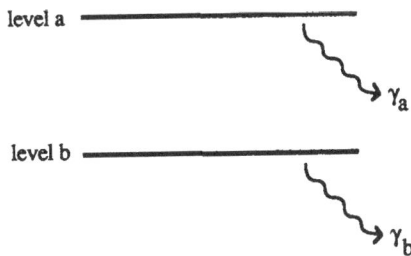

Figure 1

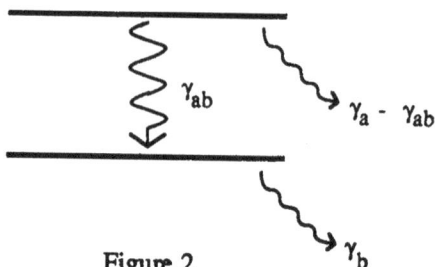

Figure 2

If we begin to consider this model of a two-level system interacting with a radiation field of density ρ tuned to frequency υ such that $h\upsilon = E_a - E_b$, then in the rate equation approximation

$$\frac{dN_a}{dt} = -\gamma_a N_a + R_a + \rho W N_b - \rho W N_a$$

and

$$\frac{dN_b}{dt} = -\gamma_b N_b + \gamma_{ab} N_a + R_b + \rho W N_a - \rho W N_b,$$

(1)

where R_a and R_b are excitation rates for the two levels, N_a and N_b are the level populations, and W is the stimulated emission (absorption) rate coefficient. If the level populations are in equilibrium with the particular field energy density ρ (which is possible if ρ is constant or slowly varying), then

$$\frac{dN_a}{dt} = \frac{dN_b}{dt} = 0 \quad \text{and the solutions of eqs. (1) are}$$

$$N_a = \frac{1}{\gamma_a} [R_a - \rho W(N_a - N_b)]$$

(2a)

and

$$N_b = \frac{1}{\gamma_b} [R_b - \rho W(N_a - N_b)] - \frac{\gamma_{ab}}{\gamma_b} N_a.$$

(2b)

These can be simplified if we first note that when $\rho = 0$, defining N_{ao} and N_{bo},

$$N_{ao} \equiv \frac{1}{\gamma_a} R_a \, , \quad \text{and} \tag{3a}$$

$$N_{bo} \equiv \frac{1}{\gamma_b} R_b + \frac{\gamma_{ab}}{\gamma_b} N_{ao} = \frac{1}{\gamma_b} R_b + \frac{\gamma_{ab}}{\gamma_a \gamma_b} R_a \, . \tag{3b}$$

Thus

$$\Delta N_o \equiv N_{ao} - N_{bo} = \frac{1}{\gamma_a} R_a \left(1 - \frac{\gamma_{ab}}{\gamma_b}\right) - \frac{1}{\gamma_b} R_b \, . \tag{4}$$

Returning to Eq. (2) we can rearrange the terms as follows:

$$N_a = \frac{1}{\gamma_a} [R_a - \rho W(N_a - N_b)] \tag{5a}$$

and

$$N_b = \frac{1}{\gamma_a} [R_b + W(N_a - N_b)] + \frac{\gamma_{ab}}{\gamma_a \gamma_b} [R_a - \rho W(N_a - N_b)] \, . \tag{5b}$$

Subtracting, we find

$$N_a - N_b = \frac{R_a}{\gamma_a} - \frac{R_b}{\gamma_b} - \frac{\gamma_{ab}}{\gamma_a \gamma_b} R_b -$$
$$\frac{\rho W}{\gamma_a} (N_a - N_b) - \frac{\rho W}{\gamma_b} (N_a - N_b) + \frac{\gamma_{ab}}{\gamma_a \gamma_b} \rho W(N_a - N_b) \, . \tag{6}$$

Solving for $N_a - N_b \equiv \Delta N$ we find

$$\Delta N = \frac{\Delta N_o}{1 + \rho W \left(\dfrac{1}{\gamma_a} + \dfrac{1}{\gamma_b} - \dfrac{\gamma_{ab}}{\gamma_a \gamma_b} \right)} \, . \tag{7}$$

For a complex system such as the one we have chosen, we see that the population difference is reduced (saturated) by interacting with the radiation field in the same manner as found by Narducci for the simpler two-level atom model. As in that case, the population difference in this case is simply described by

$$\Delta N = \Delta N_0 / (1 + I/I_s) , \qquad (8)$$

where I represents the intensity of the interacting field and I_s corresponds to that characteristic intensity given by the collection of constants in the denominator of Eq. (7).

We have defined an extremely important physical quantity, the "SATURATION INTENSITY", denoted by I_s, which is independent of the excitation processes (R_a and R_b) and depends only on the decay rates of the levels and the stimulated emission (absorption) rate coefficients.

When the intensity I reaches the saturation intensity, the population difference is reduced to one half of its zero-field or "UNSATURATED" value.

For the purposes of rate equations, the field grows or decays depending on the population difference and it appears from Eq. (8) that the population difference in multilevel systems can be well approximated as arising from a simple two-level system such as that provided as a paradigm by Narducci, particularly as the two models are saturated by an illumination intensity in the same way. Of course, if we must examine the dynamical evolution in detail, there may be important differences and one should not arbitrarily make this approximation. However, for both steady state and dynamical behavior, these two sets of equations are often indistinguishable in their effects.

B. Lineshapes, Dispersion and Bandwidths of Response

We know from the study of classical oscillators that if we stimulate a damped oscillatory system there is not just a single frequency for which the oscillator reacts but rather a spread of frequencies over which significant response can be observed, as shown schematically in Figure 3.

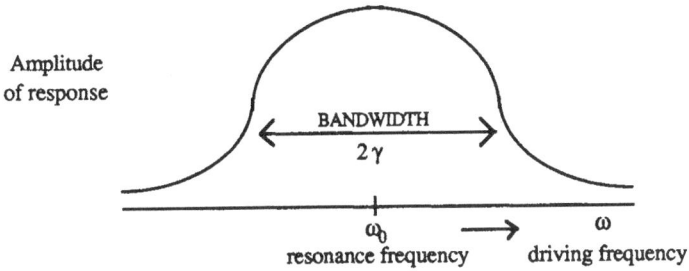

Figure 3

The bandwidth is typically given by the exponential damping rate (or decay rate) of the system if it is perturbed and allowed to freely decay. Thus it is inversely proportional to the characteristic lifetime of the oscillations of the perturbed oscillator, as shown schematically in Figure 4.

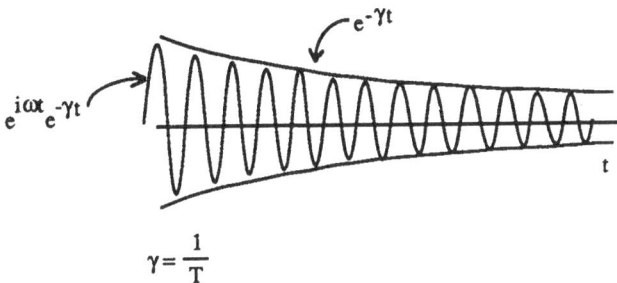

Figure 4

When absorption or emission spectroscopy is done in the optical range, one finds that materials (such as atoms or molecules) with well-defined energy levels have narrow resonant peaks. Shorter-lived states in liquids and solids (or gases at high pressure) lead to broader resonances, as shown schematically in Figure 5.

Schematic:

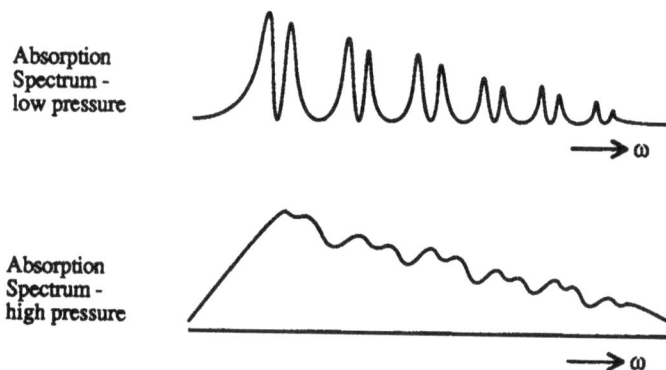

Figure 5

Thus we readily believe that there is a response bandwidth inversely proportional to the lifetime (or coherence time) of emission from the relevant transition.

If we consider what we know from only rate equations, we might be temped to concentrate on a particular excited state and to assume that its lifetime is governed by spontaneous and stimulated emission.

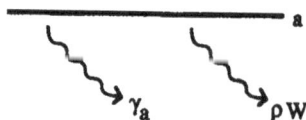

$$\frac{dN_a}{dt} = -\gamma_a N_a - \rho W N_a$$

Figure 6

As suggested in Figure 6, we might initially expect that the linewidth would be γ_a in the absence of the field and $\gamma_a + \rho W$ in the presence of the field, suggesting that stimulated emission broadens the lineshape. While such a broadening is observed and our explanation is conceptually correct, it is important to add that the explanation presented above has serious errors. First, a resonance requires two levels, while the rate equation picture is completely unable to tell us that the field must have a resonant value of frequency (given, as we know, by $h\upsilon = E_a\text{-}E_b$). Secondly, as emphasized by Narducci, the field-atom interaction as we describe it more fundamentally in quantum mechanics is not governed by a Hamiltonian of the type

$$H = I\Delta N ,$$

but rather by the form

$$H = \vec{p} \cdot \vec{E}$$

where p is the dipole moment and E is the electric field. Thus resonance theory of optical transitions must first deal with permanent or induced dipole moments in the material; by coupling them to the field we can find the resonances. It will only be in the lifetime of the material dipole moment that we will find an origin to the bandwidth of response of the medium.

If we take the simple case of the two-level atom described by quantum mechanics, we can look at the effect on the polarization of a change in the frequency of the driving field. We begin with the material equations as derived by Narducci in Eqs. (3.16) and (3.19) of his lecture notes:

$$\frac{\partial P_0}{\partial t} = -i\left(\omega_A - \omega\right) P_0 + \frac{\mu}{\hbar} E_0 D_0 - \gamma_\perp P_0 \tag{9.a}$$

$$\frac{\partial D_0}{\partial t} = -\frac{\mu}{2\hbar}\left(P_0 E_0^* - P_0^* E_0\right) - \gamma_\|\left(D_0 - \sigma\right) \tag{9.b}$$

where

$$D_0 \equiv (N_1\text{-}N_2)/N$$

is the population difference per atom. P_0 is constructed from off-diagonal elements of the atomic density matrix and corresponds to the polarization induced in the medium by the driving

166

field. We have explicitly written the frequencies ω_A (the energy level difference divided by $\hbar$) and ω (the driving frequency of the electric field given by $E = E_0 e^{i(kz-\omega t)}$).

We write E_0, P_0, D_0 as slowly varying fuctions of space and time (NOT CONSTANTS) which are obviously mutually interdependent as shown in Eqs. (9). If the field amplitude E is constant, after a transient the material variables come into equilibrium with the driving field.

Setting

$$\frac{\partial D_0}{\partial t} = \frac{\partial P_0}{\partial t} = 0$$

we can solve for P_0 and D_0 as follows:

From (9.1) we have

$$P_0 = \frac{(\mu/\hbar)\, E_0 D_0}{\gamma_\perp + i\,(\omega_A - \omega)} \quad . \tag{10}$$

Substituting into (9.2) we obtain

$$\gamma_\| (D_0 - \sigma) = -\frac{\mu^2}{2\hbar^2}\left(\frac{|E_0|^2 D_0}{\gamma_\perp + i\,(\omega_A - \omega)} + \frac{|E_0|^2 D_0}{\gamma_\perp - i\,(\omega_A - \omega)}\right), \text{ and}$$

$$\gamma_\| D_0 - \gamma_\| \sigma + \frac{\mu^2}{2\hbar^2}\,\frac{2\gamma_\perp |E_0|^2 D_0}{\gamma_\perp^2 + (\omega_A - \omega)^2} = 0 \ .$$

Then

$$D_0 = \frac{\sigma(\gamma_\perp^2 + (\omega_A - \omega)^2)}{\gamma_\perp^2 + (\omega_A - \omega)^2 + \left(\frac{\mu^2 \gamma_\perp}{\hbar^2 \gamma_\|}\right)|E_0|^2} \quad . \tag{11}$$

Substituting Eq. (11) in Eq. (10)

$$P_0 = \frac{\mu}{\hbar} E_0 \frac{(\gamma_\perp - i(\omega_A-\omega))\sigma}{\gamma_\perp^2+(\omega_A-\omega)^2} \cdot \frac{\gamma_\perp^2 + (\omega_A-\omega)^2}{\gamma_\perp^2+(\omega-\omega_A)^2 + \left(\frac{\mu^2 \gamma_\perp}{\hbar^2 \gamma_\parallel}\right)|E_0|^2} .$$

Hence

$$P_0 = E_0 \frac{\mu\sigma}{\hbar} \frac{\gamma_\perp}{\gamma_\perp^2 + (\omega-\omega_A)^2 + \frac{\mu^2\gamma_\perp}{\hbar^2\gamma_\parallel}|E_0|^2}$$

(12)

$$+ E_0 \frac{i\mu\sigma}{\hbar} \frac{\omega-\omega_A}{\gamma_\perp^2 + (\omega-\omega_A)^2 + \frac{\mu^2\gamma_\perp}{\hbar^2\gamma_\parallel}|E_0|^2} .$$

The first thing to notice is the behavior of the population difference D_0 in Eq. (11) which can be rewritten as:

$$D_0 = \frac{\sigma}{1 + \left(\frac{\mu^2\gamma_\perp}{\hbar^2\gamma_\parallel}\right)\left(\frac{|E_0|^2}{\gamma_\perp^2 + (\omega_A-\omega)^2}\right)} .$$

(13)

which is in the form of our earlier equation for a saturated population difference, Eq. (8),

$$\Delta N = \Delta N_0 / (1 + I/I_s \{1 + [(\omega_A-\omega)^2/\gamma_\perp^2]\})$$

(8)

if we take

$$I_s = \left(\frac{\gamma_\perp \gamma_\parallel \hbar^2}{\mu^2}\right) .$$

(14)

168

We see that, when the driving field is resonant ($\omega = \omega_A$), the population difference is saturated more quickly. The saturating effect of the intensity is reduced in half when $(\omega - \omega_A)^2 = \gamma_\perp^2$ or $|\omega - \omega_A| = \gamma_\perp$ and when $|\omega - \omega_A| \gg \gamma_\perp$ the saturation effect is negligible.

There is the obvious conclusion to draw which is that the reduction in the population difference by the saturating field occurs because there is increased stimulated emission of radiation. To see this effect more clearly, consider the reduced Maxwell's equation for the evolution of the electric field (see Eq. 3.17 in Narducci's notes)

$$c \frac{\partial E_0}{\partial z} = - \frac{N\mu\omega}{2\varepsilon_0} P_0 , \qquad (15a)$$

where we are considering here a traveling wave field propagating through the medium as shown in Figure 7.

$$E_0(z)\, e^{i(kz - \omega t)} \Rightarrow$$

Figure 7

As shown in Figure 8 we assume that the field can be described by a time independent amplitude $E_0(z)$ which changes only by interaction with the medium.

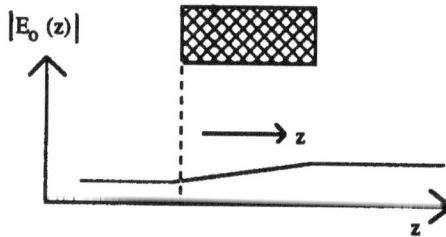

Figure 8

Equivalently we can write the evolution of the field E as

$$\frac{\partial E_o}{\partial z} = ik'E_o \qquad (15b)$$

which is equivalent to writing the following in the linear regime $(I \ll I_s)$

$$E_o(z) = E_o(0)\, e^{ik'z}$$

or

$$E(z) = E_o(0)\, e^{ik'z}\, e^{i(kz-\omega t)} \;.$$

From Eqs. (15a and 15b)

$$k' = -\frac{N\mu\omega P_o}{2\varepsilon_o i c E_o} \qquad (16)$$

and using Eq. (12)

$$k' = -\frac{N\mu^2 \sigma}{2\varepsilon_o c \hbar} \left(\frac{\omega - \omega_A}{\gamma_\perp^2 + (\omega - \omega_A)^2 + \gamma_\perp^2 \, I/I_s}\right)$$

$$+\frac{iN\mu^2 \sigma \gamma_\perp}{2\varepsilon_o c \hbar} \left(\frac{1}{\gamma_\perp^2 + (\omega - \omega_A)^2 + \gamma_\perp^2 \, I/I_s}\right) \;. \qquad (17)$$

Equation (17) gives us two very important results. Recall that σ is negative if the upper state population is greater than the lower state population. Then the real part of k' gives a dispersive effect, a change of wavelength in the medium, which is an effect we attribute to the index of refraction n. The total real wavevector in the medium is given by nk where k is the free-space wavevector. Then from Eq. (17) the index of refraction is given by

$$n(\omega) - 1 = \frac{-N\mu^2 \sigma}{2\varepsilon_o c \hbar k} \left(\frac{\omega - \omega_A}{\gamma_\perp^2 + (\omega - \omega_A)^2 + \gamma_\perp^2 \, I/I_s}\right) \qquad (18)$$

and the gain coefficient which describes the local exponential growth or decay of the field amplitude is given by $\alpha(\omega) = -(Im\, k')$,

$$\alpha(\omega) = \frac{-N\mu^2\sigma\,\gamma_\perp}{2\varepsilon_0 c\hbar} \left(\frac{1}{\gamma_\perp^2 + (\omega - \omega_A)^2 + \gamma_\perp^2\, I/I_s} \right). \qquad (19)$$

Written in the form of normalized frequencies, $(\omega - \omega_A)^2/\gamma_\perp^2$, so that we have standard functional forms, these physical quantities become

$$n(\omega)-1 = -\frac{N\mu^2\sigma}{2\varepsilon_0 c\hbar\,\gamma_\perp k} \left(\frac{(\omega - \omega_A)^2/\gamma_\perp^2}{(1 + I/I_s) + (\omega - \omega_A)^2/\gamma_\perp^2} \right) \qquad (18')$$

and

$$\alpha(\omega) = -\frac{N\mu^2\sigma}{2\varepsilon_0 c\hbar\,\gamma_\perp} \left(\frac{1}{(1 + I/I_s) + (\omega - \omega_A)^2/\gamma_\perp^2} \right). \qquad (19')$$

We see that the half width of the response function is given by

$$\Delta\omega = \gamma_\perp (1 + I/I_s)^{1/2}.$$

Let us look briefly at schematic drawings of these functions for $\sigma < 1$ (the case of population inversion) for two different strengths of the field $I(\omega)$. Note that these functions are to be evaluated only at the frequency ω at which the strong intensity $I(\omega)$ exists. These do not indicate the "residual gain" or "residual dispersion" measured at other frequencies in the presence of the strong field at some specific frequency ω.

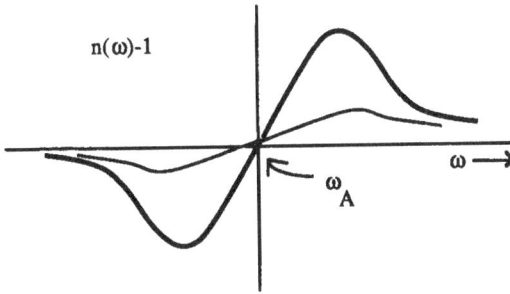

Figure 9a

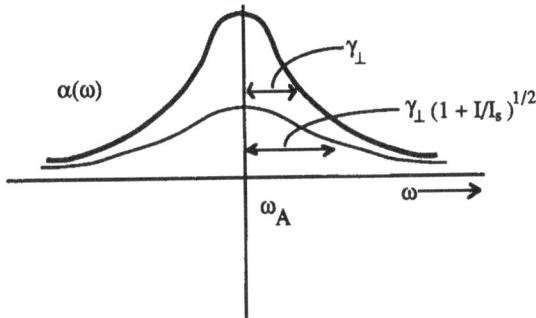

weak field (thick line); strong field (thin line)

Figure 9b

For a driving field intensity near zero we have a Lorentzian lineshape for the gain which is peaked at the resonance frequency ω_A and which has a full width at half maximum (FWHM) of $2\gamma_\perp$. The associated anomalous dispersion is positive for $\omega > \omega_A$ and negative for $\omega < \omega_A$. The extrema of the dispersion function occur at approximately $|\omega - \omega_A| = \gamma_\perp$.

For a strong saturating field that is scanned in frequency there are two noticeable effects. The gain is reduced and the resonance width is broadened, an effect known as "POWER BROADENING".

For $\sigma > 1$, the case of an absorbing medium, the functions change sign.

172

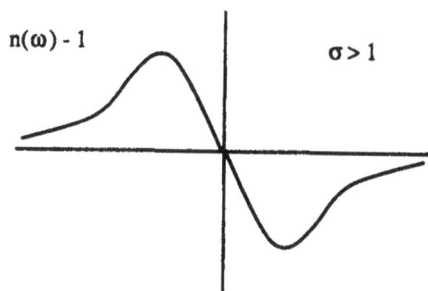

Figure 10a

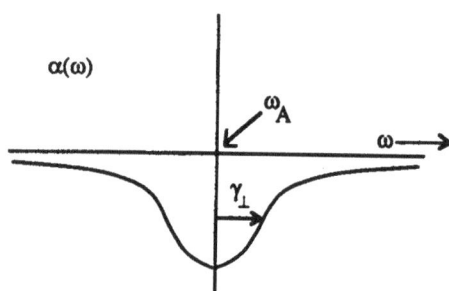

Figure 10b

The pairs of functions in Figures 9 and 10 are subtly related to each other in a manifestation of a quite general relationship which can be described qualitatively as follows:

(1) When the gain is symmetric with respect to ω_A, the dispersion function is zero at ω_A.

(2) When the gain is positive (negative) at ω_A, the slope of the dispersion function is positive (negative) at ω_A.

(3) The extrema of the dispersion function occur at approximately the frequencies where the gain is reduced to one half of its value at ω_A.

These relations are generally true for all functional forms of $k'(\omega)$ and are summarized formally by the Kramers-Kronig relations.

C. Evolution of the Intensity Upon Propagation in a Medium

We can apply these results to the evolution of the intensity of the field in the rate equation limit: since

$$\frac{dI}{dz} = 2|E_0| \frac{d|E_0|}{dz} \; ,$$

we can write

$$\frac{dI}{dz} = 2\alpha(\omega)I \; . \tag{20}$$

The intensity gain per unit length is twice the field amplitude gain per unit length because of the quadratic relationship

$$I = |E_0|^2.$$

For very weak fields the intensity grows exponentially as we can write

$$\frac{dI_\omega}{dz} = \frac{2\sigma(\omega_A)\,I_\omega}{\left(1 + \dfrac{(\omega-\omega_A)^2}{\gamma_\perp^2} + I_\omega/I_s\right)} \; , \tag{21}$$

and if $I \ll I_s$ then the multiplicative factor is an intensity-independent function of frequency

$$\frac{1}{I_\omega} \frac{dI_\omega}{dz} = g(\omega) \; .$$

The solution is then given by

$$I_\omega(z) = I_\omega(0)e^{g(\omega)z} \; .$$

Note that the total gain

$$\frac{I_\omega(z)}{I_\omega(0)} = G(\omega,z) = e^{g(\omega)z} \; , \tag{22}$$

narrows in frequency with increasing length z because the exponentiation of a peaked function g(ω) leads to an even more sharply peaked function. Under saturated conditions, the gain is still peaked and the function continues to narrow with length, though more slowly as power broadening joins the usual gain saturation.

For a particular medium of length L, one often finds experimentalists speaking in terms of "the Number of Gain Lengths" which is by definition $\ln G(0,L) = g(0) L = 2\alpha(0) L$.

For strong fields near resonance $\left(I_\omega \gg I_s, \omega-\omega_A; \text{ or } \dfrac{I}{I_s} \gg 1 + \dfrac{(\omega-\omega_A)^2}{\gamma_\perp^2}\right)$

$$\frac{dI}{dz} = 2\alpha(\omega_A) I_s, \tag{23}$$

indicating that under heavily saturated conditions, the intensity grows linearly with distance

$$I_\omega(z) = I_\omega(0) + (2\alpha(\omega_A) I_s) z. \tag{24}$$

Physically this is reasonable because under strong stimulated emission conditions essentially all of the energy supplied to the material by the pumping rate $\sigma\gamma_\parallel$ is extracted and added to the driving field so the field energy increases linearly, in proportion to the amount of materal from which the energy has been extracted.

The solutions of Eq. 21 in two extremes of intensity provided physical insight. The equations can also be solved exactly:

$$\frac{dI_\omega}{dz} = \frac{g_0 I_\omega}{\left(1 + \dfrac{(\omega-\omega_A)^2}{\gamma_\perp^2} + \dfrac{I_\omega}{I_s}\right)} \tag{25}$$

yields an implicit solution

$$\ln\left(\frac{I_\omega(z)}{I_\omega(0)}\right) + \frac{I_\omega(z) - I_\omega(0)}{I_s\left(1 + \dfrac{(\omega-\omega_A)^2}{\gamma_\perp^2}\right)} = g_0 z \left(1 + \frac{(\omega-\omega_A)^2}{\gamma_\perp^2}\right). \tag{26}$$

D.   Amplified Spontaneous Emission

A brief digression is in order to comment the effects of amplified spontaneous emission which can be very damaging to the efficiency of an "Ideal Laser." We will not provide an overly detailed discussion.

Here we consider only the incoherent contributions and so write only an equation for the intensity

$$\frac{dI}{dz} = \frac{g_0 z}{1 + I/I_s} + \beta , \tag{27}$$

where β corresponds to the average intensity of spontaneous emission per unit length. Note $g \sim (N_a - N_b)$ while $\beta \sim N_a$. Thus β also will be saturated by depletion of the upper level, a complication which should be remembered. However the spontaneous emission will not be completely diminished as $g \to 0$ when $N_a = N_b$, but in that case β remains nonzero.

A further cautionary note is that neglect of coherent interactions between the field and atom can be a serious error if the material is heavily saturated.

In the linear region ($I \ll I_s$)

$$I(z) = I(0)e^{g_0 z} + (\beta/g_0)(e^{g_0 z} - 1) . \tag{28}$$

| OUTPUT | AMPLIFIED INPUT | AMPLIFIED SPONTANEOUS |
| SIGNAL | SIGNAL | EMISSION |

This result indicates that spontaneous emission noise limits the linear amplification regime:

The amplified signal is accompanied by noise which also grows with increasing length. The amplified spontaneous emission provides a fundamental effective input noise of magnitude β/g_0. This is the spontaneous emission in the first gain length of the medium (if the gain length l is defined by $g_0 l = 1$, $\beta/g_0 = \beta l$). For most laser systems

176

$$\beta/g_0 \sim (10^{-7} - 10^{-9})I_s \qquad (29)$$

so the amplification of input signals weaker than this level will be masked by the noise. The effect is less for very short lengths $gz \leq 1$ where the full effect of the noise buildup will not have been reached.

In the nonlinear regime, $I \geq I_s$, we solve first the case $I(0) = 0$. Then, as shown by Casperson, the intensity grows as shown in Figure 11.

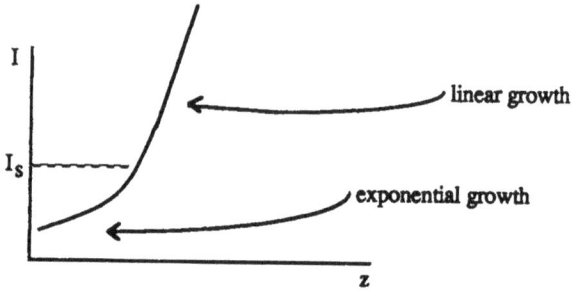

Figure 11

Because of saturation, the gain of the system as a function of total length L in the presence of spontaneous emission can be found from

$$G(L) \equiv \frac{I(L)}{I(0)} = \exp(\int_0^L g(z)dz) \qquad (30)$$

and is shown schematically in Figure 12.

Figure 12

The Amplified Spontaneous Emission (ASE) is a parasite which depletes the gain, stealing energy from the intended amplification process.

The actual experimental situation is made even worse by two additional effects:

1) ASE is bidirectional. We show in Figure 13 the two intensities.

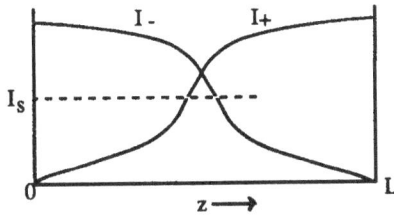

Figure 13

In Figure 14 we show the remaining local differential gain g(z)

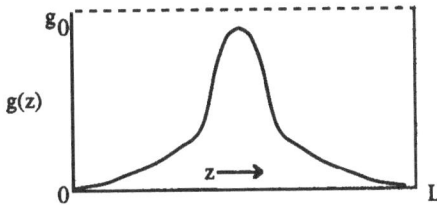

Figure 14

and in Figure 15 we show the residual total gain as a function of material length L.

178

$$G(L) = \int_0^L g(z)dz$$

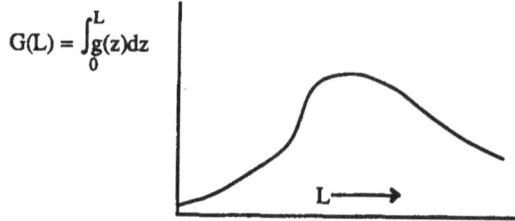

Figure 15

This differs from Fig. 12. The bidirectional effects included in Figure 15 cause the total gain to decrease with increasing length, indicating that a material that raises ASE to of order the saturation intensity has reached maximum efficiency. No increase in excitation (g_0) or length (L) will enhance the total gain G(L) further; in fact, increasing either will decrease G(L). Because β/g is typically $\sim 10^{-7}(I_s)$, ASE gives a practical limit on $g_0 L$ of $e^{g_0 L} \sim 10^7 - 10^9$

$$g_0 L \sim (9\text{-}7)\ln 10 \sim 15 - 20 \tag{31}$$

This sets a limit on $\alpha_0 L$ of 7 - 10.

2)   The gain and emission coefficients are frequency dependent. However, typically both $\alpha(\omega)$ and $g(\omega)$ have the same frequency dependence as the lineshape. Thus the ASE grows (for I(0) = 0 and I < I_s) as

$$I(z,\omega) = [\beta_0/g_0](e^{g_0(\omega)z} - 1) \tag{32}$$

creating a source that is both intense and narrow in its frequency spread. For a homogenously broadened medium, this depletes the gain throughout the spectrum. For an inhomogeneously broadened medium a hole is burned in the spectral profile near the center of the resonance line. This leaves more gain for frequencies further away from resonance, but with increasing amplifier length the ASE spectrum broadens to deplete even that part of the remaining gain in the medium.

E.   Factors influencing gain and dispersion

Returning now to our main results which are the existence of frequency-dependent gain and dispersion, let us examine briefly the physical quantities which affect the size and shape of the gain and dispersion.

1)   <u>Number density of excited atoms</u> -- N

Both gain and dispersion are linearly proportional to the number of excited atoms.

2)   <u>Dipole matrix element</u> -- $|\mu|^2$

Both gain and dispersion are linearly proportional to the absolute square of the dipole matrix element between the two states as one might expect from Fermi's Golden Rule for transition probabilities.

3)   <u>Pumping factor</u> -- σ

Both gain and dispersion are linearly proportional to the pumping rate per atom $\gamma_{//}\sigma$ which has its efficiency governed by σ which gives the unsaturated population difference per atom.

4)   <u>Polarization decay rate</u> -- $\gamma_\perp$

The polarization decay rate determines the width of the interesting region of variation of both the gain and dispersion. The Full Width at Half Maximum (FWHM) of the uinsaturated gain is $2\gamma_\perp$. The slope of the dispersion near $\omega=\omega_A$ is inversely proportional to $\gamma_\perp^2$, that is

$$\left(\frac{\partial n(\omega)}{\partial \omega} \right)_{\omega=\omega_A} \sim \frac{1}{\gamma_\perp^2} .$$

The gain at the resonance frequency ω_A , also called "the Line Center Frequency", is inversely proportional to $\gamma_\perp$. Finally the saturation intensity I_s depends linearly on $\gamma_\perp$.

180

5)   Population difference decay rate --$\gamma_{//}$

Intuitively this seems to be a crucial factor in determining our ability to achieve lasing action, but it seems remarkably absent from our equations except that the saturation intensity is found to be linearly proportional to $\gamma_{//}$. Our intuition suggests that this should give us a critical lifetime ($T_1 = 1/\gamma_{//}$) and thus a crucial bandwidth for interaction. We must here, therefore, note the role of the two important material decay rates $\gamma_{//}$ and $\gamma_\perp$. The population difference decay rate determines the effectiveness of the pumping rate $R = \gamma_{//}\sigma$ in achieving the unsaturated population difference. However, the energy extraction from the atoms by the field requires stability of the induced dipole moment (or polarization) and so it is the polarization decay rate which crucially determines the characteristics of the gain and dispersion in the manner noted above.

For two-level atoms, $\gamma_{//}$ is determined by the inverse of the lifetime of the upper level. As the lower level is the ground state, all decays (both radiative and nonradiative) contribute to $\gamma_{//}$. Any hard (inelastic) collisions or processes which deexcite the upper level are included.

For two-level atoms, $\gamma_\perp$ at its minimum (without collisions) can be shown from the Schrödinger Equation to be given by $\gamma_{//}/2$. For such systems $\gamma_\perp \geq \gamma_{//}/2$. However, for multi-level systems in which the lower lasing level is not the ground state, the condition $\gamma_\perp \geq \gamma_{//}$ applies. The minimum values are often refered to as the 'radiative limit' because the decay of the induced dipole moment due to unavoidable spontaneous radiation is given by this expression. In addition, if no other decay processes are present, then this is the limiting linewidth set by the spontaneous decay of the upper level. Many other kinds of processes may cause interruptions of the coherent field-atom interaction and thus increase $\gamma_\perp$. These include phase changing elastic collisions in gases which are pressure dependent, typically on the order of 10^7 sec^{-1} per Torr, and weak long-range and short-range interactions in solids and liquids which because of their high density may lead to rates on the order of 10^{12} sec^{-1}.

Thus in almost all practical materials, $\gamma_\perp$ is greater than $\gamma_{//}$ and it is not uncommon to find $\gamma_\perp$ much greater than $\gamma_{//}$. We will see that these factors have extremely important consequences for laser dynamics. All of the factors discussed above which change the lineshape, do so for all atoms and thus are referred to as HOMOGENEOUS BROADENING.

F.   Coherent Dynamical Response of the Medium

For multilevel atoms the models are much more complex and are called the Lamb equations. One must keep track of the populations of the levels separately instead of only monitoring the population difference, However, if the problem is reduced to an effective two-level model, we should note that the assumption that E_0 is constant does not tell us about the sensitivity of the system to perturbations.

Returning briefly to Eqs. (9.1) and (9.2), we can write

$$\frac{\partial^2 D}{\partial t^2} = \frac{\mu}{2\hbar} (-i(\omega_A-\omega)P_0{E_0}^* + \frac{\mu}{\hbar}|E_0|^2 D_0 - \gamma_\perp P_0 {E_0}^*)$$

$$\frac{-\mu}{2\hbar}(+i(\omega_A-\omega){P_0}^* E_0 + \frac{\mu}{\hbar}|E_0|^2 D_0 - \gamma_\perp {P_0}^* E_0) \tag{33}$$

$$-\gamma_{//}(-\frac{\mu}{2\hbar} (P_0 {E_0}^* + {P_0}^* E_0) - \gamma_{//}(D_0 - \sigma)).$$

As an approximation for strong fields in resonance, consider $\gamma_\perp \to 0$ and $\gamma_{//} \to 0$ (assuming spontaneous processes are much slower than stimulated processes) and $\omega \to \omega_A$. The result is that

$$\frac{\partial^2 D_0}{\partial t^2} = \frac{-\mu^2}{\hbar^2}|E_0|^2 D_0. \tag{34}$$

Even for a strong but constant E_0, the solution for D_0 and correspondingly for P_0 is predominantly oscillatory with a characteristic frequency

$$\Omega^2 = \frac{\mu^2}{\hbar^2} |E_0|^2 , \tag{35}$$

to be compared with the saturation intensity

$$I_s = \gamma_\perp \gamma_{//} \hbar^2. \tag{36}$$

We can rewrite the driven oscillation frequency as

$$\Omega^2 = (|E_0|^2/I_s) \; (\gamma_{//} \gamma_\perp). \tag{37}$$

Ω is called the Rabi frequency and represents oscillatory time dependence in the material variables induced by a strong driving field. This provides dynamical corrections to $n(\omega)$ and $\alpha(\omega)$ which will not be quantitatively considered here, but it should be noted that those corrections are extremely important for the study of dynamical behavior in lasers.

The equations of the Bloch Model for the two-level atom have damping of both the polarization and population difference which damp the coherent oscillations. The oscillations can only be seen in transients of an oscillatory nature in the polarization and population in response to perturbations. Writing $P_0 = \langle P_0 \rangle + P$, and similarly for D_0, and assuming exponential evolution $P(t) = P_1 e^{\lambda t}$ we find $P = P_0 e^{\lambda \tau}$, where

$$\lambda = -\frac{(\gamma_\perp + \gamma_{//})}{2} \pm \sqrt{\left[\frac{(\gamma_\perp + \gamma_{//})}{2}\right]^2 - \left(\Omega^2 + \gamma_\perp \gamma_{//}\right)}$$

or $\quad \lambda = -\frac{(\gamma_\perp + \gamma_{//})}{2} \pm \sqrt{\left[\frac{(\gamma_\perp + \gamma_{//})}{2}\right]^2 - \gamma_\perp \gamma_{//}\left(1 + \frac{I_0}{I_s}\right)} . \tag{38}$

These damped oscillations suggest that the system will be susceptible to perturbations at the frequency given by $Im \, \lambda$ (which is approximately Ω for large Ω).

G. Inhomogeneous Broadening

One other form of broadening requires discussion because it is important for the gas laser systems which will be discussed shortly. INHOMOGENEOUS BROADENING refers to the fact that the overall gain lineshape is made wider because of a mixture of atomic resonant frequencies ω_A which differ significantly on the scale of $\gamma_\perp$. There are many important causes of inhomogeneous broadening including (1) fine structure or level splittings of the resonance transition, (2) differences in the sites of the excited atoms in a solid material with the changes in local fields causing related changes in the energy level spacing, (3) differences among the excited atoms caused by isotopic shifts in the energy level spacing, (4) different energy positions in resonance bands as may occur in solids or liquids, and (5) Doppler shifts in the absorption or emission resonance frequency because of motion of the atoms or molecules relative to the laboratory reference frame. For now we will concentrate on the Doppler shifts as they are most important for our low pressure gas lasers.

Typically the gas is made up of atoms or molecules in random thermal motion resulting in a Gaussian velocity distribution which follows from the thermodynamics of the Maxwell-Boltzmann distribution. The width of the distribution increases as the temperature increases (which for gas discharges typically can range from room temperature, 300K, to 500K) and decreases as the mass increases. When we convert velocity distributions into resonant frequency distributions using the Doppler shift we find that the spread in frequencies varies as the spread in velocities and also varies inversely as the wavelength corresponding to ω_A. If we write our distribution of resonance frequencies as a standard Gaussian

$$P(\omega_A) = \frac{1}{\sqrt{2\pi\sigma_D^2}} e^{-(\omega_A - \omega_{Ao})^2 / 2\sigma_D^2} , \tag{39}$$

we find that

$$\sigma_D \sim (T\,m)^{1/2} (\lambda_{Ao})^{-1} , \tag{40}$$

where T is the temperature, m is the mass and λ_{Ao} is the wavelength of the resonant transition when the atom is at rest. Doppler broadening is less important in the infrared than in the visible and that it is less important for heavy atoms.

This spread of frequencies can also be characterized spectoscopically by a Half Width at Half Maximum given by

$$\Delta \omega_D = (2 \ln 2)^{1/2} \ \sigma_D. \tag{41}$$

The formal details of deriving this Doppler distribution are provided in Narducci's lecture notes in the chapter on inhomogeneous broadening.

There are important differences between the Doppler (Gaussian) and Homogeneous (Lorentzian) lineshapes. Relative to their half widths a Lorentzian lineshape falls to 20% of its maximum at two half widths from resonance and 1% of its maximum at seven half widths from its maximum, while a Gaussian falls to 6.25% of its maximum at two half widths and 10^{-15} at seven half widths. Thus we find that the Lorentzian profile is much larger in the wings that is a Gaussian profile of the same half width. This is very important because the Gaussian distribution is a form of broadening and is not a lineshape itself. Properly a Doppler-broadened lineshape is written by combining the Lorentzian lineshapes for the different velocity groups of atoms with the Gaussian distribution giving the relative probability of the different resonant frequencies. The resulting lineshape is a convolution of the Gaussian and Lorentzian functions as

$$P(\omega) = \int_{-\infty}^{\infty} \frac{1}{\left(2\pi\sigma_D\right)^{1/2}} \exp\left(-\left(\omega_A - \omega_{A_0}\right)^2 / 2\sigma_D^2\right) \left(\frac{\gamma_\perp/\pi}{\left(\omega - \omega_A\right)^2 + \gamma_\perp^2}\right) d\omega_A , \tag{42}$$

where we have written a distribution normalized so that $\int P(\omega)\, d\omega = 1$, and we have taken the unsaturated limit for the Lorentzian. The distribution can be derived formally from the Maxwell-Bloch equations as is shown in Narducci's notes. It is characterized by two widths, σ_D and $\gamma_\perp$. It is Gaussian near the center and Lorentzian in the wings.

The 'change-over frequency' is governed roughly by where

$$e^{-\Delta\omega^2/2\sigma_D^2} \quad \sim \quad \frac{\gamma_\perp^2}{\Delta\omega^2 + \gamma_\perp^2} . \tag{43}$$

For $\sigma_D = 10\gamma_\perp$, this changeover occurs approximately for

$$| \omega - \omega_{Ao}| \equiv \Delta\omega \sim 3\sigma_D . \tag{44}$$

The convolution given in Eq. (42) is called a Voigt profile and is one of the most common spectral profiles found in gas laser spectroscopy. For our purposes it is also important to note that, even when there is large Doppler broadening, atoms many $\gamma_\perp$ away in resonance frequencies contribute in a non-negligible way to response at any particular frequency because of their Lorentzian response functions. We cannot consider the Doppler distribution to be a spread of delta function responses.

Often to emphasize this point we illustrate a Voigt profile as shown in Figure 16 below.

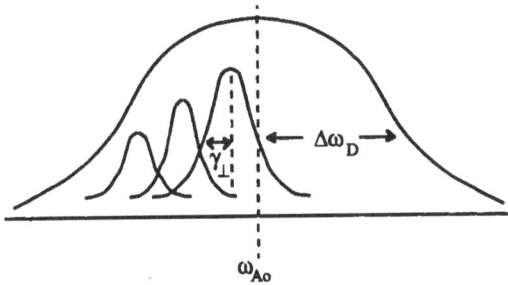

Figure 16

H. Laser Cavities and Decay Rates

Our next important considerations are with the design of laser cavities which provide feedback and resonant structures. Such processes use the amplification process to build a strong source by repeated amplification, use gain saturation to stabilize the intensity output and use a resonant structure to select a particular narrow band of frequencies from the wide range of frequencies where gain is possible in order to generate nearly monochromatic and highly coherent light.

CAVITY DECAY RATE - CAVITY LIFETIME

     We begin with the notion of a cavity decay time. Assume either a unidirectional ring laser of length L or a Fabry Perot laser of round trip length 2L, as shown in Fig.17,

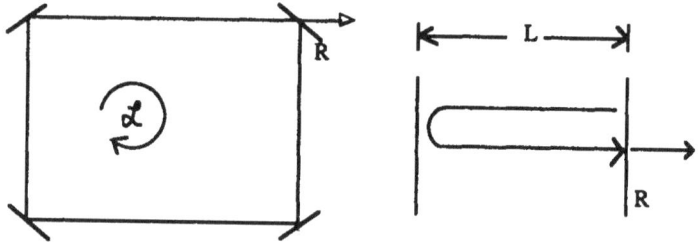

Figure 17

with one mirror of reflectivity R and other mirror(s) of 100% reflectivity. Suppose the field inside the cavity is initially uniform, as shown by the broad line in Figure 18.

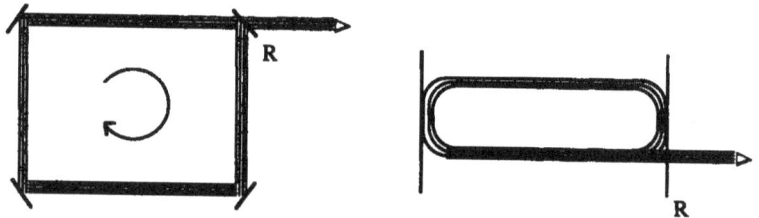

Figure 18

After a short time the part of the field which has passed by the mirror is weaker in intensity as shown by the dotted line in Fig. 19.

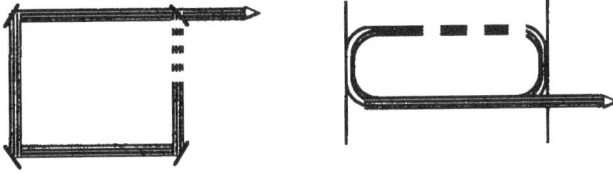

Figure 19

After a time given by the roundtrip time in the cavity t_R ($t_R = L/c$ for the ring, $t_R=2L/c$ for the Fabry Perot) the intensity everywhere has taken on the reduced value:

$$I(z,t_R)=RI(z,0). \qquad (45)$$

If we think of this as having occured as a slow, continuous process everywhere in space we can write

$$I(t)=e^{-t/t_c} I(0) . \qquad (46)$$

In this case

$$I(t_R)=e^{-t_R/t_c} I(0) = RI(0) , \quad \text{so that} \qquad (47)$$

$$e^{-t_R/t_c} = R, \qquad \text{or}$$

$$-t_R/t_c = lnR . \qquad (48)$$

$$t_c = \frac{-t_R}{lnR} = \begin{cases} \dfrac{-L/c}{lnR} & \text{RING} \qquad (49a) \\[2em] \dfrac{-2L/c}{lnR} & \text{FABRY PEROT} \qquad (49b) \end{cases}$$

where t_c is called the cavity lifetime and refers to the characteristic decay time for the intensity (or more properly, for the energy) inside of the cavity,

$$\frac{dI}{dt} = - \frac{I}{2t_c} \tag{50}$$

It is somewhat problematical, however, when R is not very close to unity. Then there is a significant step change in the longitudinal spatial pattern of the field. The notion of a cavity decay time makes good physical sense only if $R \sim 1$ which means that $t_c \gg t_R$. However, even when $R<1$, the characteristic decay time of the space-averaged energy density in the cavity is given by t_c. Even when the reflectivity causes an abrupt change in the field pattern we can speak of a decay rate which, strictly speaking, is a conversion of boundary conditions to a kind of distributed loss within an otherwise perfect cavity.

Noting that $I=|E|^2$ one can write

$$\frac{dE}{dt} = - \frac{E}{2t_c} \tag{51}$$

and using $1/2t_c$ to define an electric field amplitude decay rate κ, we can write

$$\frac{dE}{dt} = - \kappa E . \tag{52}$$

It is important to remember that t_c is defined for intensity and κ is defined for the field amplitude and that they are related by

$$\kappa = \frac{1}{2t_c} . \tag{53}$$

I.   Cavity Resonances - Mode Structure

The boundary value problem for modes of a perfect cavity can be solved by well established techniques learned by students in courses on electromagnetic theory. For cavities of small transverse dimensions it is useful to breakup the wave equation into transverse and longitudinal parts. An eigenmode of the cavity typically is identified by indices which enumerate the longitudinal and transverse mode eigenvalues.

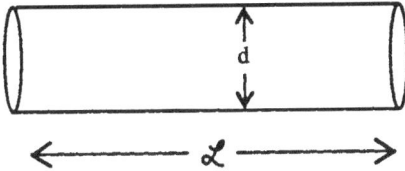

$$d\sin\theta = 1.22\lambda$$
$$\text{if } \theta \sim d/\mathscr{L} \text{ then } d^2 = 1.22\lambda\mathscr{L} \tag{54}$$

Figure 20

For our geometries we will consider long cylindrical systems which are many wavelengths long and only the lowest transverse order dimension wide as shown in Figure 20. As there are diffraction limits to single transverse mode analysis, we can estimate the transverse dimension by assuming a transverse Gaussian spatial profile that is diffraction limited.

Then $d \sim \sqrt{1.22\,\lambda\mathscr{L}}$.   (55)

For a 30 cm HeNe laser at .6 μm,
$$d \sim .5mm.$$
For a 1m Ar$^+$ laser at .5 μm,

$$d \sim .8 \text{ mm.}$$

For a 1m CO_2 laser at 10 μm,

$$d \sim 3.5 \text{ mm.}$$

In the longitudinal direction, modes can be identified by the number of wavelengths in the roundtrip of the path, numbers of order 200,000 are typical for visible lasers of order 1 meter in length. For a perfect resonator we expect the approximate relations

$$L = m\lambda \quad \text{(RING)}$$
$$\text{and}$$
$$2L = m\lambda \quad \text{(FABRY PEROT)}. \tag{56}$$

With the result in free space that the frequencies are given by $\omega = 2\pi c/\lambda$, the resonant frequencies are:

$$\omega_m = m\,(2\pi c/L) \quad \text{(RING)}$$
$$\omega_m = m\,[2\pi(c/2L)] \quad \text{(FABRY PEROT)}. \tag{57}$$

These relations are exact for plane waves in perfect cavities. We expect that transverse field distributions and imperfect cavities may to lowest order simply shift the absolute values of the optical frequency resonances by small amounts. However, because the relevant mode numbers are $m \sim 10^5$, the spacing between modes is essentially unchanged by these corrections, thus

$$\omega_m - \omega_{m+1} \sim 2\pi(c/L) \quad \text{(RING)}$$
$$\text{or}$$
$$\sim 2\pi(c/2L) \quad \text{(FABRY PEROT)}. \tag{58}$$

Recalling the definition of the cavity round trip time, t_R, we see that these spacings are given by

$$\omega_{m+1} - \omega_m \sim 2\pi/t_R. \tag{59}$$

Because the field in the cavity decays, these resonances are not perfectly sharp. The decay rate for the field inside the cavity gives an effective linewidth for the cavity resonances

$$\kappa = 1/2 \ t_c \ .$$

The cavity resonances provide a good reference basis set to describe the laser if the resonances are relatively narrow compared with their separation $(\kappa \ll \omega_{m+1} - \omega_m)$, which requires that

$$1/2 \ t_c \ll 2\pi/t_R \ ,$$

or

$$t_c \gg t_R/4\pi \ , \tag{60}$$

which is achieved if the reflectivity of the mirrors approaches unity. Then the picture shown in Figure 21 is helpful.

We can see the picture of single mode laser or multimode laser more clearly by comparing Figure 21 to the plot of the gain function shown in Figure 22.

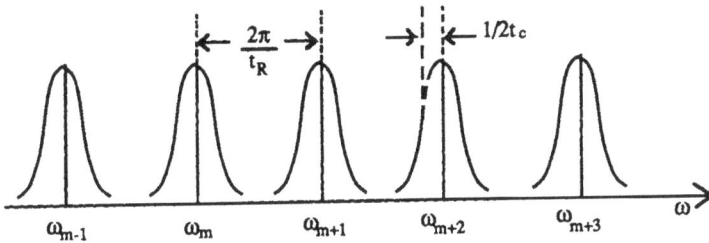

Figure 21

192

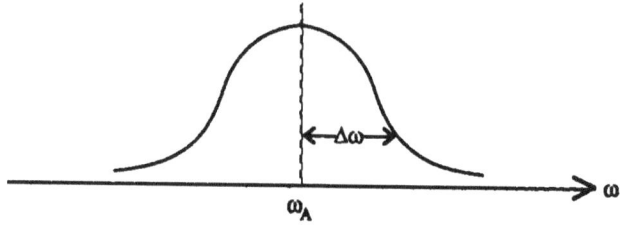

Figure 22

If we could know how many different cavity modes see significant amplification, we could guess whether the system will function on only one (or more than one) mode. "Significant" gain can be made more precise. As shown before, the cavity losses caused by the reflectivity can be lumped into a distributed loss κ. The equation for the field amplitude is then given approximately by

$$\frac{dE}{dt} = - \kappa E + \frac{cl}{L}\, \alpha(\omega,E)E ,\qquad (61)$$

where l/L is the fraction of the cavity filled by the medium. We can speak of the "above threshold" range in frequency as that range for which $(cl/L)\, \alpha(\omega,0) > \kappa$. Graphically this can be displayed by plotting both $(cl/L)\, \alpha(\omega,0)$ and κ on the same picture as in Figure 23.

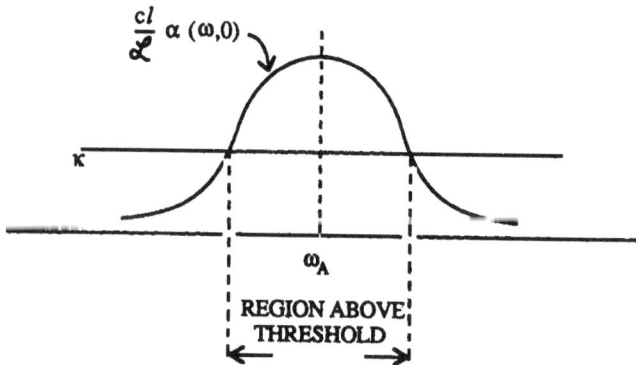

Figure 23

Roughly speaking, any cavity mode which falls within the above threshold region has the possibility to oscillate. This notion is made more precise in Narducci's notes where it will be shown that any cavity mode falling in this region has a corresponding steady-state solution. (Mode pulling effects modify this. More precisely, a mode whose operating frequency falls within the above threshold range will have a corresponding steady state solution.)

Approximately speaking, if the spacing between cavity modes $\omega_{m+1}-\omega_m \equiv \alpha_1$ is much larger than the width of the above-threshold region, then the laser is likely to operate on only one mode at a time. This is called "single-mode operation." If α_1 is much less that the width of the above-threshold region, then it may be possible for more than one mode to oscillate. This is called "multimode oscillation".

If $(cl/L)\,\alpha(\omega,0) > \kappa$, then an initial weak field at frequency ω which is resonant with the cavity will grow in amplitude, until the gain is saturated. Equilibrium is established by $(cl/L)\,\alpha(\omega,E) = \kappa$. For the homogeneously broadened case this can be written as

$$(cl/L)\,\alpha(\omega,0)\,\frac{1}{1 + I(\omega)/I_s + \dfrac{(\omega-\omega_A)^2}{\gamma_\perp^2}} = \kappa . \qquad (62)$$

Solving for the intensity

$$I(\omega)/I_s = \left(\frac{(cl/L)\,\alpha(\omega,0)}{\kappa} - 1 - \frac{(\omega-\omega_A)^2}{\gamma_\perp^2}\right) . \qquad (63)$$

194

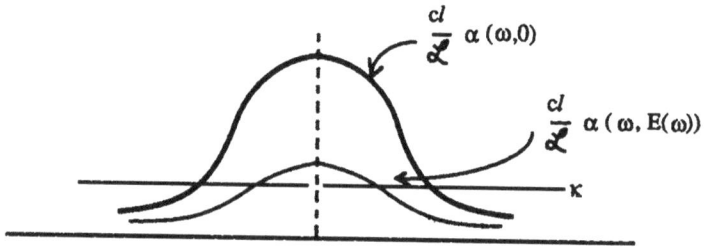

Figure 24

We see that in the steady state solution for the gain at frequency ω, the field amplitude $E(\omega)$ grows until the quantity $[(cl/L)\,\alpha(\omega,E)]$ is "clamped" to the value which is the threshold gain for laser action (where gain balances the loss) as shown, for example in Fig. 24.

From Eq. (63) we immediately see two charateristics of the parametric dependence of the laser output which are shown in Fig. 25. They are the linear dependence of the power output on the unsaturated gain (or excitation) $(cl/L)\,\alpha(\omega,0)$, shown in Fig. 25a, and the parabolic dependence of the laser power output on frequency detuning from the resonant frequency ω_A, shown in Fig. 25b.

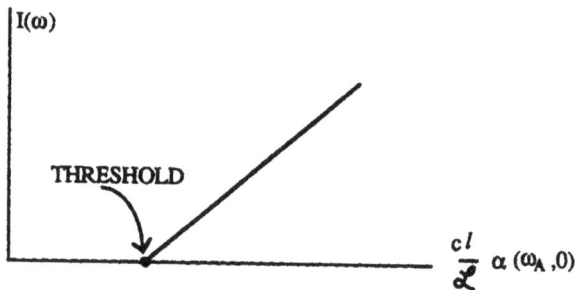

Figure 25a

For fixed ω, and varying the amount of gain, the threshold gain at ω is specified by the value yielding $I(\omega) = 0$.

$$\left[\left(\frac{cl}{L}\right)\alpha(\omega_A,0)\right]_{\text{THRESHOLD FOR }\omega} = \kappa\left(1 + \frac{(\omega-\omega_A)^2}{\gamma_\perp^{\,2}}\right) \qquad (64)$$

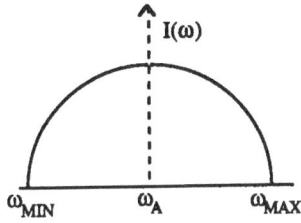

Figure 25b

For fixed $(lc/L)\alpha(\omega_A,0)$ and varying ω, the tuning range is given by extremes ω_{MIN} and ω_{MAX} which are solutions of:

$$\frac{(\omega - \omega_A)^2}{\gamma_\perp^2} = \frac{(cl/L)\alpha(\omega_A,0)}{\kappa} - 1 \ . \tag{65}$$

ω_{MAX} and ω_{MIN} are the frequencies at which the unsaturated gain $(cl/L)\,\alpha(\omega_A,0)$ just equals the loss κ.

Finally we consider the question of cavity resonance in the presence of a non-linear dispersive medium which partially fills the laser cavity. Consider the set-up as shown in Figure 26.

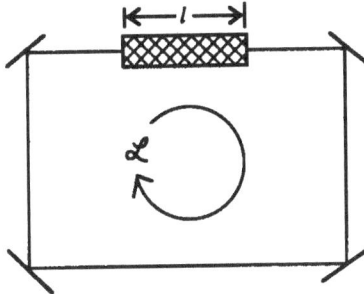

Figure 26

Here the medium has a length l and the cavity has a length L. The resonance condition may be established by observing that the medium has an index of refraction $n(\omega)$ for the length l, and an index of refraction of unity for the remaining part of the roundtrip path. Requiring an integral number of wavelengths in a roundtrip leads to Eq. (66).

$$\frac{(L-l)}{\lambda} + \frac{l}{\lambda n(\omega)} = m \quad , \tag{66}$$

which can be rearranged as

$$L - l + l\,n(\omega) = m\lambda = \frac{m2\pi c}{\omega} \quad . \tag{67}$$

Writing $n(\omega)$ as $1 + (n(\omega)-1)$ we have

$$L + l\,(n(\omega)-1) = \frac{m2\pi c}{\omega} \quad , \quad \text{or} \tag{68}$$

$$\omega + \frac{l}{L}\,(n(\omega)-1) = \frac{m2\pi c}{L} \quad . \tag{69}$$

Using the mth free cavity resonance,

$$\omega_c^{\,m} = \frac{m2\pi c}{L} \quad , \quad \text{we find}$$

$$\frac{l}{L}\,(n(\omega)-1) = \omega_m - \omega \quad . \tag{70}$$

The solution of this equation gives the resonant frequencies. Of course a solution must be found that is self-consistent with the solution of Eq. (62) because $n(\omega)$ is quite generally $n(\omega, E)$ and can be saturated or distorted by the field amplitude. If $n(\omega)$ is independent of E, then it is helpful to first solve Eq. (70) graphically as shown in Fig. 27.

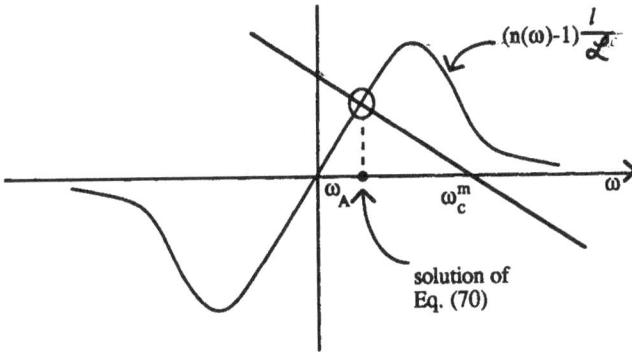

Figure 27

This approach is nearly exact for an extremely inhomogeneously broadened medium. In this case the solution is generally closer to the atomic frequency ω_A than was the empty cavity resonance ω_c^m. This effect is called *mode-pulling* as the atomic resonance "pulls" the solution away from the cavity resonance and toward the atomic resonance.

These notions are again made more precise in Narducci's lecture notes. For now it is important to also note the results for the homogeneously broadened case. There

$$(n(\omega,E)-1) = \left(\frac{\omega-\omega_A}{\gamma_\perp}\right) \alpha\,(\omega,E) \tag{71}$$

and since $(cl/L)\alpha\,(\omega,E) = \kappa$, then

$$(n(\omega,E)-1) = (L/cl)\,\kappa\left(\frac{\omega-\omega_A}{\gamma_\perp}\right) \tag{72}$$

with the result that (substituting Eq. (72) into Eq. (70))

$$\frac{\omega-\omega_A}{\gamma_\perp} = \frac{\omega_c^m-\omega}{\kappa}\,, \tag{73}$$

or rewriting

198

$$\omega - \omega_A = \frac{\omega_c^{\,m} - \omega_A}{(1 + \kappa/\gamma_\perp)} \,. \tag{74}$$

These are expressions for the mode pulling in a homogeneously broadened medium. This relation is independent of the intensity of the solution. In this case the graphical picture of Fig. 18 can be highly misleading because of the strong dependence of $n(\omega,E)$ on E.

Returning to the solutions for different modes

$$\omega_c^{\,m} = 2\pi mc/L \,,$$

where m is an integer, we find that these modes may be either densely packed on the scale of $\gamma_\perp$, in which case we speak of a "multimode situation," or they may be well separated, in which case we speak of a "single mode situation" as illustrated in Fig. 28.

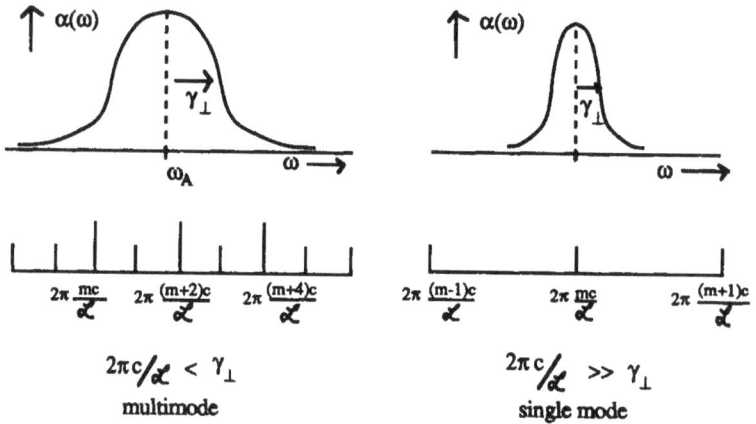

Figure 28

To make these notions more precise we must include the frequency pulling and we must compare the frequency spread to the range of frequency over which the gain exceeds the loss as indicated in Fig. 29.

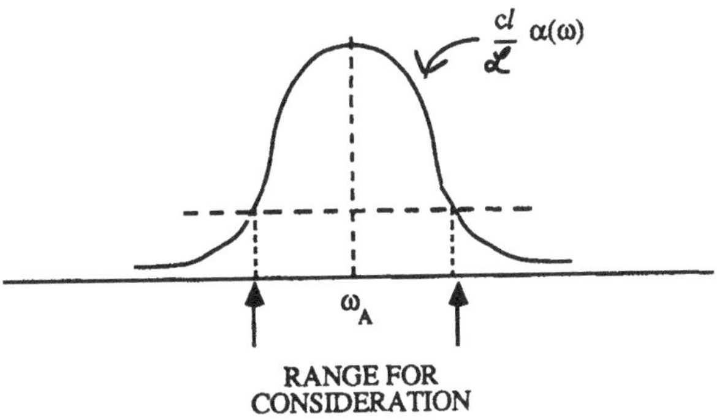

Figure 29a

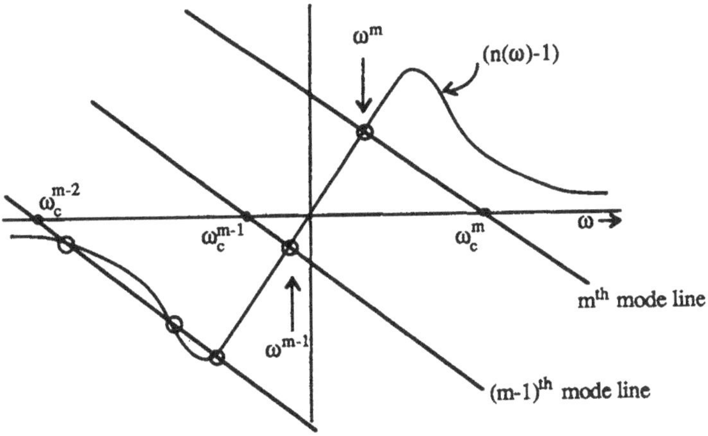

Figure 29b

The ω_c^{m-2} case shows an extraordinary richness. Each circle (o) corresponds to a possible steady state solution from the frequency resonance taken alone. If the gain at the corresponding frequency exceeds the loss then a steady state solution is assured. The 'solution' suggested by the circles in Fig. 29b will be valid only if the dispersion $(n(\omega)-1)$ is not changed by the strength of the field solution E_0 which is controlled mainly by Fig. 29a.

Achieving the multiple solutions implied for ω_c^{m-2} requires extremely high gain and therefore very high intensities which introduce nonlinear effects. In theoretical analyses we have found conditions for these solutions but it is unlikely that the necessary conditons for observing the case of three solutions will ever be found in a functioning laser.

J. Comparisons of Various Laser Materials and their associated parameters

We now have several different key parameters by which we can classify lasers. While the two-level atom model will not always be appropriate, it will be sufficiently correct qualitatively if the following parameters are specifically defined for that msyrtisl:

Decay Rates:

$\gamma_\perp$ - decay of polarization

$\gamma_\parallel$ - decay of population difference

κ - decay of field in the cavity (Depends on: t_R - roundtrip time and R - reflectivity) Gain: $\alpha(\omega,0)$ - (unsaturated gain for the field)

$g(\omega,E)$ - intensity gain per unit length

Inhomogeneous Half Linewidth: $\Delta\omega_I$ (a subclass is the DOPPLER half linewidth $\Delta\omega_D$)

Cavity Mode Spacing: α_1

Classifications of Materials

We can classify materials by their dominant broadening mechanisms. A few comments on known or suspected features of lasers of each type will be given.

1. HOMOGENEOUSLY BROADENED

Commonly believed to be the most stable kind of single mode laser. Mode-pulling of the laser frequency is power independent. Single mode instabilities require a $\kappa > \gamma_\perp + \gamma_\parallel$ (a "bad cavity") and high excitation. It is commonly believed that these lasers function only on the single cavity mode which gives oscillation closest to the atomic resonance. Recent work suggests this is not always correct. Transitions to multimode pulsations require high excitation if the laser is resonantly tuned; lower excitation is required for a detuned laser. Such materials are dye lasers, room temperature solid state lasers, semiconductor lasers and high pressure gas lasers.

2. INHOMOGENEOUSLY BROADENED

It is commonly believed that these lasers can easily support multiple modes simultaneously and that the modes function quasi-independently. This might be called "mode coexistence". Mode pulling of the laser frequency depends on the power level; pulling decreases with increasing power. Single mode instabilities happen very easily if $\kappa > \gamma_\perp$. The full details of the multimode operation are not well explained yet by formally developed theory. Common lasers of this type are low pressure gas lasers and low temperature solid state lasers.

3. MIXED BROADENED

$$3 < \frac{\Delta\omega_I}{\gamma_\perp} < 3 \ .$$

This is a very common type of laser system. Small, but non-negligible, amounts of inhomogenous broadening make these very complicated cases. Generally single mode instablilities and multimode operation are more possible here than in the homogeneously broadened case. Such materials include middle pressure gas lasers and middle temperature solid state lasers.

202

Classification of Lasers

When the laser cavity characteristics are included there are several important relationships between the cavity linewidth and the material linewidths which can be used to classify different lasers. For plane-wave ring lasers, the following results are known:

A. $\gamma_\perp, \gamma_\parallel \gg \kappa$

DYE, ATMOSPHERIC
PRESSURE LASERS

The single mode states are stable with real eigenvalues. Single mode instabilities are impossible. Cavity linewidths are narrow so laser frequencies are very close to cavity resonance frequencies.

B. $\gamma_\perp \gg \kappa \gg \gamma_\perp, \gamma_\parallel$

RUBY,
Nd:YAG; Nd:Glass

SEMICONDUCTOR
MID-PRESSURE CO_2

Damped relaxation oscillations are common. These lasers are often sensitive to harmonic perturbations at

$\sim \sqrt{2\kappa\,\gamma_\parallel(A-1)}$, where A is the threshold parameter. Laser frequencies are very close to cavity resonance frequencies (little mode pulling).

C. $\kappa \sim \gamma_\perp \sim \gamma_\parallel$

HeNe, Ar*, HeXe, FIR

Unstable oscillations from a single mode steady state are possible for $\kappa > \gamma_\perp + \gamma_\parallel$. Laser frequencies may be strongly pulled from cavity resonances towards the atomic frequency if $\kappa \gg \gamma_\perp$.

Summary of Specific Lasers

Material characteristics for linewidths are given and the cavity parameters for a typical specified reflectivity R and characteristic size (giving t_R).

1. Dye

$$\gamma_\perp \sim 10^{13} \ sec^{-1}$$

$$\gamma_\| \sim 10^9 \text{-} 10^{10} \ sec^{-1}$$

$$\Delta\omega_D << \gamma_\perp, \gamma_\|$$

$\alpha_1 = 6 \times 10^9 \ sec^{-1}$, $R = 80\%$, $t_R = 1$ ns, $\kappa = 10^8 \ sec^{-1}$

2. CO_2

$$\gamma_\| \sim 3 \times 10^4 \ sec^{-1}$$

$$\gamma_\perp \sim 3 \times 10^8 \ sec^{-1} \text{ at 10 Torr}$$

$$\Delta\omega_D = 1.9 \times 10^8 \ sec^{-1}$$

$\alpha_1 = 7.5 \times 10^8 \ sec^{-1}$, $R = 80\%$, $t_R = 8$ ns, $\kappa \sim 10^8 \ sec^{-1}$

3. Ruby

$$\gamma_\perp \sim 10^{12} \ sec^{-1} \qquad\qquad \text{room temperature}$$

$$\gamma_\| \sim 3 \times 10^3 \ sec^{-1}$$

$$\Delta\omega_T \sim 10^5 \ sec^{-1}$$

$\alpha_1 = 1.5 \times 10^9 \ sec^{-1}$, $R = 80\%$, $t_R = 4$ ns, $\kappa \sim 2.5 \times 10^7 \ sec^{-1}$

4. Nd:YAG $\qquad$ $\gamma_\perp \sim 3.8 \times 10^{10}$ sec^{-1} $\qquad$ room temperature

$\gamma_\parallel \sim 6 \times 10^5$ sec^{-1}

$\Delta\omega_T \sim 10^5$ sec^{-1}

$\alpha_1 = 6 \times 10^9$ sec^{-1}, R = 95%, t_R = 1 ns, $\kappa \sim 2.5 \times 10^7$ sec^{-1}

5. Nd:Glass $\qquad$ $\gamma_\perp \sim 9 \times 10^{12}$ sec^{-1}

$\gamma_\parallel \sim 10^4$ - 10^3 sec^{-1}

$\Delta\omega_T \sim 10^5$ sec^{-1}

$\alpha_1 = 6 \times 10^9$ sec^{-1}, R = 98%, t_R = 1 ns, $\kappa \sim 10^7$ sec^{-1}

6. He-Ne .6328μm $\quad$ $\gamma_\perp \sim 10^8$ sec^{-1}

$\gamma_\parallel \sim 10^7$ sec^{-1}

$\Delta\omega_D \sim 4.7 \times 10^9$ sec^{-1}

$\alpha_1 = 9 \times 10^9$ sec^{-1}, R = 99%, t_R = 2 ns, $\kappa \sim 2.5 \times 10^6$ sec^{-1}

7. He-Ne 3.39μm $\quad$ $\gamma_\perp \sim 2 \times 10^8$ sec^{-1}
   at 5 Torr

$\gamma_\parallel \sim 1 \times 10^8$ sec^{-1}

$\Delta\omega_D \sim 10^9$ sec^{-1}

$\alpha_1 = 3 \times 10^9$ sec^{-1}, R = 90%, t_R = 2 ns, $\kappa \sim 2.5 \times 10^7$ sec^{-1}

8. He-Xe 3.51μm $\gamma_\perp \sim (1.2 - 25) \times 10^7 \text{ sec}^{-1}$

$\gamma_\parallel \sim 1.2 \times 10^7 \text{ sec}^{-1}$

$\Delta\omega_D \sim 3 \times 10^8 \text{ sec}^{-1}$

$\alpha_1 = 3 \times 10^9 \text{ sec}^{-1}$, $R = 50\%$, $t_R = 2$ ns, $\kappa \sim 1.6 \times 10^8 \text{ sec}^{-1}$

9. Ar$^+$ 5440 $\gamma_\perp \sim 7 \times 10^8 \text{ sec}^{-1}$

$\gamma_\parallel \sim 10^8 \text{ sec}^{-1}$

$\Delta\omega_D \sim 3 \times 10^9 \text{ sec}^{-1}$

$\alpha_1 = 7.5 \times 10^8 \text{ sec}^{-1}$, $R = 95\%$, $t_R = 8$ ns, $\kappa \sim 3 \times 10^6 \text{ sec}^{-1}$

10. Semiconductor $\gamma_\perp \sim 10^{14} \text{ sec}^{-1}$

$\gamma_\parallel \sim 10^9 \text{ sec}^{-1}$

$\Delta\omega_D \sim ? < \gamma_\perp$ (in dispute)

$\alpha_1 = 10^{13} \text{ sec}^{-1}$, $R = 40\%$, $t_R = 6 \times 10^{-13}$ sec, $\kappa \sim 3 \times 10^6 \text{ sec}^{-1}$

Notes: 1) It is argued that carrier diffusion removes any spatial patterns in semiconductors formed saturation by standing wave fields. However, semiconductors retain a strong intensity dependent dispersion (not modeled by rate equations or Maxwell Bloch equations) even though the polarization is often neglected because of its rapid relaxation. 2) Molecular lasers such as dye and CO_2 lasers typically need a more complex model to describe the interband relaxation processes and thus two-level models often fail to give quantitatively correct results.

Sources of high gain

A brief comment is also in order about various sources of high gain. The most important are a high dipole matrix element, μ, and a high density of excited atoms, N. If the response is a narrower function ($\gamma_\perp$, $\Delta\omega_I$ as small as possible) then the gain will be more sharply peaked for the same number of atoms. Also, for fixed $\Delta\omega_I$ a larger $\gamma_\perp$ (up to $\gamma_\perp \sim \Delta\omega_I$) permits a single frequency probe to draw gain from a wider range of the otherwise unused atoms spread away from resonance by the inhomogeneous broadening.

The most notable differences between different wavelength regimes come from considering the ratio of the spontaneous emission rate to the simulated emission rate. In terms of the Einstein A and B coefficients, this can be written

$$\frac{A}{B} = \frac{8\pi n^3 h}{\lambda^3} \ .$$

The wavelength dependence results primarily from the fact that there are more spatial modes for spontaneous emission at the higher frequencies.

As the wavelength gets longer, the excited level is less likely to decay spontaneously making it easier to store energy in the upper level. Also, for transitions of the same spontaneous decay rate (similar linewidths), the stimulated emission rate increases with λ as λ^3. Thus it is not surprising that the HeNe transitions at .6328μm and 3.39μm (which share a common upper level and common level decay scheme) differ so much in their gain. The .6328 transition has a gain of about .2 - .4 decibels per meter (2% - 4% gain per meter) while the 3.39 transition has a gain of about 20 - 50 dB per meter (10^2 - 10^5 per meter). Based on wavelength alone, the gain is expected to differ by a factor of $(5.4)^3 = 160$ which accounts for most of the difference.

This also helps to explain the extremely high gain and relatively narrow homogeneous linewidths found in far infrared lasers with wavelengths in the 18μm to 700μm range which have been used with great success by Weiss, Glorieux, Harrison and Lawandy in the study of laser instabilities. Recall also that Doppler broadening depends inversely on wavelength. Thus the far infrared lasers are better candidates to be nearly fully homogeneously broadened.

K. References

Rate Equations

A. Yariv, *Introduction to Optical Electronics* (Holt-Rinehart, NY, 1987).

Kramers-Kronig Relations

A. Yariv, *Quantum Electronics* (Wiley, NY, 1975).

Gain Narrowing and ASE

L.W. Casperson and A. Yariv, IEEE J. Quantum Electron. *QE-8*, 69 (1972).

L.W. Casperson, J. Appl. Phys. *48*, 256 (1977)

Modepulling by graphical solution

L.W. Casperson and A. Yariv, Appl. Phys. Lett. *17*, 259 (1970).

Modepulling details for Inhomogeneous Broadening

N.B. Abraham, L.A. Lugiato, P. Mandel, L.M. Narducci and D.K. Bandy, J. Opt. Soc. Am. B *2*, 56 (1985).

Material Decay Constants

A. Yariv, *Introduction to Optical Electronics* (Holt-Rinehart, NY, 1987).

A. Yariv, *Quantum Electronics* (Wiley, NY, 1975).

A. Siegman, *Lasers* (University Science Books, CA, 1986).

Semiconductors: C. Henry, IEEE J. Quantum Electron. **QE-19**, 1391 (1983).

CO_2 lasers: E. Arimondo, F. Casagrande, L.A. Lugiato and P. Glorieux, Appl. Phys. B *30*, 57 (1983); F.T. Arecchi, W. Gadomski, R. Meucci and J.A. Roversi, Opt. Commun. *65*, 47 (1988).

Dye Lasers: F. Hong and H. Haken, J. Opt. Soc. Am. B *5* (to be published 1988).

III. Experimental Measurements of Dynamical Instabilities and their Analysis in Single Mode Inhomogeneously Broadened Lasers

A. Introduction

Inhomogeneously broadened lasers offer a complex problem for detailed theoretical modelling, particularly when the broadening arises from Doppler shifts of the resonant frequencies because of the thermal spread of velocities of individual atoms, as is true for gas lasers. Nevertheless there are several high-gain transitions in gas lasers (3.39 μm in neon, and 3.51 μm and 5.57 μm in xenon) which are particularly easy to use in constructing laboratory lasers. We are fortunate that, despite the formal difficulties of developing exact models, these lasers often lend themselves to straight-forward pictorial representations in terms of gain and dispersion for many forms of steady-state (and some simple time-dependent) behavior in the interaction between fields and the material. This makes qualitative discussions of the physics of these lasers both possible and particularly instructive[1-20].

Among the interesting characteristics of such lasers are the following:

1) Pressure increases tend to increase the power output.

2) Unlike the homogeneously broadened laser, the inhomogeneously broadened laser when detuned has an operating frequency for steady state conditions which depends on the intensity.

3) For Doppler broadening, spectral hole-burning for Fabry-Perot lasers leads to a dip (the "Lamb dip"[5]) in the power output as the frequency of the cavity is tuned to resonance with the medium (such that the counter-propagating fields interact with the same velocity class(es) of atoms).

4) The anomalous dispersion of high gain transitions and the hole-burning at the Lamb dip can cause complex variations in the laser frequency as a function of the cavity frequency. Three values for the laser frequency for a fixed cavity frequency have been predicted in the vicinity of the Lamb dip[7-8] and for detunings of about one Doppler linewidth[1].

5) Spectral hole-burning (the saturating of only the part of the medium that is close to being resonant with the single frequency lasing field) seems to be hardly apparent or relevant to the steady state operation of single-mode, unidirectional ring lasers, but it indirectly offers explanations for key features of dynamical instabilities[1-4].

These are only some of the interesting phenomena which have motivated our studies of the high gain gas laser transitions. They have provided us with a rich diversity of steady-state and time-dependent behavior.

We began our studies of the 3.51 μm xenon lasers because we wanted stable reference lasers for spectroscopic and heterodyne applications. We discovered that a relatively high buffer gas pressure was necessary to achieve stable laser output. Later we undertook a study of mode-pulling in a very lossy resonator in order to see if we could find the predicted condition of three different frequencies of single mode lasing for the same mode pattern and the same laser length[1]. When we went to lower pressure (where the inhomogeneous broadening effects were the greatest and thus where mode-pulling effects also would be the greatest) we found the laser performance was limited by time-dependent pulsations..

These two clues in the 1970's led us along a path of discovery and exploration of unstable behavior in these lasers which has continued to the present. Although our early studies were made with Fabry-Perot lasers[9-15] and although we have come to understand their unstable behavior, the theoretical analyses of this problem have remained daunting (for all but Casperson[16,17], whose preliminary work in 1978 has been followed by recent detailed work, and Englund[8]).

Hence we will turn first in our discussions of experiments on inhomogeneously broadened gas lasers to our more recent studies of unidirectional ring lasers[18-22]. In later sections we will return to discuss some of our observations of the single mode instabilities of Fabry-Perot lasers and our interpretations of those results.

210

B. Basic Results for Unidirectional Ring Lasers

The basic design of a ring laser is quite simple as shown schematically in Figure III-1. We wanted to have as small a cavity length as possible to ensure single longitudinal mode operation. The length chosen for these experiments was approximately 0.8m, giving a free cavity mode-spacing of 375 MHz. With a Doppler-broadened line width of 110 MHz for the 3.51μm transition we can expect single mode operation for excitation up to four times above the laser threshold in the absence of mode pulling effects. An aperture is used to select against transverse modes. A piezoelectric crystal permits tuning of the cavity length. A pair of 45° Faraday rotators and two linear polarizers set at 0° and 45° form an isolator which permits only a single direction of propagation and only the vertically polarized light to pass unattenuated around the cavity. The extinction ratio of modes differing in polarization or direction of propagation is 40:1, ensuring unidirectional operation for all accessible excitations.

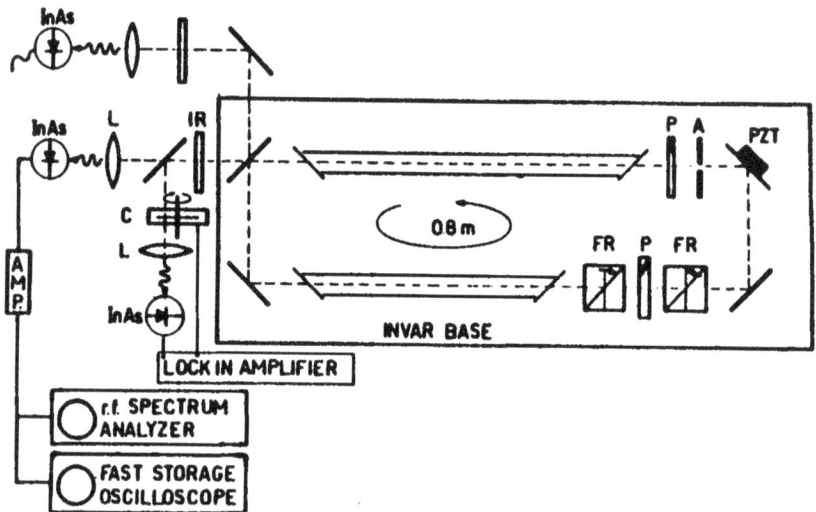

Figure III-1. Schematic of a unidirectional ring laser InAs, high-speed infrared detector; P, wire-grid polarizer; A, aperture; FR, 45-degree Faraday rotator element; PZT, piezoelectric mirror translator; C, chopper; L, quartz lens; IR, 3.51 μm dielectric filter.

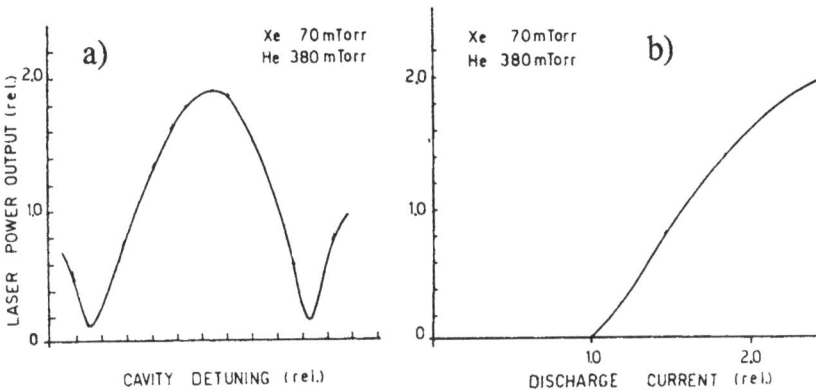

Figure III-2. Average power output for ring laser versus excitation (a) and detuning (b). Data taken in this case for 70 mTorr of xenon and 380 mTorr of helium (from ref. 19).

Measuring only average power output, we see curves as in Figure III-2 for different values of the discharge current and the cavity tuning. For low currents the population inversion is proportional to the current while for higher currents some pumping of the lower level of the transition leads to less efficient creation of the inversion and ultimately to a reduction in the inversion and a decrease in the output. When the cavity detuning is varied we see the resonance peak as the lasing mode frequency is brought into coincidence with the atomic frequency. The absence of side peaks indicates we are working on a single longitudinal mode. The slight asymmetry in the output versus detuning is a result of minor alignment changes which occur with the movement of the mirror and of frequency-dependent gain and dispersion focussing which change the mode size and thus the total power for a given intensity. Nonlinearities in the variation of the PZT length with applied voltage have been calibrated away by interferometric techniques using a HeNe .6328μm laser.

The fast detector (1 kHz-100 MHz) reveals that there are frequently pulsations in the intensity output. Samples are shown in Fig. III-3 and Fig. III-4 for different pressures.

212

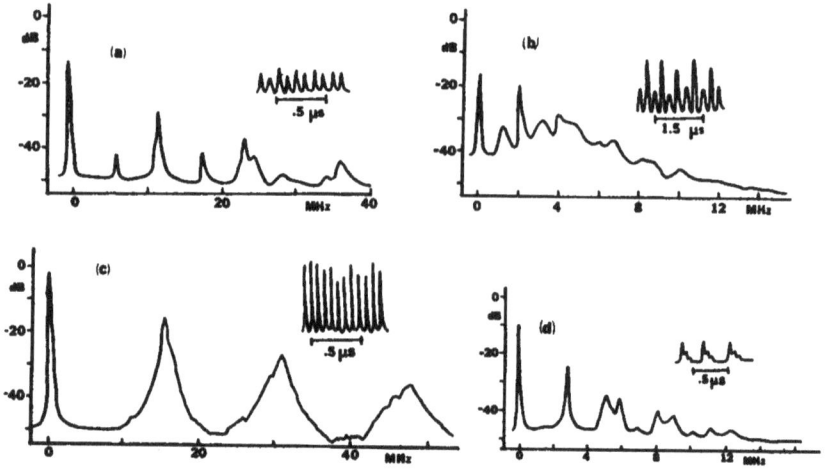

Figure III-3. Sample power spectra and real-time snapshots for single-mode pulsations from ring laser operation at 170 mTorr of xenon (from ref. 18) for different excitation levels.

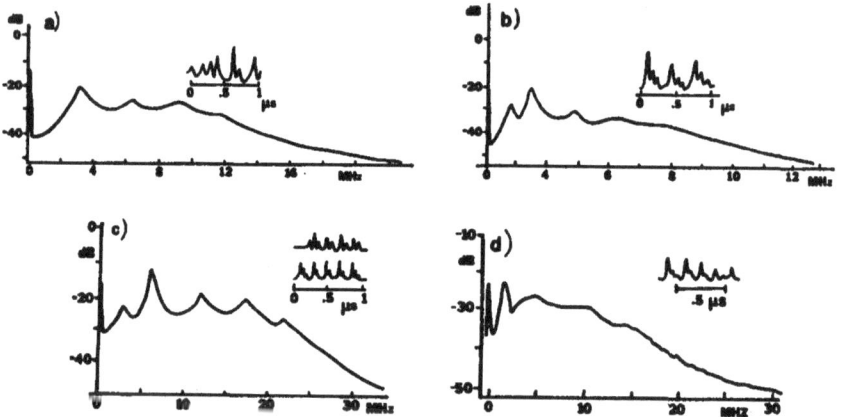

Figure III-4. Sample power spectra and real-time snapshots of single mode ring laser pulsations (from ref. 18) for more chaotic (low pressure) operating conditions of 16 mTorr xenon and 27 mTorr helium at different excitation levels.

Some are dominantly periodic (Figs. III-3a, and perhaps III-3b and III-3c) defined by a single frequency and its harmonics and subharmonics hence we call these PERIODIC and label them P1, P2, P4,... indicating the subharmonics which are observed (e.g., P2 means with the fundamental and the first subharmonic). Others are called QUASIPERIODIC, such as Fig. III-3d, when two relatively incommensurate frequencies appear. These are labelled 2F. Various spectra show a broadband spectrum underlying a few peaks and these cases are called CHAOTIC (e.g. Fig. III-4 and maybe Figs. III-3b and III-3c). (This name will be justified in later sections as the irregular pulsation can be shown to be a form of appropriate for deterministic chaos.) These may be labelled C1, C2, or C depending on whether there is a frequency peak and its harmonics above the broadband spectrum (C1), a subharmonic of the fundamental peak (C2) or merely the smoothly varying broadband chaos (C).

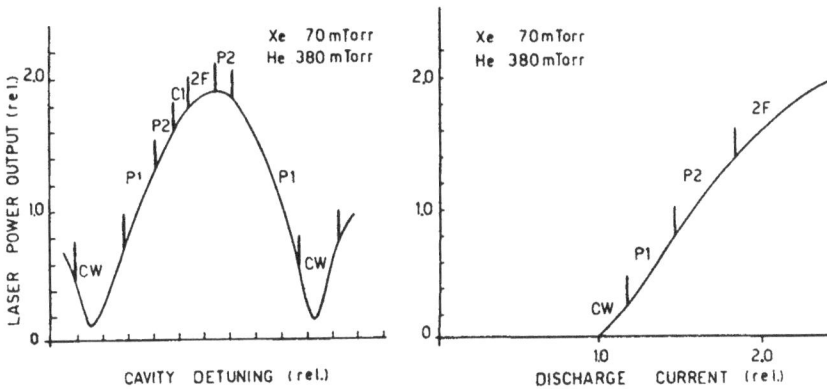

Figure III-5. Domains of different dynamical behavior from ref. 19.

Using this nomenclature we can label regions on Figs III-2 by the type of pulsations observed. Figure III-5 shows such labelling (cw indicates "continuous wave" operation -- no pulsations). We can see that detuning of the laser can be thought of as having the same effect as reducing the excitation of the laser relative to the lasing threshold for that detuning. There are many subtleties, but for these parameters this simple description works well.

The theoretical analyses[16,20,23-29] of inhomogeneously broadened ring lasers indicate that there are several key parameters in addition to the excitation and the detuning. These are the relative linewidths (decay rates) of the cavity resonance (κ, decay of the field) and of the atomic response

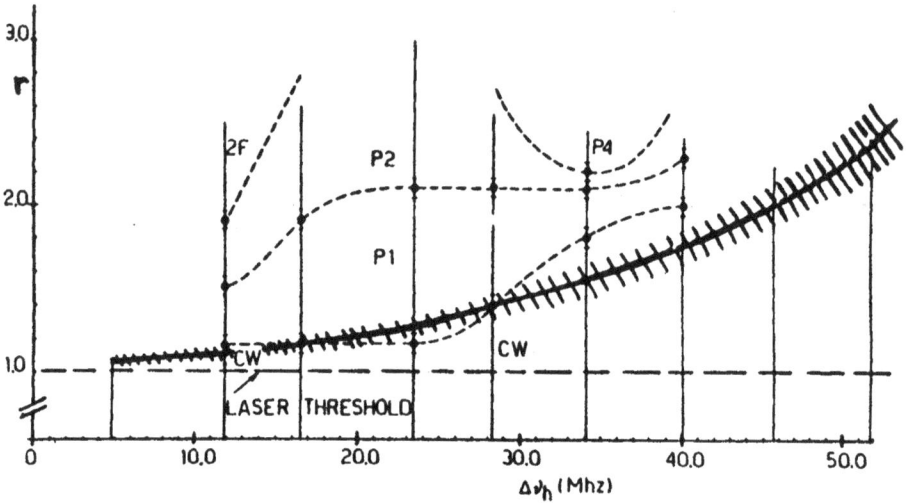

Figure III-6.   Domains of different dynamical behavior as the pressure of the laser is increased, increasing the homogeneous linewidth. r is the ratio of the excitation to the excitation at the threshold for laser action at that pressure. Operating conditions were a Doppler linewidth of 110 MHz, a natural linewidth of 4.6 MHz, and a cavity linewidth of 83 MHz. Hatched region shows theoretical uncertainty associated with uncertainties in the physical parameters.

($\gamma_\perp$, decay rate of the atomic polarization; $\gamma_{||}$, decay rate of the population inversion). For our four-level laser system, we will always have $\gamma_\perp \geq \gamma_{||}$. The instability condition in the models requires $\kappa \geq \gamma_\perp$ and the gain threshold for instabilities is lowered as κ becomes much larger than $\gamma_\perp$. Also crucial is that the inhomogeneous linewidth, σ_D, be much larger than $\gamma_\perp$; then the gain threshold for instabilities is lowered to very close to the threshold for the onset of lasing action.

For our lasers, κ, σ_D and $\gamma_{||}$ are held fixed and it is easiest to change the gas pressure which changes $\gamma_\perp$. As $\gamma_\perp$ increases the laser becomes more stable because both $\sigma_D/\gamma_\perp$ and $\kappa/\gamma_\perp$ decrease. In Fig. III-6 we tabulate results from curves such as Fig. III-5a for different pressures. We see that increasing the pressure increases the stability of the laser. Each of the data sets is compiled for line center tuning of the cavity.

The solid line indicates a theoretical prediction of the boundary of instability with a width given by the uncertainties in the experimental parameters. This curve was generated from the theory described in Refs. 26-27 by Prof. D.K. Bandy.

C. Origins of Single Mode Instabilities in Inhomogeneously Broadened Lasers

Casperson[16] developed an ingenious picture to give physical insight on the origin of these single mode pulsations. The technique is an extension of the procedure introduced by Casperson and Yariv[1] which we used in section II-G to find steady state solutions by separately satisfying the conditions of 1) gain = loss (which gives the intensity) and 2) matching an integer number of wavelengths in the dispersive medium to the cavity length (which gives the frequency). In general the two steps are coupled and must be satisfied self-consistently. However, for many inhomogeneously broadened lasers where σ_D is relatively much larger than $\gamma_\perp$, these two conditions become more nearly independent and hence the simple graphical pictures become very accurate.

Consider a situation such as that shown in Fig. III-7 for a resonantly tuned cavity and an inhomogeneous medium.

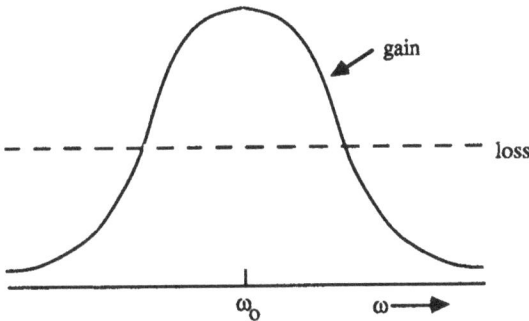

Figure III-7a. Gain and loss levels.

216

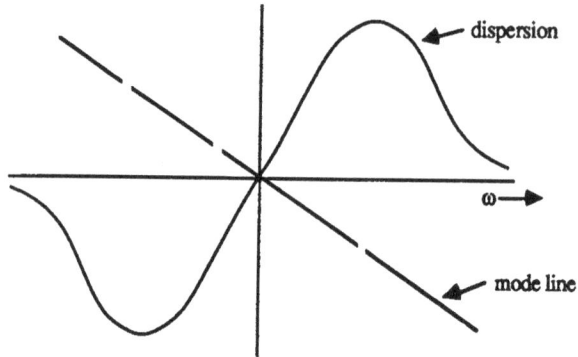

Figure III-7b. Dispersion curve and mode line.

Suppose that the resonance condition is satisfied by the intersecton at $\omega_L = \omega_0$. The intensity of the steady- state solution is determined by the intensity needed at frequency ω_L that is sufficient to reduce the gain at ω_L to the loss level. We now try to plot and picture the residual gain--that found by a weak probe field (or a pair of symmetrically detuned probes) in the presence of the lasing field and its saturation effects. Because the medium is inhomogeneously broadened the gain will not be uniformly reduced; rather a hole will be burned with a width given approximately by the power broadened homogeneous line width (hole half width $\cong \gamma_\perp (1 + I/I_s)$) as shown schematically in Fig. III-8.

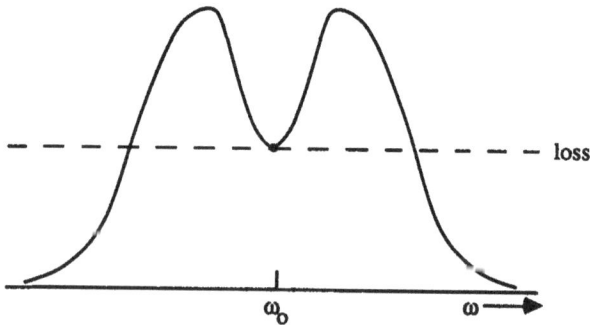

Figure III-8a. Saturated gain profile.

To a first approximation, Fig. III-8a is the "residual gain" whereby we mean that it is the gain that a weak probe at frequency ω would see in the presence of the strong saturating field at the laser solution ω_L. Sargent and Hendow[25] have pointed out that this curve is not precisely determined and that one must make a careful calculation of the residual gain for quantitative accuracy. However, for large inhomogeneous broadening and near threshold operation, this curve is suitably accurate.

As a consequence of the saturation of atoms leading to the hole in the gain profile, we must also correct the dispersion for the absence of these atoms. This is equivalent to adding the dispersive effect of a narrow absorber (the missing atoms) to the unsaturated dispersion of the original gain profile. The result is shown in Fig. III-8b.

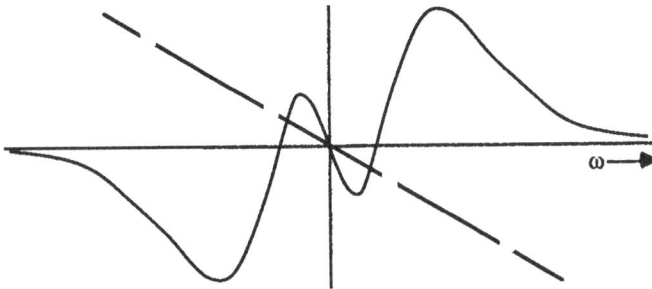

Figure III-8b. Residual dispersion and mode line.

The argument for instabilities is made based on the two new intersections of the mode line with the residual dispersion curve. At these two new frequencies there is both a) gain (note that in Fig. III-8a the gain at these frequencies exceeds the loss) and b) resonance (guaranteed by the intersections in Fig III-8b). Thus sidebands should grow at these frequencies leading to an amplitude modulation instability of the single mode steady state solution. The intensity pulsation frequency should be the frequency difference between the laser frequency, ω_L, and the sideband frequency which should be about the width of the hole. Thus we would expect the characteristic pulsing frequency of single mode instabilities to be of order the homogeneous linewidth and to increase with increasing pressure or with increasing laser intensity. Both of these effects are observed experimentally.

This phenomenon has been termed "mode splitting" by Casperson[16] or "dynamic modesplitting"[10] to distinguish it from the "passive modesplitting"[10] of multiple steady state

218

solutions for a single cavity mode as first discussed by Casperson and Yariv.[1] The mode-splitting procedure works so well that it was bound to have its detractors. Formal arguments concerning its validity[17,24-26] seem to be resolved as follows:

1) The exact residual gain and dispersion must be calculated by formally adding two weak sidebands to the laser equations in the presence of the strong mode solution and calculating the response of the system.

2) At the instability threshold (where the sidebands have gain just equal to loss and where they are resonant) the weak sideband or "mode-splitting" approach gives identical results to those of the linear stability analysis procedure.[26]

3) Above or below the instability threshold the gain or loss coefficient for the sidebands must be properly incorporated into a complex frequency.

4) Hence the instabilities can be satisfactorily viewed as arising from the dispersive effects of hole burning.

Several further comments are in order:

First, the presence of other sideband frequencies (either as weak probes of stability or as part of a pulsing solution) causes a modulation of the intensity and thus also a modulation of the atomic variables. The modulation frequency is the beat frequency and higher harmonics may also develop. The modulation of the atomic variables modulates the dispersion if the laser cavity is detuned and this can cause frequency modulation of the laser field as well. In both resonant and detuned cases the depth and width of the hole and the distortions of the dispersion are all modulated in time. For this reason the spectral hole-burning picture for the instability rapidly breaks down.

Second, a schematic drawing of the gain found by a pair of symmetrically placed sideband probe frequencies is in order. In Fig. III8c such a schematic is drawn. It has the following features:

a) At the lasing frequency the gain is below the loss - this is because the laser is stable with respect to amplitude fluctuations at the lasing frequency. (The usual picture of the gain balancing the loss for a steady state solution is still correct. We are now plotting the "residual gain" for an increment in the intensity. It is this residual gain which is negative.

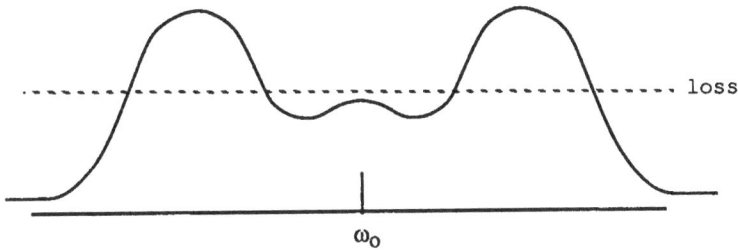

Figure III-8c. Schematic of the residual gain for a pair of weak side band frequencies placed at $\pm\omega$, in the presence of a laser field at ω_0 for an inhomogeneously broadened laser.

b) The gain for frequencies near the lasing frequency is typically even less than the gain at the lasing frequency. Only for larger detunings is the gain again positive.

c) The spectral width of the "hole" in the gain is still approximately that of the power broadened homogeneous linewidth as in the simpler picture of Fig. III-8a.

d) Because the gain is less than the loss for frequencies near the laser frequency, unstable sidebands will appear only for sidebands with a suitable non-zero detuning. For this reason the laser instability threshold is somewhat above the lasing threshold and the first instability frequency is a finite (non-zero) value of order one half of the power broadened homogeneous linewidth.

Third, there is no reason why a weak sideband approach could not be used for instabilities in single-mode homogeneously broadened lasers, a procedure explored in the graphical pictures of Hendow and Sargent[25] which are more detailed versions of the earlier pictures of Risken and Nummedal (who drew only the sideband gain curves). In this case, however, the sideband "residual gain" and dispersion are so modified from their unsaturated forms that it is hard to develop a simple intuitive picture. A schematic of the sideband gain for a homogeneously broadened laser is shown in Fig. III-8d for different strengths of the lasing field.

220

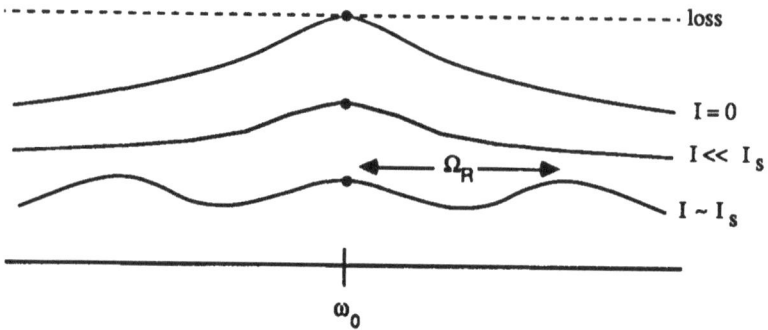

Figure III-8d. Schematic of sideband gain for a homogeneously broadened laser with laser frequency ω_0 for different laser intensities.

The instability will arise in this case because of parametric coupling of the sidebands to the laser mode when they are spaced by approximately the Rabi frequency ($\pm\Omega_R$) from the lasing frequency. Under suitable conditions the emerging bumps lead to net gain. The dispersive effects of this parametric resonance can also induce new intersections with the cavity modeline giving the "mode-splitting" resonance condition for the instability. Because a strong lasing field is needed to create strong parametric coupling and to achieve gain which exceeds the loss, the homogeneously broadened laser has a relatively high instability threshold and its pulsing frequency occurs when the Rabi frequency is a few times the homogeneously broadened half width.

D. Further Subtleties Revealed by Heterodyne Detection: Measuring the Electric Field Amplitude of Pulsing Lasers

A nagging worry in studies of nonlinear systems is whether the effects that one is measuring result from nonlinearities in the system or from nonlinearities in the detection process. In the study of optical systems this is a particular worry when the relevant field variable is the electric field amplitude while the measured variable is the intensity. [This is not always a worry in optical systems as in some cases the dynamics depend only on the intensity; e.g., the hybrid bistable systems[30,31] and the single mode "class B" laser systems discussed by Arecchi[32]. However most multimode systems (involving resonant cavities) and "coherent" single mode systems (where the polarization dynamics are important) depend on the field amplitude as one of the fundamental dynamical variables.]

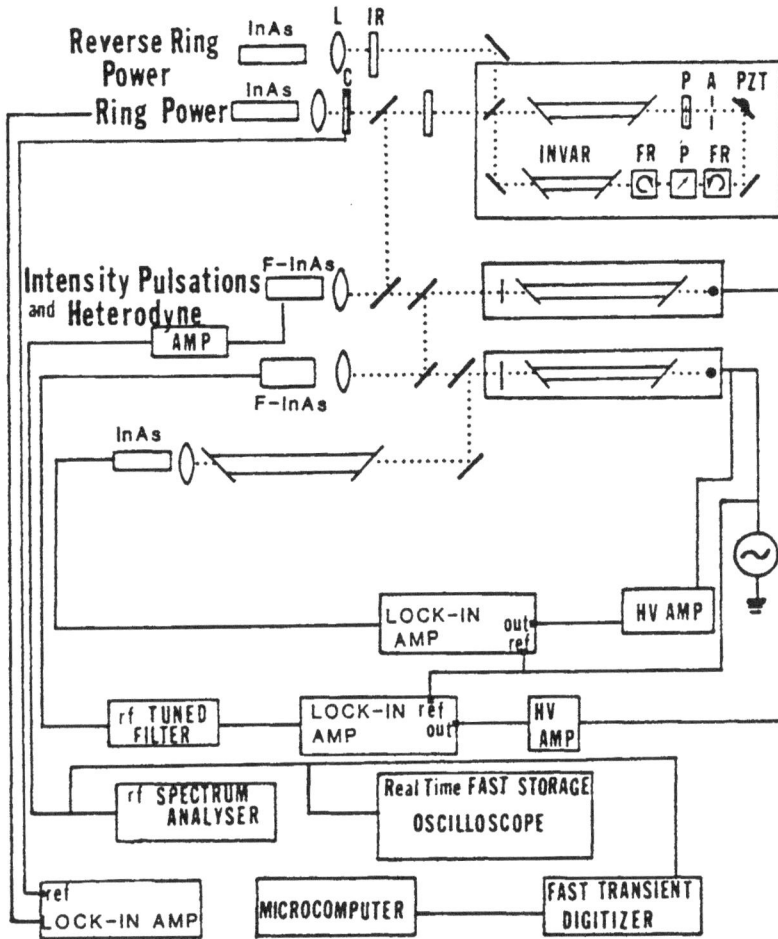

Figure III-9. Stabilization and heterodyne reference setup. One laser is feedback stabilized to the peak of the gain cell response, the other laser is stabilized to a selected frequency offset.

In order to fully detect the effects of the field dynamics we must measure the field amplitude. This cannot be done directly because of the high frequency of the optical carrier wave. However, by heterodyning with a stabilized single-frequency reference laser [that is, illuminating the detector with spatially coherent light from the reference laser and the unstable

222

laser] we find that the resulting signal from the photodetector includes signals proportional to the amplitudes of the two lasers times a carrier wave oscillating at the frequency difference of the carriers of the two lasers. If the reference laser frequency is adjusted to a convenient offset from the frequency of the unstable laser, the 'heterodyne spectrum' is clearly separated from the 'homodyne spectrum' (the intensity pulsation spectrum) and directly corresponds to the electric field amplitude spectrum of the unstable laser. If the reference laser has a large enough amplitude and adequate phase stability, the time-dependent electric field amplitude of the unstable laser can be extracted from the heterodyne signal, otherwise only the norm of the amplitude spectrum can be recovered (but this is very useful additional information).

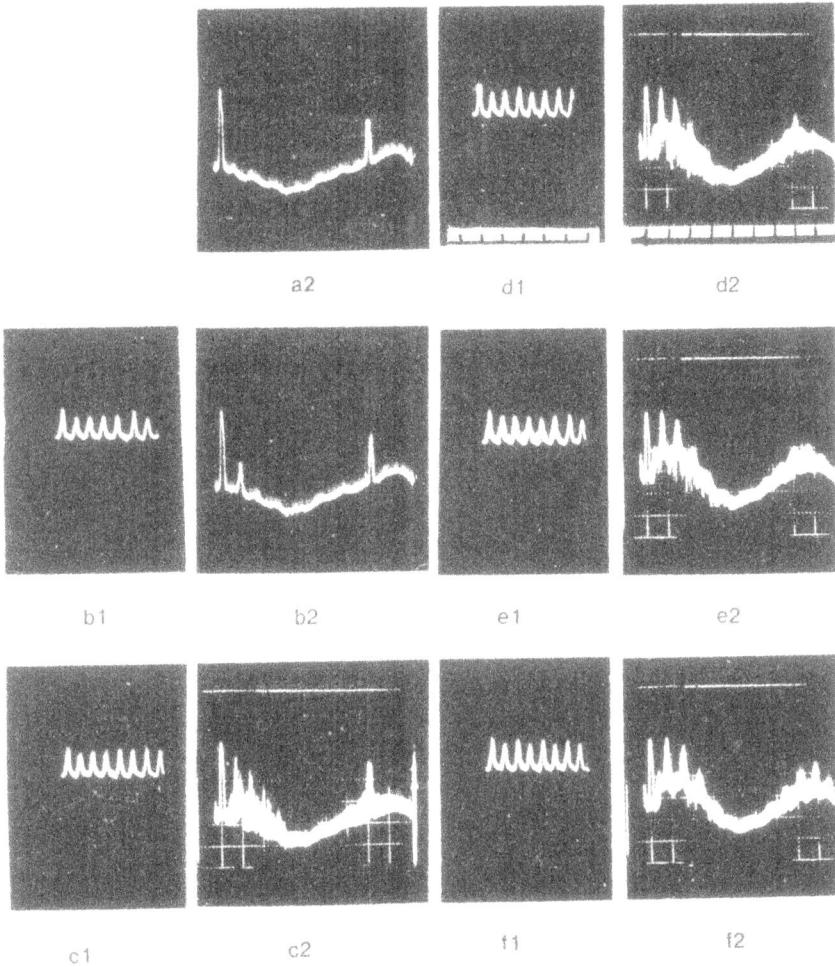

Figure III-10. Time dependent intensity pulsations, and homodyne and heterodyne spectra for laser tuned to resonance with increasing gain from a) to f) (from ref. 22). Laser pressures are 172 mTorr Xe-136 and 1.0 Torr helium. Time snapshots are 50 ns long. Spectra cover a 90 MHz range at 10 dB/div. on the vertical scale.

We achieve the measurements of the heterodyne spectrum with a stabilized reference laser system as shown in Fig. III-9. Two lasers are used in the stabilization circuit to achieve a frequency offset from the peak output of the laser and to obtain a stabilized, but unmodulated, laser for use as the reference.

Selected examples of the time-dependent intensity pulsations and of the homodyne and heterodyne power spectra are shown in Figure III-10. Stable output is shown in Fig. III-10a. An example of periodic output is shown in Figs. III-10b. Examples of C1 and C2 output are shown in Figs. III-10c and III-10d-III-10f, respectively. This information allows us to make two other types of summaries of the data as a function of the parameters (gain and detuning) which we can easily vary. Figure III-11 shows the frequencies of the peaks in the spectra as the cavity length is scanned.

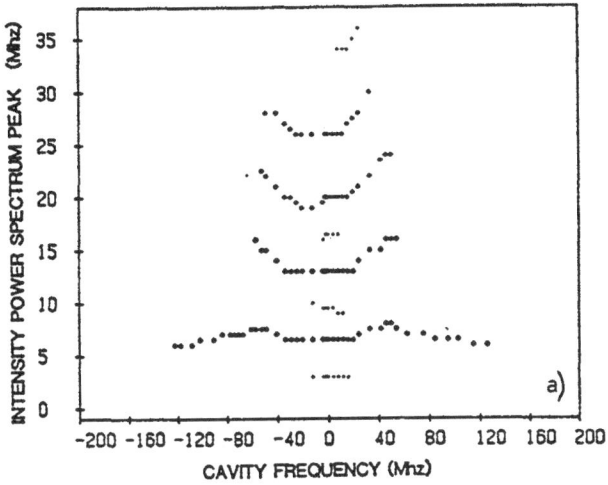

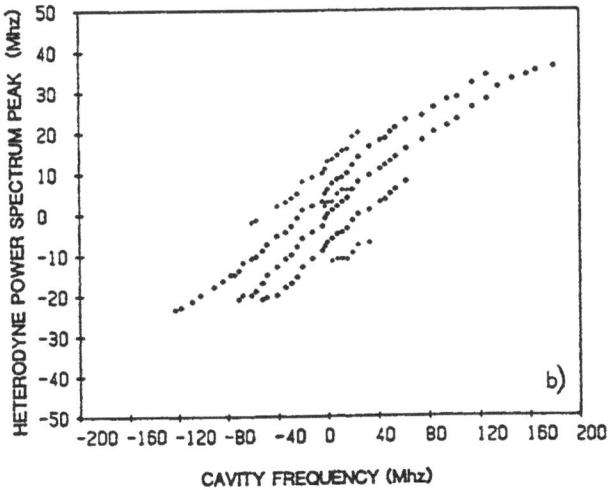

Figure III-11. Peaks in the homodyne (a) and heterodyne (b) spectra as the cavity length is detuned (from ref. 22). The intensity pulsing frequency increases with increased excitation near line center with a discontinuity where the heterodyne spectrum breaks from one dominant peak to two nearly equally strong peaks. The intensity pulsing frequency seems to decrease slightly in the period doubling region.

Figure III-12 shows the frequencies of the peaks in the spectra as the gain is scanned for resonant detuning.

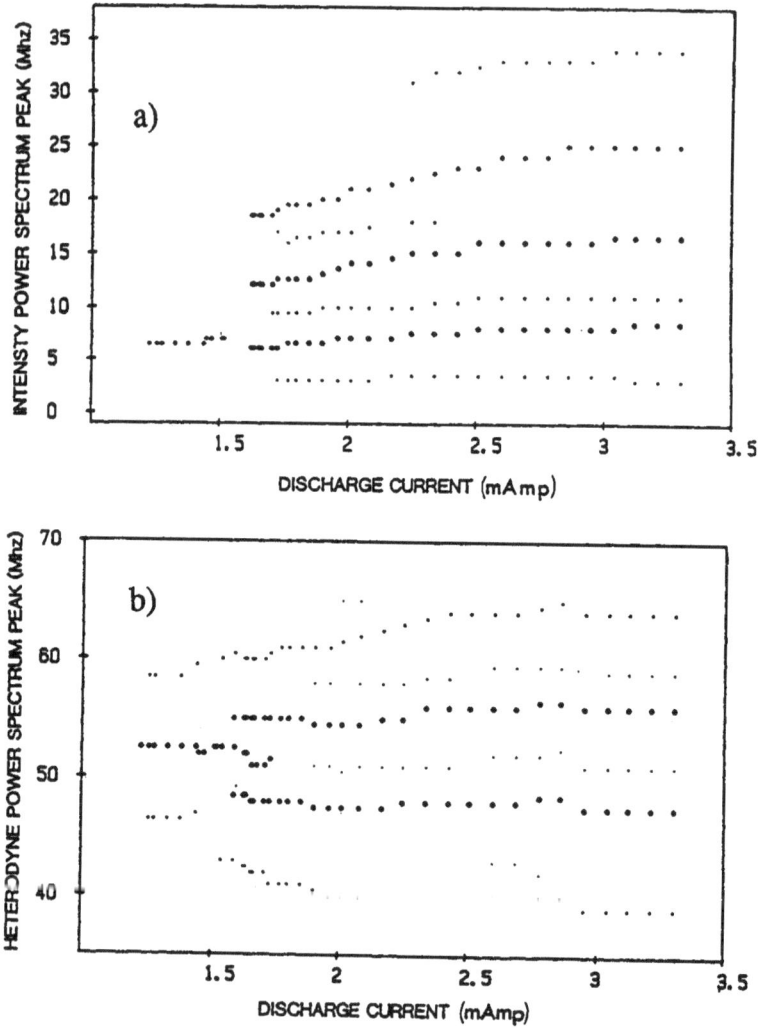

Figure III-12. Peaks in the homodyne (a) and heterodyne (b) spectra as the gain is scanned. Intensity pulsing frequency increases slowly with excitation with a dip arising from transformation of one strong peak to two strong peaks.

Figure III-11b shows that the laser frequency varies more slowly than the cavity frequency, an effect called "mode pulling". In most lasers this is a small effect of a few percent or less. Here, however, we see an effect which ranges from factors of four to six over the detuning range.

Figures III-11 and III-12 permit us to better understand the transitions from stable to unstable behavior in this kind of laser. We can picture the solutions in the multivariable phase space of the system. The stable laser of finite intensity corresponds to a fixed point in the variable space. Here "fixed point" means that small perturbations relax back to this solution. Considering the projection of the variable space onto the two-dimensional plane of ReE versus ReE we see the figure plotted in Fig. III-13a.

Increasing the excitation leads to small amplitude modulation of the laser as shown in Figure III-10b. This is shown schematically in Figure III-13b. As the trajectory grows in phase space (with increasing excitation) it crosses from ReE>0 to ReE<0 as shown in the transition from Fig. III-13c to Fig. III-13d. For the intensity in Fig. III-13d there are two peaks per cycle around the closed trajectory and it appears to be a "period-doubled" solution a name often applied when the intensity peaks alternate regularly in height, particularly when there seems to be no natural reference frequency to compare with). A fully developed periodic solution which is symmetric in ±ReE (such as seen in Fig. III-10f) would appear as in Fig. III-13e while the chaotic equivalent is shown in Fig. III-13f.

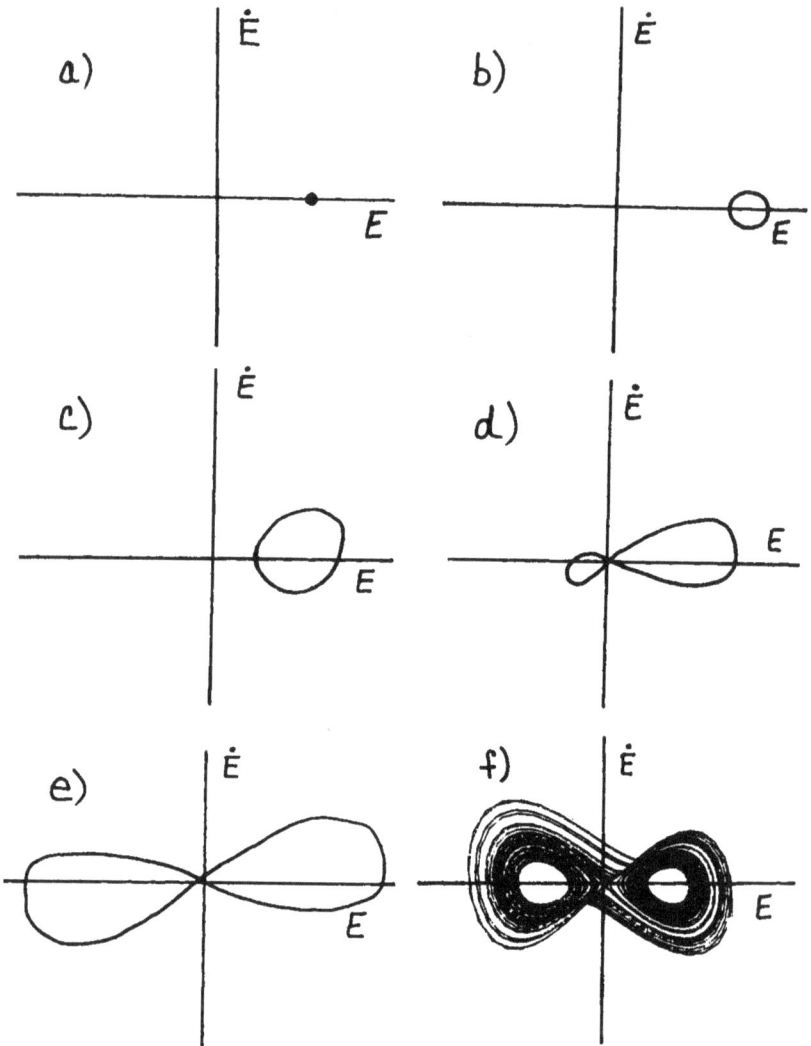

Figure III-13. Schematic phase space portraits of ReĖ vs. ReE for stable operation (a), weak pulsing (b,c), stronger asymmetric pulsing (d), and fully symmetric pulsations (e,f).

For resonant tuning the equations are symmetric under the interchange E→-E, P→-P, D→D (where E, P and D are the field amplitude, polarization amplitude and population inversion, respectively). Under this condition we refer to solutions as being either "SYMMETRIC" or

"ASYMMETRIC". Figs. III-13a-d are "ASYMMETRIC" and Figs. III-13e,f are "SYMMETRIC". The heterodyne spectra, with a central peak, represent "ASYMMETRIC" solutions (Figs. III-10a-e). The "subharmonics" seen in the intensity (homodyne) spectrum of Figs. III-10c-e and recorded in Figs. III-11a and III-12a are revealed by the field (heterodyne) spectra to not represent period-doublings, but rather slightly asymmetric solutions.

This provides useful insight into the physical processes in these lasers and reminds us that systems with the inversion symmetry in their variables cannot undergo period doublings of symmetric solutions[33]. Symmetric solutions must undergo symmetry breaking if they are to follow period doubling sequences. A symmetric solution may make a direct transition to chaos or it may follow a symmetry breaking followed by glueing together of the two asymmetric solutions to form a new (more complex) symmetric solution followed by further symmetry breakings and glueings[34]. These many results for the unidirectional ring laser are in excellent agreement with the results from quite complicated models (which have sometimes required as many as 302 equations to fully model the inhomogeneous lineshape)[20].

When there is detuning the symmetries in the equations is broken. The symmetry properties of the solution are also changed. Period-doublings are now possible and are frequently observed. Phase and frequency modulation also become common. It is often possible, however, to use the approximate symmetry for effective descriptions.

E. Measurements to Confirm Chaotic Laser Behavior

We have chosen to call "CHAOTIC" that behavior which has a broadband component to its power spectrum. However, the term "chaotic" is used here to distinguish deterministic aperiodic behavior from random (or stochastic) behavior. Low dimensional chaos in continuously varying dissipative systems[35] has such properties as: 1) at least one positive Lyapunov exponent (corresponding to the exponential divergence of neighboring trajectories in phase space), 2) a finite (but non-zero) entropy, and 3) a dimension for its asymptotically attracting subset of the phase space that is a finite number (typically non-integer) that is greater than two.

As stochastic noise (suitably filtered) can have the same spectrum as deterministic chaos, we cannot immediately infer chaotic behavior from broadband spectra or finite entropy. Rather we must analyze the time-dependent signals in detail to establish that chaos underlies the observed erratic behavior.

What we wish to do is to describe the pulsing signal as the result of motion on an attractor in the variable space. Consider the three cases shown below in Fig. III-14. In Fig. III-14a the solution is a fixed point; in Fig. III-14b the solution is a closed loop (called a "limit cycle"); in Fig. III-14c, the solution is a two-torus (involving two incommensurate frequencies); and in Fig. III-14d, the solution is the strange attractor of the Lorenz-Haken model.

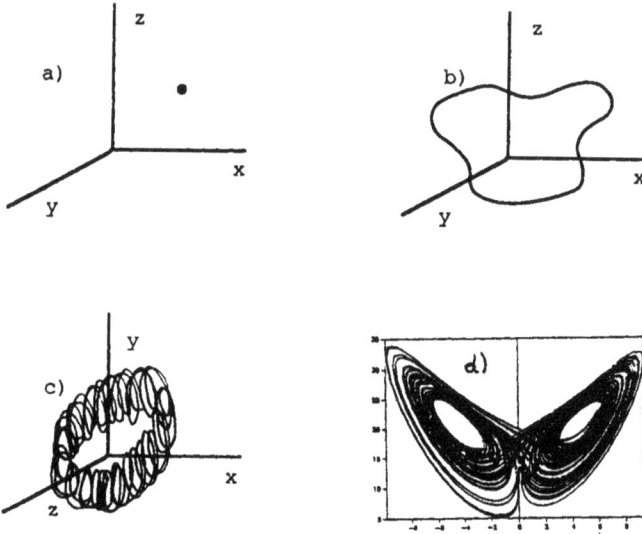

Figure 14. Schematics of different kinds of solutions in a variable space.

If these solutions are digitized then they become sequences of points instead of lines. We measure the dimension by examining small neighborhoods of radius ϵ near each point. If we find that the number of points in a ball of radius ϵ grows as $N(\epsilon) = \epsilon^D$, then we say the local dimension is D.. This examination must be done at length scales as small as possible and after the attracting set has been fully reconstructed by a sufficiently long time series. A strange attractor (chaos) typically has a fractal dimensionality.

The methods for this analysis have been outlined in detail over the last four years[36] and refined in various ways[37]. We have recently demonstrated that dimensions can be accurately calculated in some cases with as few as several hundred points with 10-bit accuracy[33]. Of course, when digitizing capabilities permit, it is useful to take larger numbers of data samples. However, more than ten thousand data points can seriously tax the computer used for calculating the results.

The basic method for calculating the dimension is outlined in the sequence of Figures III-15-17. The data selected had intensity spectra as shown in Fig. III-15a and time series as shown in Fig. III-15b.

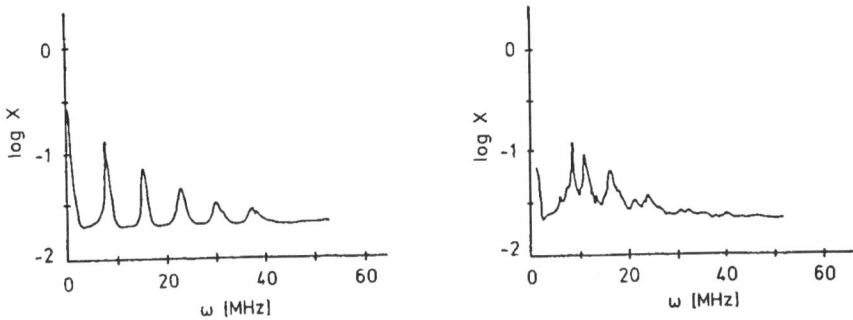

Figure III-15a. Intensity spectra of selected data.

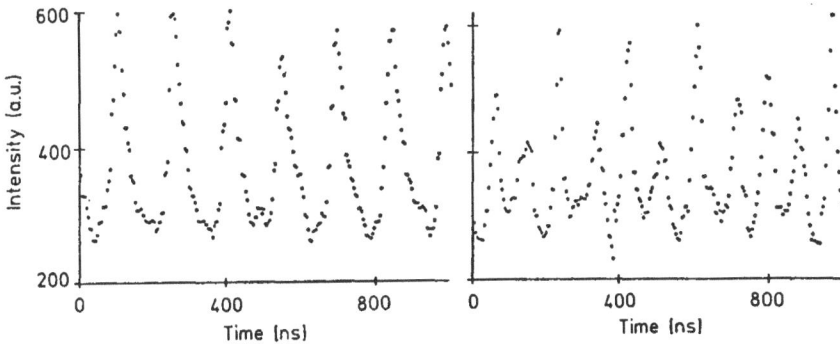

Figure III-15b. Intensity time series of selected data.

One is periodic, the other is quasi-periodic or perhaps chaotic. Figure III-15b shows a portion of the digitized time series of equally spaced points for each case. We embed the data to reconstruct the topology of the attractor by forming vectors which have components which are successively delayed values from the time series. For example, a two-dimensional embedding would be made up of vectors $(I(t_n),I(t_{n+q}))$, where q is a suitably chosen delay. The number of components in a vector is called the embedding dimension. An attractor dimension for each embedding is calculated by examining the number of points in the neighborhood of each point as a function of the neighborhood size. If the attractor has a well-defined dimension then the number of points, $N(\varepsilon)$ should scale as ε^D over some range of ε. If we take ε too large it exceeds the boundaries of the attractor and scaling breaks down. For ε too small we may not have enough occurrences of these interpoint distances or experimental noise or digitizing accuracy becomes a distorting factor. We hope to find a scaling region for intermediate valves of ε after a finite number of embedding dimensions. A topology theorem tells us that the topology of the attractor (for example, its dimension) in the variable space can be fully recovered by the embedding technique by taking a large enough dimension for the embedding of the time series of any single one of the interacting variables or from a time series of any nonlinear function of such a variable. Practically, the embedding dimension needs to be only a little larger than 2D+1 but the time between successive components of an embedding vector must be chosen to give suitably correlated but independent information. In Fig. III-16 we give a log-log plot of the number of interpoint distances less than ε versus ε for embedding dimensions 10-20. If the reconstructed attractor has a uniform dimension, there will be a range of ε (or, correspondingly, a range of $(N(\varepsilon))$) for which the log-log plots have the same slopes. Plotting the slopes of the curves in Fig. 16 we look for flat regions as found in the two cases at 1.1 ± 0.1 and 2.6 ± 0.3, respectively. Also from the vertical separation of the curves in Fig. III-16 we can calculate an estimator of the entropy which should be zero for periodic signals, finite and non-zero for chaotic signals, and infinite for white-noise random signals.

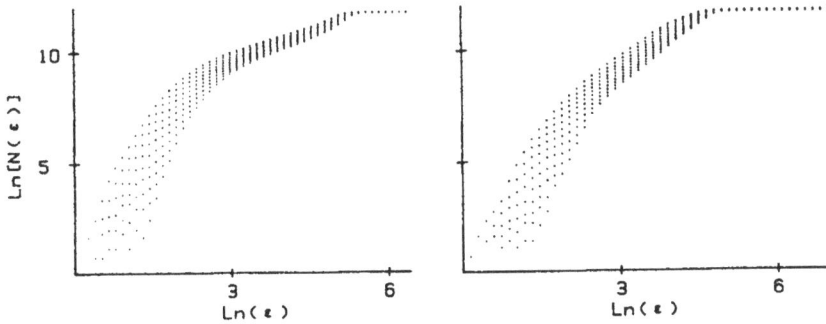

Figure III-16. Ln.of the correlation sum (N(ε) normalized to N(∞)) vs. ln.ε for the two data sets.

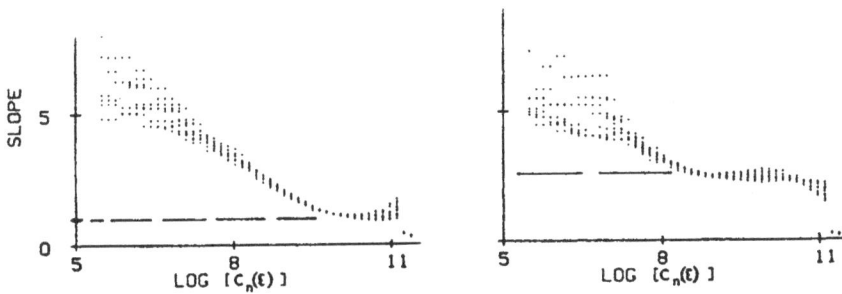

Figure III-17. Slope of curves in Fig. III-16 vs. ln.of correlation sum.

Applying these methods to many different spectra for different detunings of the ring laser we obtained the results shown in Figure III-18. There are clear regions of dimension of order 1 at large detunings and a region near line center where the dimension is in the range of 2-3. This range corresponds to broad peaks in the power spectrum and to higher entropy values, all evidence that the broadband spectra arise from chaotic behavior. Numerical results for chaotic pulsing from the theoretical model for this laser have been analyzed for fractal dimension and that point is plotted on the curve.

234

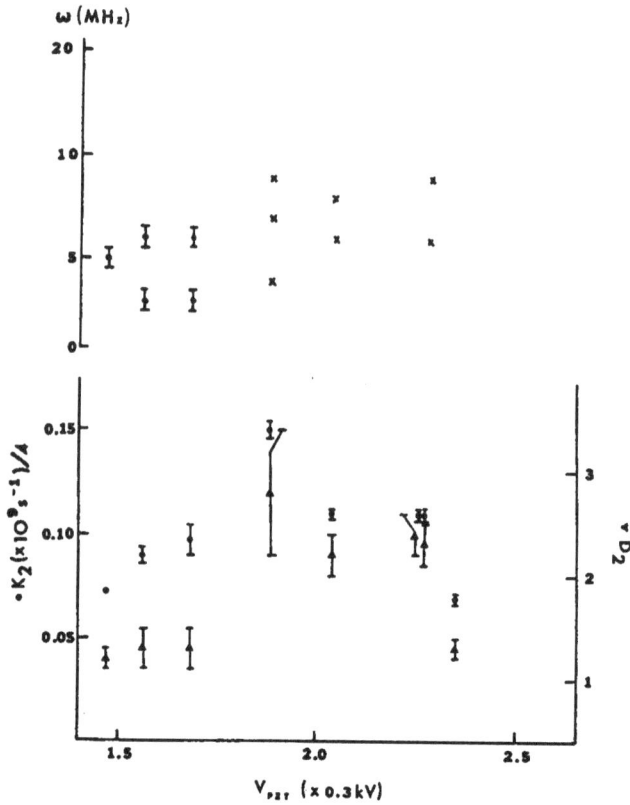

Figure III-18. Results for dimension and entropy for different cavity detunings and corresponding data on peaks in the homodyne spectrum.

Data is limited in most cases in terms of number of data points and digitizing accuracy. How then should one go about applying the method? First one must pick the embedding time so that the components in an embedding vector are somewhat correlated with each other. The total time spanned by the components of a single vector (sometimes called "the window") must be less than the time over which the envelope of the autocorrelation function of the signal falls to near its asymptotic value. However, if the data points used as components are too closely spaced in time, they do not contain new information. A compromise is necessary. When the spectrum shows a strong periodic component (peak), a good rule of thumb is to take data with about 20 points per period and to use an embedding time of 2 or 3 sampling intervals.

The total span of the time series (sometimes called "the epoch") must far exceed the time over which the autocorrelation function differs from its asymptotic value to give full visitation of the attractor. One must also avoid taking too many data points per period unless a suitably large number of periods is digitized. Too many points per period gives emphasis to each particular trajectory of the attractor and may bias the calculated dimension to the value D=1 characteristic of an isolated filamentary structure. We have found such problems when 50 or 100 points are taken per period even when the total data record is a few hundred periods. As a good rule of thumb, the number of points per period should be an order of magnitude smaller than $N^{1/D}$ where N is the total number of data points and D is the dimension of the attractor.

As a further caution we note that an attractor may not have its fractal structure as large in some portions of the attractor as in others. Hence the algorithm averaged over the attractor for any particular length scale may average accurate fractal dimensions with distorted values from other regions of the attractor which are more compressed or expanded. The best way to avoid this is to calculate the dimension locally around only a single point (pointwise dimension) or in a small region of the reconstructed attractor.

Other systematic errors also occur from the following:

a) Limited precision - the estimated attractor dimension decreases as the number of bits of precision decreases.

b) Excessive embedding dimension - the estimation of the dimension increases with increasing the embedding dimension so care should be used when selecting the embedding time so that convergence is obtained for dimensions less than about 20.

c) Noise - A systematic increase in the dimension at all length scales results from noise with a distortion greater at small length scales. The error scales as σ^2/ϵ^2 where σ is the standard deviation of the noise. The error also increases linearly with the embedding dimension.

d) Different requirements for good estimates of Entropy and Dimension: Several studies show that the time spanned by a single vector must be several times the inverse of the entropy to get a good convergence of the estimate of the entropy. Good values of the dimension are usualy obtained when the time spanned by a

236

single vector is of order a single period (if there is a partly periodic structure) or, if not, then when it is of order the autocorrelation time.

e) A good estimate of the entropy comes from the derivative at zero time of the envelope of the autocorrelation function.

f) Averaging the correlation integral over all interpoint differences assumes that the fractal structure has the same absolute length scale or a common range. If this is not true, errors result such as dimensions between one and two for period doubled and quasiperiodic attractors.

g) Two different norms are commonly used for calculating the distance between points in the embedded space. The usual Euclidean distance is more demanding of computer time and has the feature that length scales of the attractor are stretched by the square root of the number of embedding dimensions. When using the Euclidean norm it is often convenient to compare slopes of the correlation sum at the same values of the correlation sum to offset the rescaling of ϵ with embedding dimension. The other norm is sometimes called "the maximum norm" and is the maximum of the set of the absolute differences of the components of two vectors. This is much faster to calculate, particularly for integer data such as is generated by experiments and the length scale of the attractor is held constant. However, in both cases the influence of noise added to the pure data increasingly interferes as the embedding dimension is increased.

F. Results for Standing-Wave Lasers

We can now turn to the standing-wave lasers where instabilities, period doublings and chaos[9-15] were first seen in single-mode lasers. The simplest results have been found in HeNe 3.39 μm lasers. Several selected results are shown in Figs. III-19 and III-20. Figure III-19 shows the power output curves and pulsing frequencies versus detuning and the laser frequency versus detuning for different levels of excitation.

We see both strong mode pulling and the Lamb dip in the power output for the stable laser in Fig. III-19a. However, in both III-19a and III-19b we notice a "kink" in the mode pulling line as if near resonance the mode pulling effect disappeared. This reduced mode pulling in the Lamb dip because of "hole repulsion" has been predicted by Bennett[5,7]. For this laser the instability remains located only in the Lamb dip, even for relatively large excitaiton. We also see that the heterodyne spectrum is dominated by two large peaks although the intensity spectrum shows complex variations that depend sensitively on the cavity detuning. Figure III-20 shows some samples of the periodic, period doubled and chaotic spectra observed in this laser.

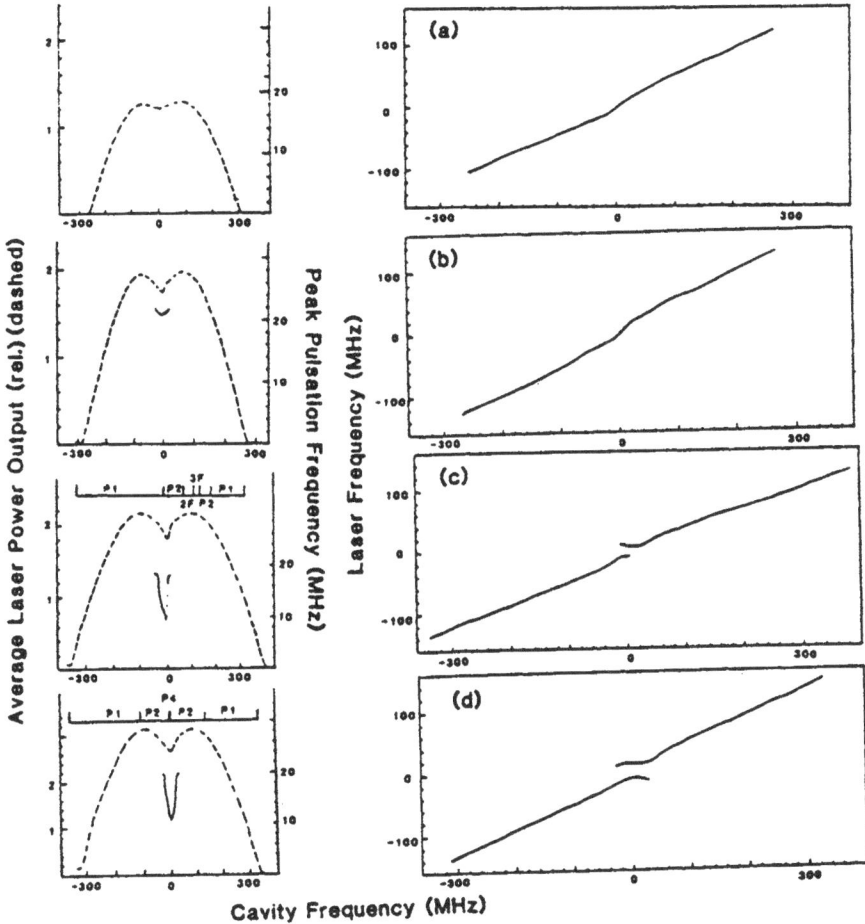

Figure III-19. HeNe laser of ref. 13b with power output, pulsing frequencies, and laser frequency versus cavity detuning.

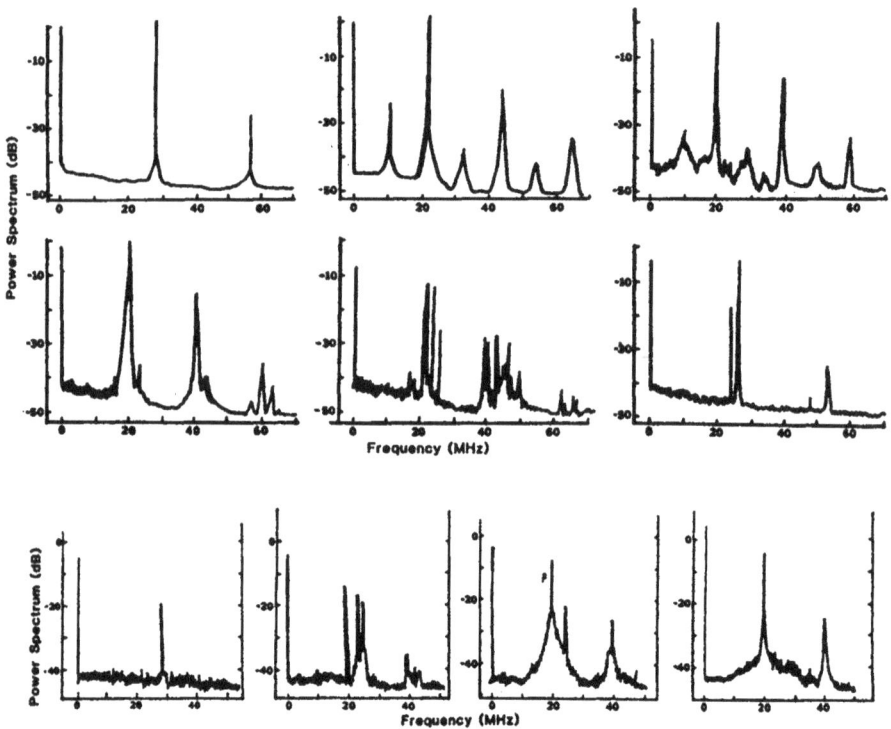

Figure III-20. Selected spectra of HeNe laser from ref. 13b.

More complex results were found for the HeXe 3.51 μm standing wave lasers which are more unstable than the 3.39 μm lasers. Results similar to Figs. III-19 and III-20 are shown in Figs. III-21-23. The instabilities appear over a much larger range and well outside the Lamb dip region.

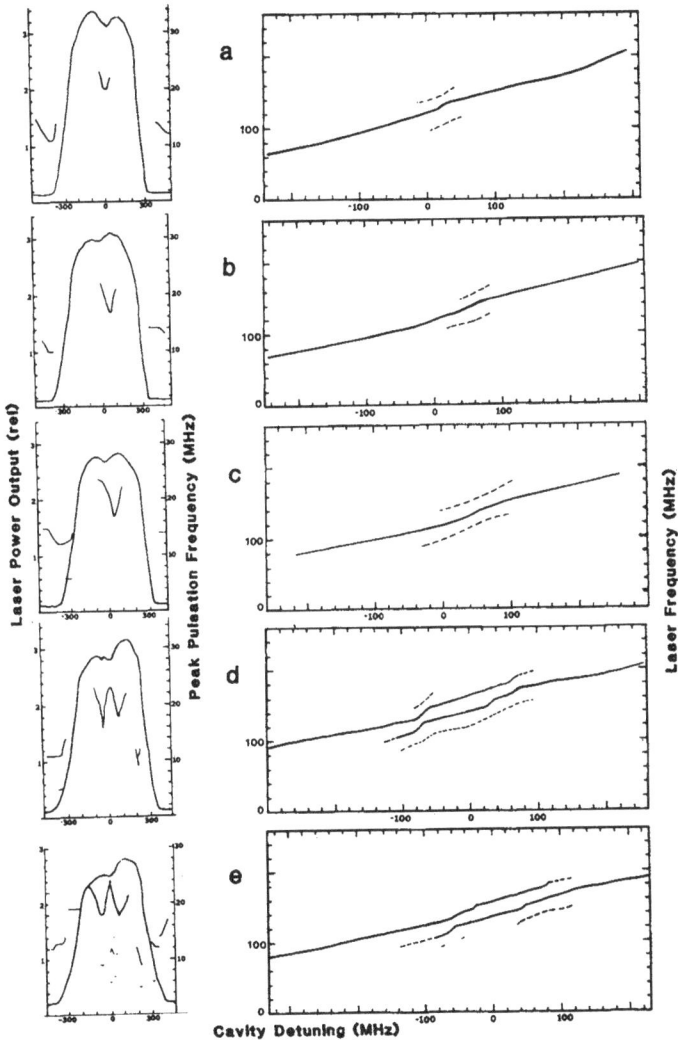

Figure III-21. HeXe laser instabilities from ref. 12b showing average power, pulsing frequencies, and laser output frequency.

For large detunings, as shown in Fig. III-22a, the pulsing frequency is roughly proportional to the output power, More complex phenomena are observed for large $\kappa/\gamma_\perp$ and large $\sigma_D/\gamma_\perp$ as seen in Fig. III-21. Some of the sequences of spectra are captured in Fig. III-24 where we can clearly see period doubling to chaos and periodic→quasiperiodic→locking→chaos. Perhaps most interesting is the plot shown in Fig. III-23 of the amplitudes of the spectral components as they are tuned through resonance. Near resonance each shows a kink as in Fig. III-21d and each shows a

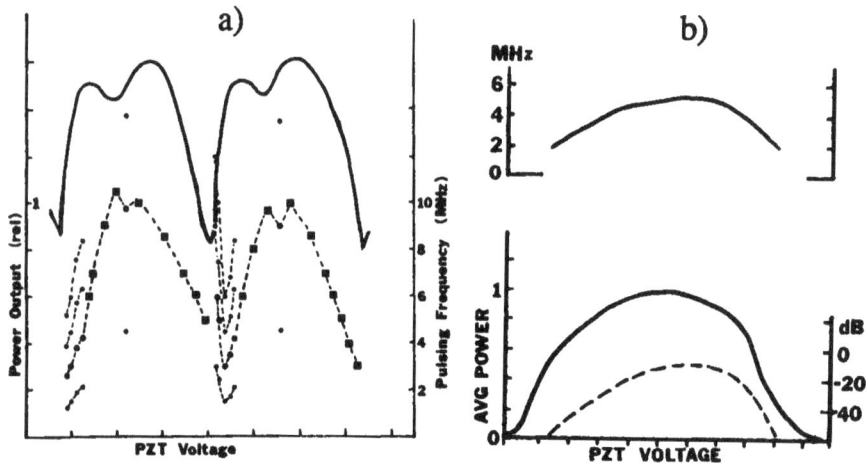

Figure III-22. Pulsing frequency and power output versus detuning for selected HeXe data. a) Fabry Perot laser data from ref. 10 (solid line is power output and dotted line is principal pulsing frequency); b) ring laser data from ref. 18 (solid lines give pulsing frequency and power output, dotted line gives strength of spectral peak). Note that for these pressures the pulsing was of small amplitude hence the pulsing for the ring did not show the dip near line center seen in earlier ring laser data in Fig. III-11 where the pulsing became more complex.

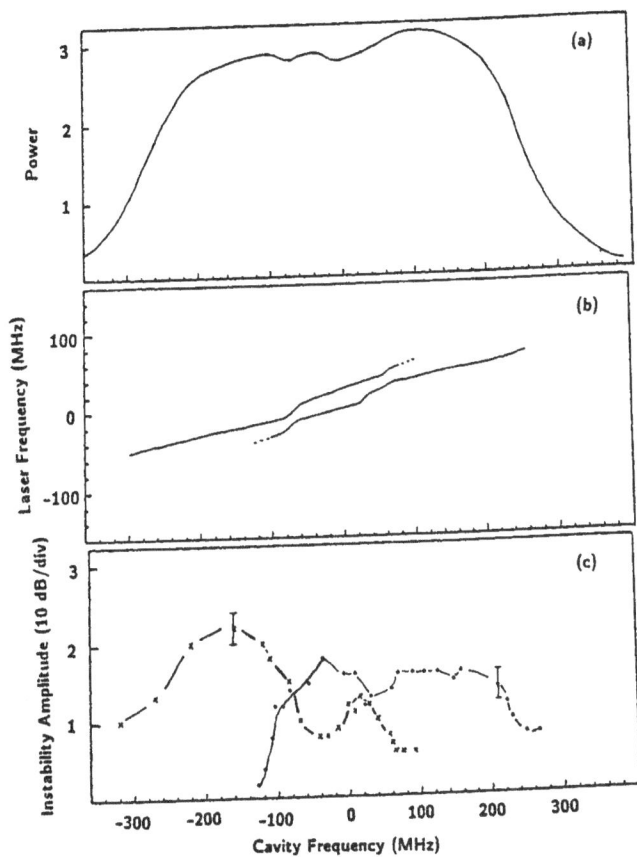

Figure III-23. Variation of heterodyne spectral peak amplitude corresponding to Fig. III-21d showing dips as each component passes through resonance with the zero-velocity atoms (from ref. 39).

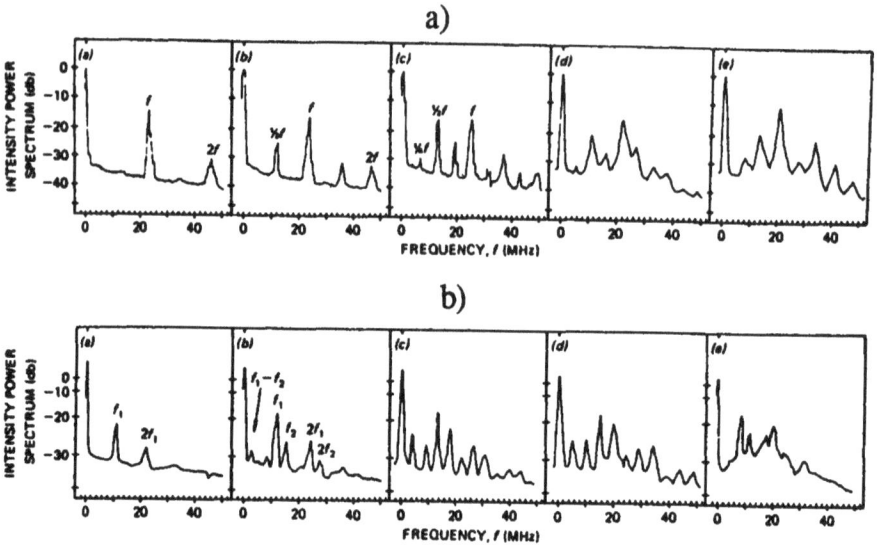

Figure III-24. Progressions of spectra from ref. 13a showing period doubling (a) and quasiperiodic (b) routes to chaos.

dip in output[39]. This is rather curious because the spectral components are not independent but coupled by the dynamical time evolution of the system as a whole. It is fascinating to see that the steady state Lamb dip behavior is also found in each spectral component in each spectral component in the time-dependent case.

G. A Careful Look at Field and Intensity Spectra and Their Correlation Functions

In this section we wish to address the question of the appearance of chaos as a modulation of the intensity and of the electric field for the same system[40,41]. In part, we wish to address the fact that "fully developed chaos" (a smooth broadband spectrum) is relatively rarely seen in the intensity power spectra of our lasers. Is this because our lasers are only weakly chaotic? Or is it because the intensity is a nonlinear distortion of the more fundamental electric field variable? Specifically, if we turn to the best signal-to-noise spectra we have obtained, those from the HeNe laser as shown in Fig. III-20, we ask if spectra shown in Figs. III-20c and III-20d are chaotic.

To address this question we look first to the results of the numerical model for inhomogeneously broadened ring lasers. Figure III-25 shows the evolution of the electric field and of the intensity for unstable operation. [The results are extraordinarily similar to those for the homogeneously broadened single mode laser[42] ("Haken-Lorenz model"), which we atribute to an overall low dimensional evolution of the inhomogeneously broadened system despite its 302 equations.]

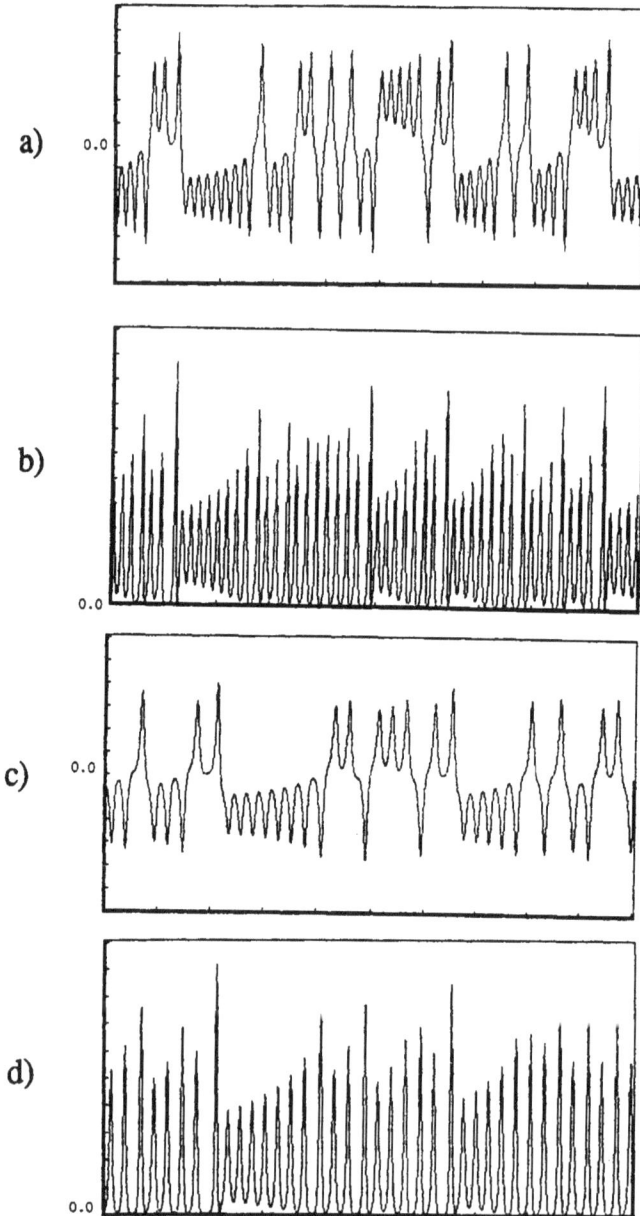

Figure III-25. Electric field (a) and intensity (b) versus time for inhomogeneously broadened ring laser model and (c) and (d) for the field and intensity of the Haken-Lorenz homogeneously broadened laser model.

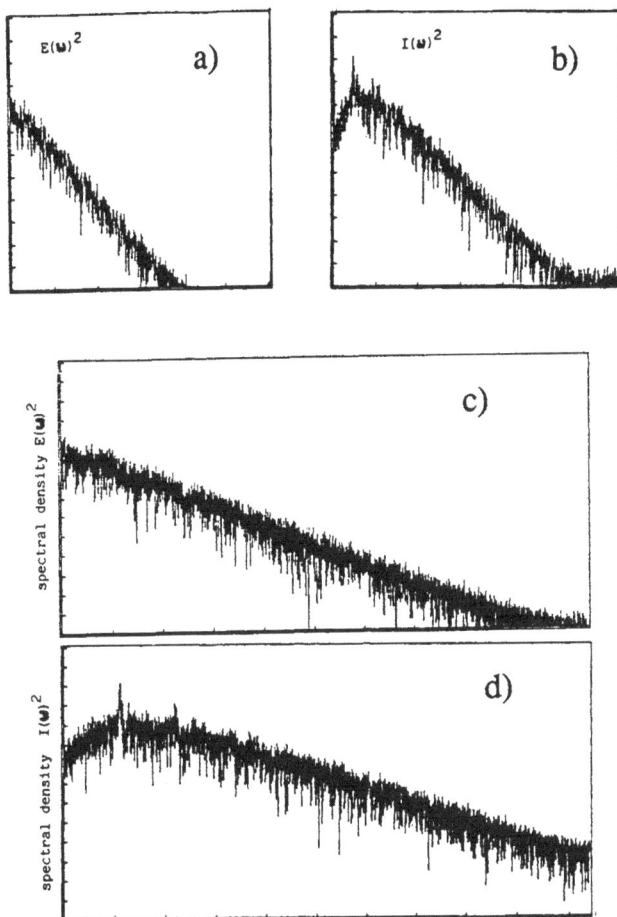

Figure III-26. Power spectra of field (a) and intensity (b) for inhomogeneously broadened ring laser and (c), (d) corresponding results for the homogeneously broadened laser.

In Fig. III-26 we show the power spectra of these two signals. The electric field is fully developed in its chaos while the intensity has a narrow peak that rises 20 dB above the broad background. The broadband field spectrum is peaked at zero frequency while the broadband portion of the intensity spectrum is 20 dB lower at zero frequency than its maximum values. In these characteristics the spectra of Fig. III-26 agree quite well with Figs. III-20c, 20d indicating that they may well be "fully-developed chaos." It is an important reminder to us of the phase correlations hidden in the amplitude spectrum for chaotic signals which are essential to giving the peaks in the intensity spectrum.

It is useful to know that the "topology" of the chaos in these two cases is the same. Calculations of the dimensions in both cases yield a value of order 2.0 even though the chaotic structure in the attractor is a very different fraction of the overall size in the two cases[41].

It is also useful to look at the higher order correlation functions for the field and intensity as shown in Fig. III-27 which have their own distinctive features. Special among them is the evidence for two time scales in the decay of the envelope functions. One time scale is short,

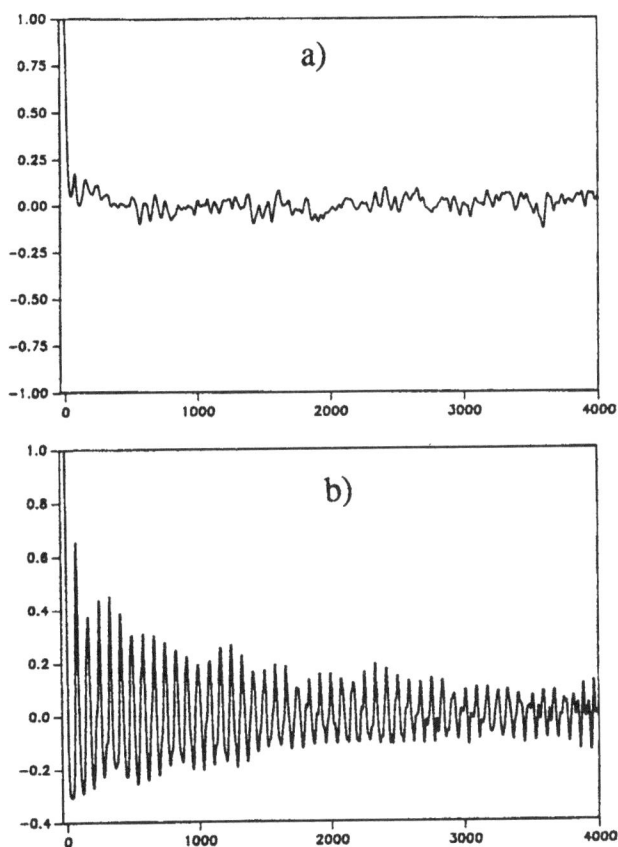

Figure III-27a. Auto-correlation functions C_{11} for the field (a) and intensity (b) outputs of the inhomogeneously broadened ring laser model.

248

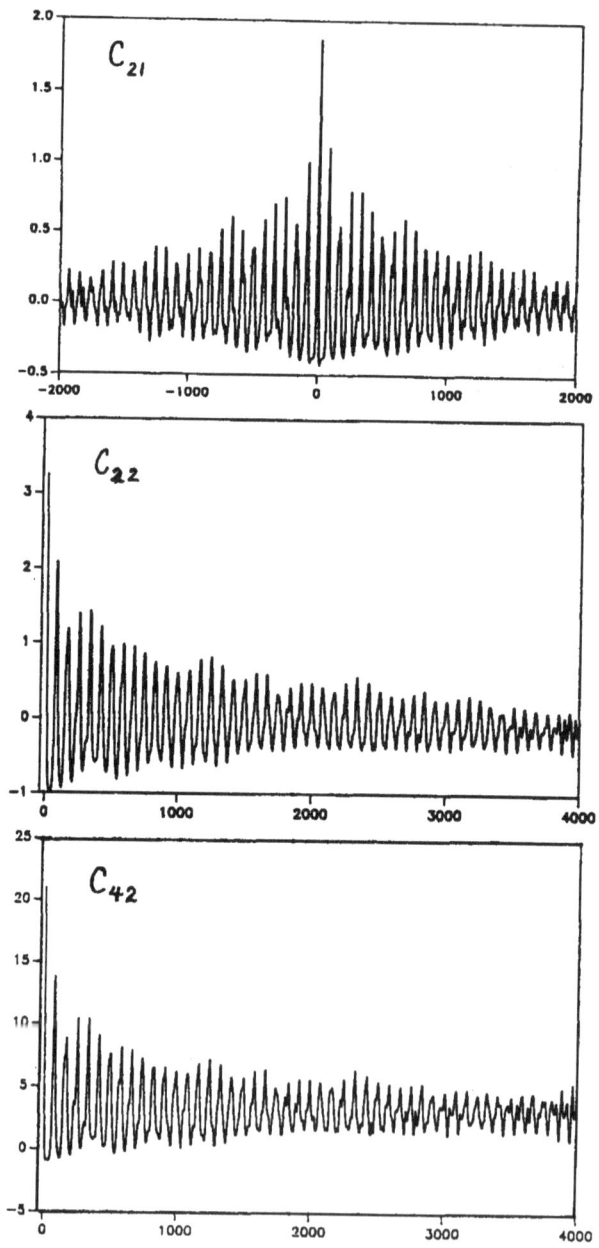

Figure III-27b. Higher order auto-correlation functions of field and intensity for the IHB laser numerical data.

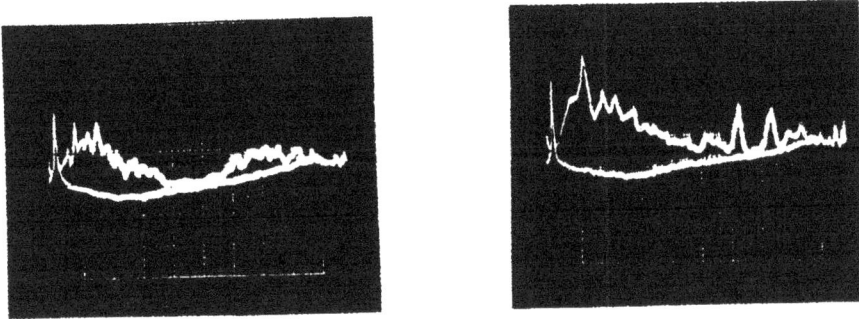

Figure III-28. Samples of homodyne and heterodyne spectra for experimental ring laser
     data from ref. 41.

of order the phase switching time of the field. The other time scale is longer, of order the inverse
bandwidth of the broadband portion of the spectrum. (The various correlation functions C_{ij}
shown here are the time averages of the product of the signal to the i-th power at time t times the
signal to the j-th power at time t+τ, where signals are AC-coupled [zero-mean].)

Returning to the ring laser experiments we show two interesting cases of chaotic behavior in
Fig. III-28. Figure III-28 shows the intensity spectra and the corresponding amplitude
(heterodyne spectra) for two selected data sets. Figure 29 shows the higher order correlation
functions with the long memory characteristic of chaos, the highly periodic nature characteristic
of the intensity in "Lorenz-like" chaos, and the two different decay times in the envelopes of C_{11}
and C_{21}.

The dimensions show similar values in these two cases as well.

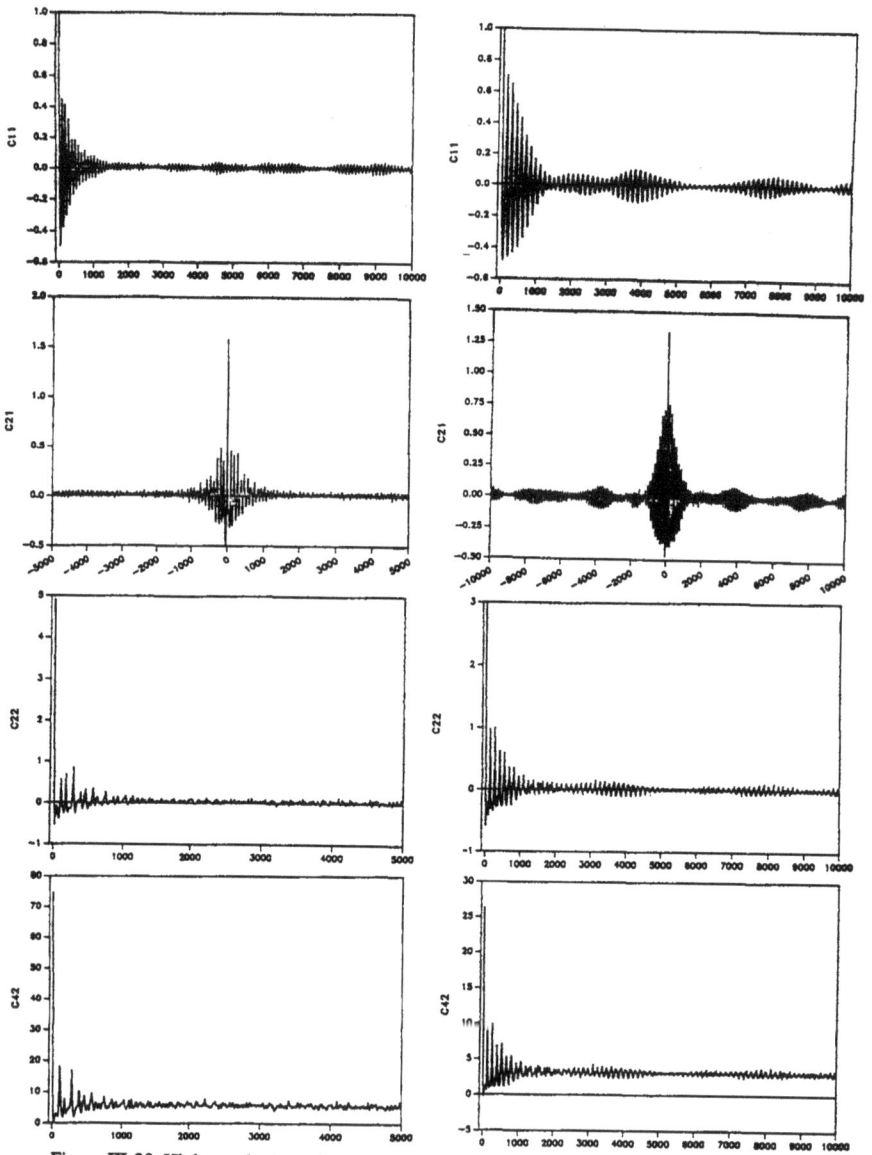

Figure III-29. Higher order intensity correlation functions for the two experimental data sets.

The differences in these two signals lie mainly in their phase portraits as can be infered from their heterodyne spectra. Both are symmetric in the electric field, but the broadband one must be of the type of the complex spiraling and looping figure-8's of the Lorenz model while the one with two prominent peaks is of the form of a nearly periodic figure-8 structure as has been reported elsewhere in our numerical studies [see Ref. 20].

There is a rich reservoir of information describing the chaotic behavior of our lasers that can be extracted from the many correlation functions that can be calculated. Instead of a single dimension and entropy, it is well-known that there is a series of dimensions and entropies which can be used to move fully characterize a chaotic system. Further progress in this area can be expected.

H. Conclusions

Inhomogeneously-broadened, high-gain noble gas lasers provide many interesting dynamical features to aid in the study of optical instabilities and chaos. The dynamical behavior may require many equations for suitable modelling but the resulting chaotic behavior is low dimensional (of order 3). Many of the apparently simple results inferred from intensity power spectra are, instead, more subtle and complex as can be revealed in part by measuring the heterodyne spectra. Standing-wave lasers are not much more complicated in their fundamental behavior than ring lasers except in the narrow region of the Lamb dip. Intensity and electric field spectra together with their correlation functions offer many helpful insights into the characteristics of chaotic signals.

I. References

1. L.W. Casperson and A. Yariv, Appl. Phys. Lett., 17, 259 (1970).
2. L.W. Casperson and A. Yariv, Appl. Opt., 11, 462 (1972).
3. A. Yariv, Introduction to Optical Electronics (Holt-Rinehart, NY, 1987); Quantum Electronics (Wiley, NY, 1975).
4. M. Sargent, III, M. Scully, and W.E. Lamb, Jr., Laser Physics, (Addison-Wesley, Reading, 1974).
5. W.E. Bennett, Jr., Phys. Rev., 126, 580 (1962); Appl. Opt., Suppl. #1, 573 (1963).
6. W.E. Lamb, Jr., Phys. Rev., 134, A1429 (1964).
7. W.E. Bennett, Jr., The Physics of Gas Lasers (Gordon and Breach, NY, 1977) p. 128.
8. J.C. Englund and W.C. Schieve, J. Opt. Soc. Am. B, 2, 81 (1985); J.C. Englund, in Optical Instabilities, R.W. Boyd, M.G. Raymer and L.M. Narducci, eds., (Cambridge U. Press, Cambridge, 1986) p. 235; in Optical Bistability III, H.M. Gibbs, P. Mandel, N. Peyghambarian, and S.D. Smith, eds., (Springer, Berlin, 1986) p. 356; Phys. Rev. A, 33, 3606 (1986).
9. J. Bentley and N.B. Abraham, Opt. Commun., 41, 52 (1982).
10. M. Maeda and N.B. Abraham, Phys. Rev. A, 26, 3395 (1982).
11. N.B. Abraham, T. Chyba, M. Coleman, R.S. Gioggia, N.J. Halas, L.M. Hoffer, S.-N. Liu, M. Maeda, and J.C. Wesson, in Laser Physics, J.D. Harvey and D.F. Walls, eds., Lecture Notes in Physics, vol. 182, (Springer-Verlag, Berlin, 1983) pp. 107-131.

12. R.S. Gioggia and N.B. Abraham, Phys. Rev. Lett., 51, 650 (1983); in Coherence and Quantum Optics V, L. Mandel and E. Wolf, eds., (Plenum, NY, 1984) pp. 563-570.

13. R.S. Gioggia and N.B. Abraham, Opt. Commun., 47, 278 (1983); Phys. Rev. A, 29, 1304 (1984).

14. A.M. Albano, J. Abounadi, T.H. Chyba, C.E. Searle, S. Yong, R.S. Gioggia, N.B. Abraham, J. Opt. Soc. Am. B, 2, 47 (1985).

15. R.S. Gioggia and N.B. Abraham, Acta Physica Austriaca, 57, 127 (1985).

16. L.W. Casperson, IEEE J. Quantum Electron., QE-14, 756 (1978); in Laser Physics, J.D. Harvey and D.F. Walls, eds., Lecture Notes in Physics, vol. 182, (Springer-Verlag, Berlin, 1983) pp. 87-106.

17. L.W. Casperson, presented at the International Workshop on Instabilities and Chaos in Nonlinear Optical Systems, July 1987; in Instabilities and Chaos in Quantum Optics II, N.B. Abraham, F.T. Arecchi, and L.A. Lugiato, eds., (Plenum, NY, 1988).

18. L.E. Urbach, S.-N. Liu, and N.B. Abraham, in Coherence and Quantum Optics V, L. Mandel and E. Wolf, eds., (Plenum, NY, 1984) pp. 593-600.

19. L.M. Hoffer, T.H. Chyba and N.B. Abraham, J. Opt. Soc. Am. B, 2, 102 (1985).

20. M.F.H. Tarroja, N.B. Abraham, D.K. Bandy and L.M. Narducci, Phys. Rev. A, 34, 3148 (1986).

21. N.B. Abraham, A.M. Albano, T.H. Chyba, R.S. Gioggia, L.M. Hoffer, and C.E. Searle, in Atti Del IV Congresso Nazionale di Elettronica Quantistica e Plasmi, (ENEA, Serie Simposi, 1984) pp. 235-242.

22. N.B. Abraham, A.M. Albano, T.H. Chyba, L.M. Hoffer, M.F.H. Tarroja, S.P. Adams and R.S. Gioggia, in Instabilities and Chaos in Quantum Optics, F.T. Arecchi and R.G. Harrison, eds., (Springer, Berlin, 1987) pp. 49-71.

23. L.W. Casperson, Phys. Rev. A, 21, 911 (1980); 23, 248 (1981); M.L. Minden and L.W. Casperson, IEEE J. Quantum Electron., QE-18, 1952 (1982).

24. P. Mandel, Opt. Commun., 44, 404 (1983); 45, 269 (1983) in Coherence and Quantum Optics V, L. Mandel and E. Wolf, eds., (Plenum, NY, 1984) pp. 579.

25. S. Hendow and M. Sargent III, Opt. Commun., 40, 385 (1982); 43, 59 (1982); J. Opt. Soc. Am. B, 2, 84 (1985).

26. L.A. Lugiato, L.M. Narducci, D.K. Bandy and N.B. Abraham, Opt. Commun., 45, 115 (1984); in Coherence and Quantum Optics V, L. Mandel and E. Wolf, eds., (Plenum, NY, 1984) pp. 217; N.B. Abraham, L.A. Lugiato, P. Mandel, L.M. Narducci and D.K. Bandy, J. Opt. Soc. Am. B, 2, 35 (1985).

27. D.K. Bandy, L.M. Narducci, L.A. Lugiato, and N.B. Abraham, J. Opt. Soc. Am. B, 2, 56 (1985).

28. L.W. Casperson, J. Opt. Soc. Am. B, 2, 62 (1985); 2, 73 (1985), in Optical Instabilities, R.W. Boyd, M.G. Raymer and L.M. Narducci, eds., (Cambridge U. Press, Cambridge, 1986).

29. J.Y. Zhang, H. Haken and H. Ohno, J. Opt. Soc. Am. B, 2, 141 (1985).

30. H.M. Gibbs, F.A. Hopf, D.L. Kaplan and R.L. Shoemaker, Phys. Rev. Lett., 46, 474 (1981).

31. R. Vallée, C. Delisle, and J. Chrostowski, Phys. Rev. A, 30, 336 (1984); R. Vallée and C. Delisle, IEEE J. Quantum Electron., QE-21, 1423 (1985); Phys. Rev. A, 31, 2390 (1985); 34, 309 (1986).

32. F.T. Arecchi in Instabilities and Chaos in Quantum Optics II, N.B. Abraham, F.T. Arecchi and L.A. Lugiato, eds., (Plenum, NY, 1988); and in Instabilities and Chaos in Quantum Optics, F.T. Arecchi and R. Harrison, eds., (Springer, NY, 1987) pp. 9-48.

33. J.W. Swift, and K. Wiesenfeld, Phys. Rev. Lett., 52, 705 (1984).

34. P. Coullet and C. Vanneste, Hel. Phys. Acta, 56, 813 (1983) and in the Synergetics Workshop at Schloss Elmau, 1984; Y. Kuramoto and S. Koga, Phys. Lett., 92A, 1 (1982). A. Arneodo, P. Coullet and C. Tresser, Phys. Lett., 81 A, 197 (1981); P. Coullet, in Chaos and Statistical Methods, Y. Kuramoto, ed., (Springer-Verlag, Berlin, 1984). 62; for an illustration for homogeneously broadened lasers see also H. Zeghlache and P Mandel, J. Opt. Soc. Am. B, 2, 19 (1985).

35. J.P. Eckmann, Rev. Mod. Phys., 53, 643 (1981); P. Cvitanovic (ed.), Universality in Chaos, (Hilger, London, 1984) and Acta Phys. Polonica A, 65, 203 (1984); P. Berge, Y. Pomeau, C. Vidal, L'Ordre dans le chaos: vers une approche deterministe de la turbulence (Hermann, H. Schuster, Paris, 1985).

36. P. Grassberger and I. Procaccia, Phys. Rev. A, 28, 2591 (1983); Phys. Rev. Lett., 50, 346 (1983); Physica, 9D, 189 (1983); Physica, 13D, 34 (1984); A. Ben Mizrachi, I. Procaccia, P. Grassberger, Phys. Rev. A, 29, 97 (1985); A. Cohen and I. Procaccia, Phys. Rev. A, 31, 1872 (1985); J.D. Farmer, E. Ott, and J. Yorke, Physica, 7D, 153 (1983); J.D. Farmer in Fluctuations and Sensitivity in Nonequilibrium Systems, W. Horsthemke and D.K. Kondepudi, eds., Springer Proc. in Physics, Vol. 1 (Springer, Berlin, 1984) p. 172.

37. Dimensions and Entropies in Chaotic Systems, G. Mayer-Kress, ed. (Springer-Verlag, Berlin, 1986); A.M. Albano, J. Muench, C. Schwartz, A.I. Mees, and P.E. Rapp, Phys. Rev. A, (submitted January 1988); J.G. Caputo and P. Atten, Phys. Rev. A, 35, 1311

(1987); F. Mitschke, M. Möller and W. Lange, Phys. Rev. A, $\underline{37}$, (to be published 1988); J.-P. Eckmann, S. Oliffson Kamphorst, D. Ruelle, and S. Ciliberto, Phys. Rev. A, $\underline{34}$, 4971 (1986); G. Broggi, J. Opt. Soc. Am. B, $\underline{5}$, (to be published, May 1988); R. Badii and A. Politi, J. Stat. Phys., $\underline{40}$, 725 (1985).

38. N.B. Abraham, A.M. Albano, B. Das, G.C. de Guzman, S. Yong, R.S. Gioggia, G.P. Puccioni and J.R. Tredicce, Phys. Lett., $\underline{114A}$, 217 (1986); N.B. Abraham, A.M. Albano, G.C. de Guzman, M.F.H. Tarroja, S. Yong, S.P. Adams and R.S. Gioggia, in Perspectives in Nonlinear Dynamics, M.F. Shlesinger, R.W. Cawley, A.W. Saenz and W. Zachary, eds., (World Scientific, Singapore, 1986) p. 214.

39. R.S. Gioggia, N.B. Abraham, W.E. Lange, M.F.H. Tarroja, and J.C. Wesson, J. Opt. Soc. Am. B, $\underline{5}$, to be published (1988).

40. N.B. Abraham, A.M. Albano, B. Das and M.F.H. Tarroja, in Fundamentals of Quantum Optics II, F. Ehlotsky, ed., (Springer-Verlag, Berlin, 1987).

41. M.F.H. Tarroja, Ph.D. Thesis, Bryn Mawr College 1988.

42. A.V. Uspenskii, Radio Eng. and Electron. Phys. (USSR), $\underline{8}$, 1145 (1963); $\underline{9}$, 605 (1964); V.V. Korobkin and A.V. Uspenskii, Sov. Phys. JETP, $\underline{18}$, 693 (1964); A.Z. Grazyuk and A.N. Oraevskii, in Quantum Electronics and Coherent Light, P.A. Miles, ed., (Academic Press, NY, 1964); H. Haken, Z. Phys., $\underline{190}$, 327 (1966); H. Risken, C. Schmidt and W. Weidlich, Z. Phys., $\underline{194}$, 337 (1966); H. Haken, Phys. Lett., $\underline{53A}$, 77 (1975); E.N. Lorenz, J. Atmos. Sci., $\underline{20}$, 130 (1963); C. Sparrow, in Optical Instabilities, R.W. Boyd, M.G. Raymer and L.M. Narducci, eds., (Cambridge U. Press, Cambridge, 1986) p. 72; C. Sparrow, The Lorenz Equations: Bifurcations, Chaos and Strange Attractors (Springer-Verlag, Berlin, 1982).

256

IV. Other Examples of Dynamical Instabilities and Chaos in Lasers

A. Introduction

More can be learned about dynamical behavior of lasers by considering recent results from several other special examples in this general field. More comprehensive references and discussions of examples of laser dynamics can be found, for example, in N.B. Abraham, P. Mandel, and L.M. Narducci, in Progress in Optics XXV, ed. E. Wolf (Elsevier, Amsterdam, 1988) and in R.G. Harrison and D.J. Biswas, in Progress in Quantum Electronics 11, 127 (1985); and in C.O. Weiss and R. Vilaseca, Laser Dynamics (Physics-Verlag, Weinheim, to be published). In section B some of the results are given for the single-mode lasers which are almost as unstable as the inhomogeneously broadened lasers considered in section III--namely, the optically pumped far-infrared lasers. In section C the pulsations of lasers with saturable absorbers are reviewed. In section D results are reviewed from modulated single-mode lasers where the external forcing frequency can excite complex periodic and chaotic behavior in lasers which remain stable in the absence of modulation. Another form of modulation is achieved when an injected signal has a frequency that differs from that of the free-running laser, a case that is briefly reviewed in section E. In section F the simplest example of the interaction of two modes is considered as it occurs in a bidirectional ring laser. The multimode instability of a unidirectional homogeneously broadened ring laser is reviewed in section G. As a final example, in section H, the case of a laser with three modes at different optical frequencies is used to illustrate how modes can interact to generate modulation frequencies that are much lower than the intermode beat frequencies.

B. Single Mode Pulsations in FIR Lasers

Optically-pumped far-infrared lasers use an excitation scheme such as shown in Figure IV-1.

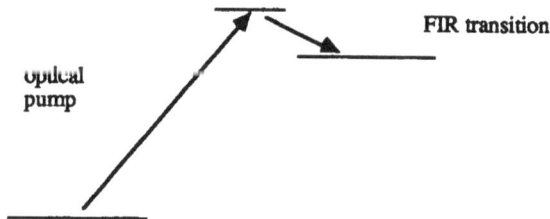

Figure IV-1.

The gain on the FIR transition can be very high and the homogeneous linewidth of the transition depends on pressure with a value of about one megahertz typical for moderate pressures of the FIR gas. Because of the high gain and low homogeneous linewidth, Weiss and Klische[1] proposed that it might easily be possible to achieve the "bad cavity" ($\kappa > \gamma_{\perp}$) conditions necessary for pulsations in the simple theoretical model for the "Lorenz-like" single mode homogeneously broadened laser. The Doppler broadening in the FIR gas is effectively eliminated for the FIR transition because of velocity-selective pumping and the reduced Doppler broadening at the longer FIR wavelength for the range of excited velocities. Suitable tuning (or detuning) of the pump transition and the FIR laser cavity can make the FIR laser in a ring geometry operate in only one direction.

The simple picture of FIR lasers matching the Lorenz model has been challenged because of the inherent 3-level nature of the FIR laser, the AC-Stark broadening of the optically pumped level, and the additional detunings and coherences that appear. However, two simple experimental setups seem to give results very much like that predicted for simple models and they seem to show 3-level coherence effects only for certain ranges of their operation. The physical reasons for the "simplified" behavior are not fully clear, but Weiss suggests they may arise from mixing of the different Zeeman sublevels when the pump laser and FIR laser have crossed polarizations. The backward emission of the FIR laser also seems to be more Lorenz-like, perhaps because its emission profile is less AC-stark broadened.

Unidirectional Ring Laser Operation

An example of the experimental setup of the experiments of Weiss and coworkers is shown in Figure IV-2.

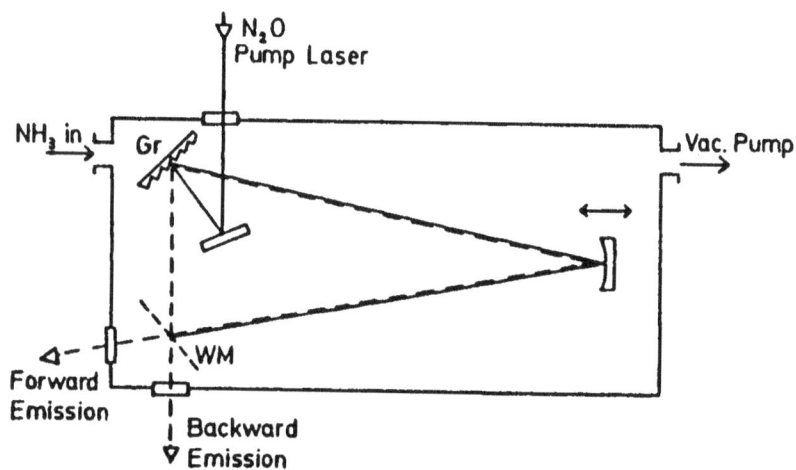

Figure IV-2. FIR ring-laser setup from ref. 3

Results from this type of laser operation have been most "Lorenz-like" on various FIR transitions of NH_3, notably those at 81.5μm in $^{14}NH_3$[4,5], and at 153μm[6] and 3763,7μm in $^{15}NH_3$. Examples which emulate those of the single mode two-level atom laser model are shown in Figures IV-3 to IV-6 for the 81.5μm laser. In Figure IV-3 we see classic "Lorenz-type" spirals for intensity pulsations expected for chaotic behavior in the simple two-level atom model on resonance.

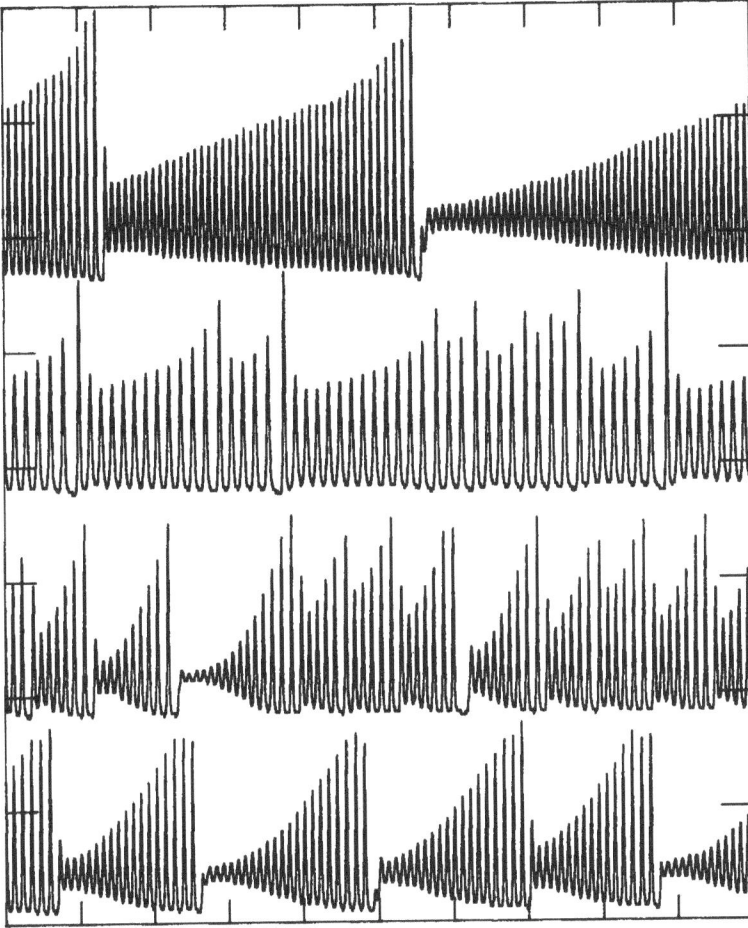

Figure IV-3. Experimental Lorenz-like (spiral) intensity pulsations at 81.5μm in an NH₃ FIR laser. Different traces are for different pressures and excitation levels.

As predicted by the single mode model the transition from stable single mode operation to the Lorenz chaos occurs abruptly for excitations about 10-15 times larger than the threshold for laser action as long as the FIR gas pressure exceeds about 8 Pa as illustrated in Fig. IV-4. Lower thresholds for pulsations and transitions from stable operation to periodic pulsations which period-double with increasing excitation were found for lower pressures as is predicted for the 3-level models. Perhaps when the pressure is high enough the collisions wash out the 3-level coherences of the optical pumping leaving a system which behaves like the two-level model.

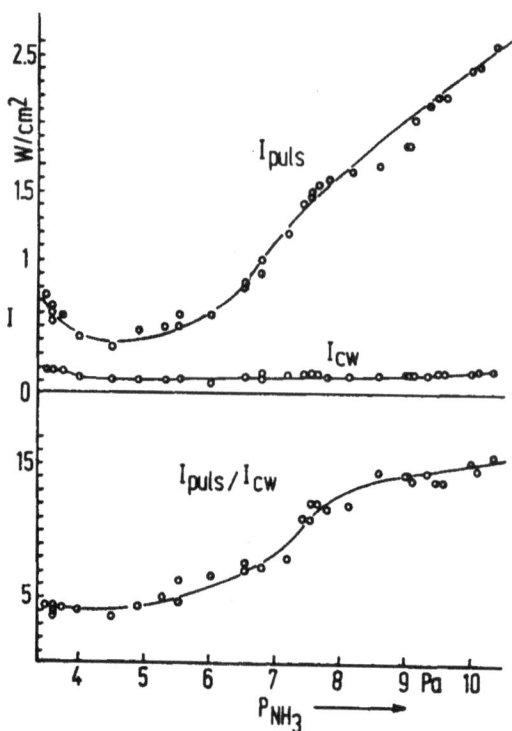

Figure IV-4. Results from Ref. 5 showing the pump laser intensity required for single mode pulsations, the pump intensity for FIR laser operation, and their ratio. The FIR laser excitation is assumed to be proportional to the pump intensity.

With cavity detuning, the laser shows behavior such as that found in the simple single-mode model showing a change from spiral chaos to more irregular behavior with periodic windows and ultimately period-doubling chaos which inverse period doubles with increasing detuning. Examples of various intensity pulsations in time are shown in Fig. IV-5 and sample intensity power spectra are shown in Fig. IV-6.

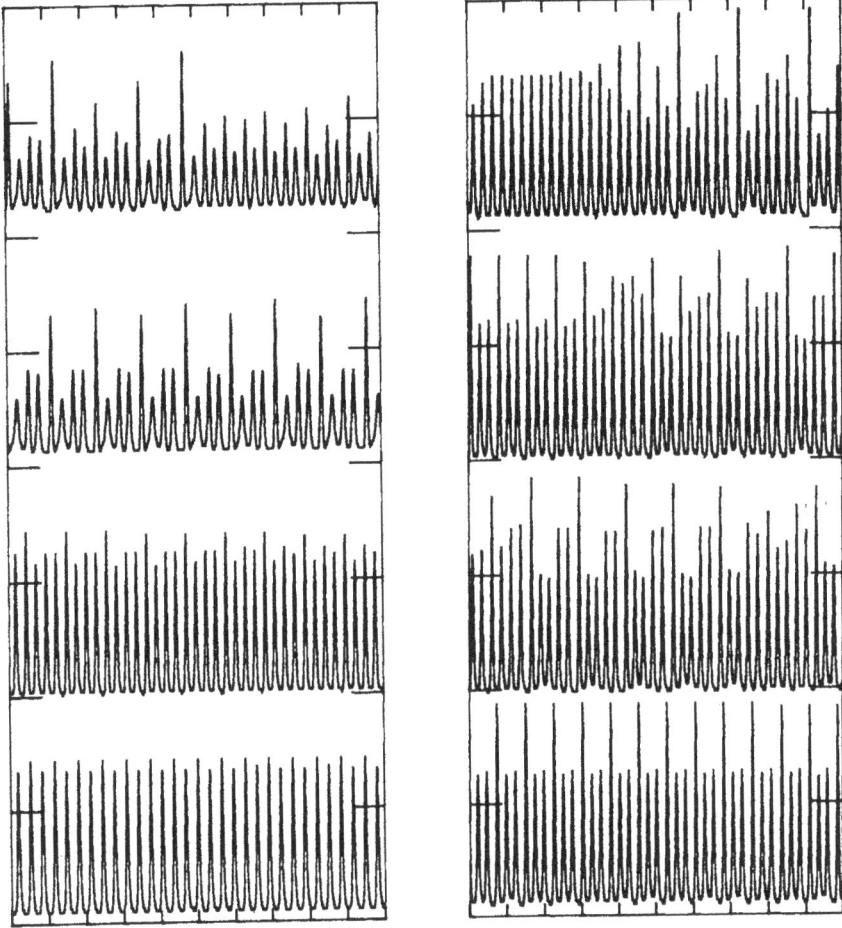

Figure IV-5. Experimental records of intensity pulsations, from an 81.5μm NH₃ FIR laser (courtesy of C.O. Weiss).

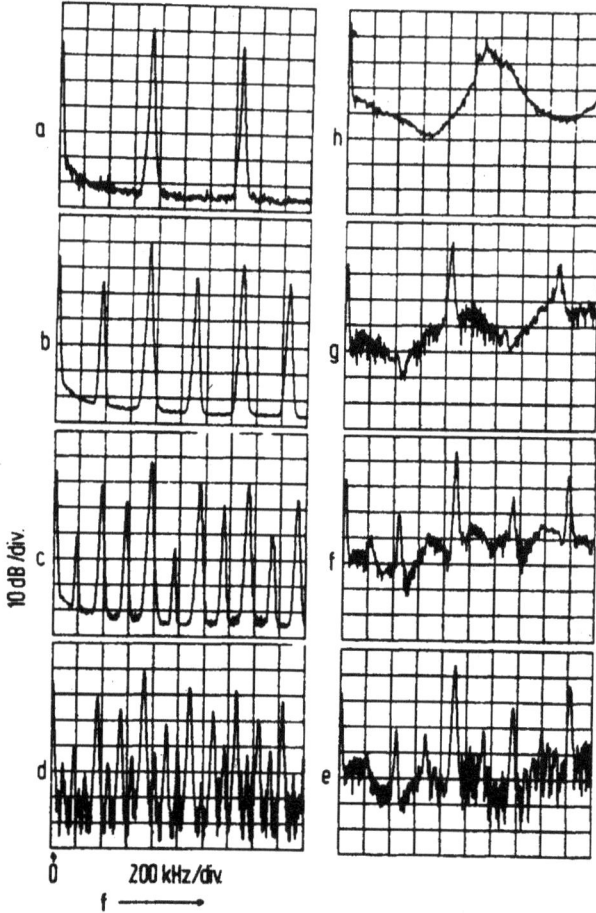

Figure IV-6. Intensity power spectra from Ref. 5. a) - h) are examples with decreasing detuning, where h) represents chaotic spirals on line center and a) is the largest detuning.

It remains an open question whether these results only "resemble" those of the simple model (retaining important, if subtle, differences) or whether the FIR systems explored here really do represent the first direct illustrations of the theoretical models that are classic in both laser physics and convective hydrodynamics[8] and thus paradigmatic for many studies of nonlinear dynamics. Recent experiments have tested other properties of these laser pulsations (dimensions, entropies, phase dynamics, attractor symmetries), in order to make more precise comparisons with the theoretical models, but the results are not yet conclusive. [Abraham, Weiss, Hübner, work in progress 1988.]

Standing Wave Operation

Another simple laser model, that of a single mode laser field interacting with a medium having two simple and independent two-level atom resonances, is also nicely illustrated by optically pumped FIR lasers using a standing wave cavity for both the pump and the FIR laser field. The model was first studied by Idiatulin and Uspenskiy[9] as a first approximation to considering the effects of an inhomogeneously broadened laser medium. More recently the model has been studied in greater detail because it seems to explain many initially puzzling standing wave FIR laser experiments.[10]

Schematically the resonance characteristics of such a medium and the associated dispersion are illustrated in Fig. IV-7.

From previous considerations of the graphical solution of resonant frequencies of a laser, it is clear that the rapidly varying dispersion with a negative slope near the center of the total gain profile can lead to multiple steady state laser solutions if the cavity is sufficiently lossy. (Recall that the slope of the mode line depends on the cavity losses.) Furthermore, as the gain in this case at the center frequency is less than that for certain detunings, it will not be surprising if with increasing excitation a resonantly-tuned laser cavity reaches a threshold for *detuned* steady state operation at detuned laser field frequencies before it reaches the threshold for operation at the resonantly tuned frequency.

264

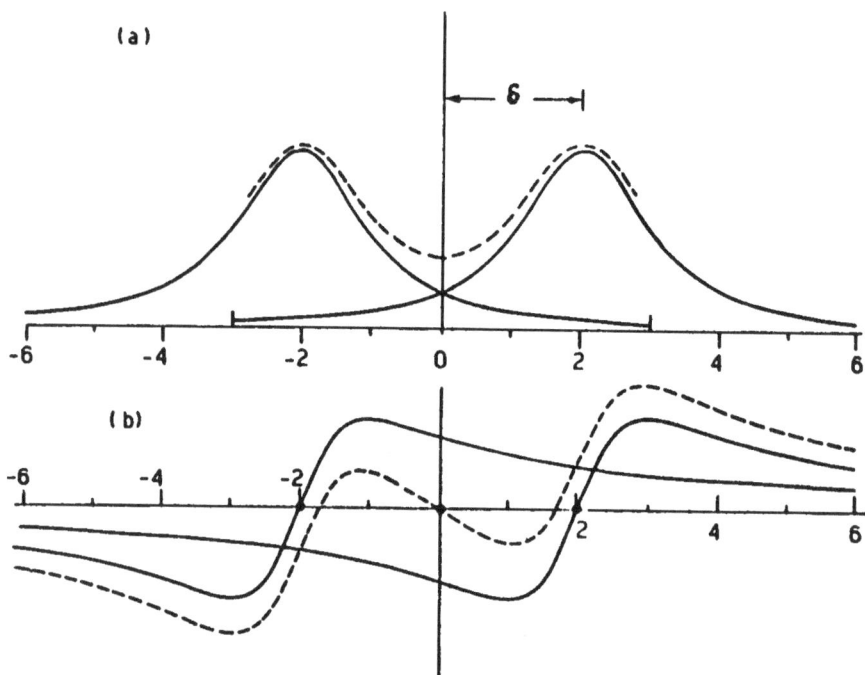

Figure IV-7. a) unsaturated gain and b) unsaturated dispersion for a medium composed of equal amounts of two level atoms with resonances separated by two linewidths (FWHM). (after ref. 9a, Fig. 7). Solid curves show effects of the two groups of atoms separately, dashed curves give totals.

This is in fact the case for suitable splittings of the gain line and suitable cavity detunings as illustrated in Fig. IV-8.

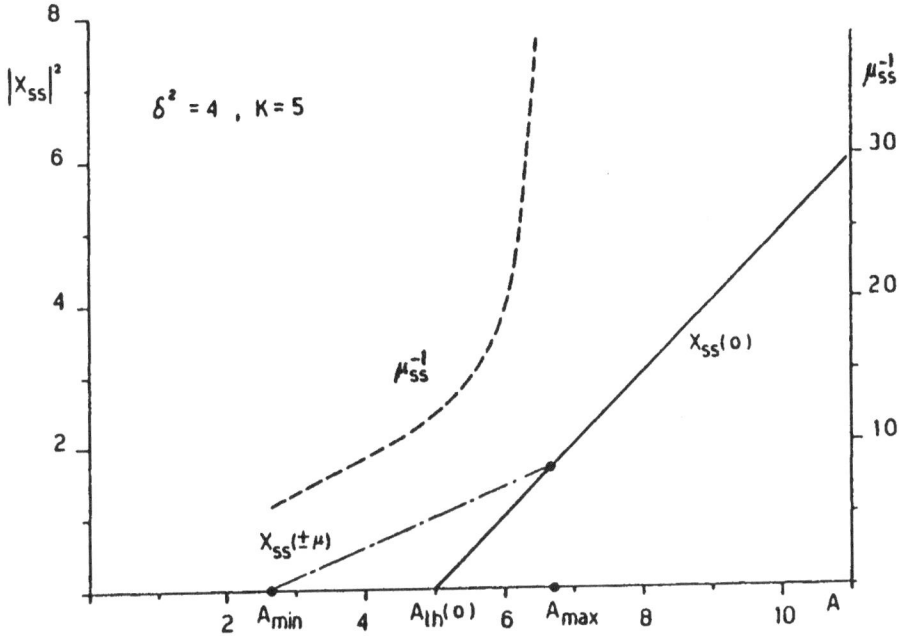

Figure IV-8. Steady state solutions of the laser model of Ref. 10 showing the frequency μ_{ss} and intensity.

The two detuned solutions with laser field frequencies $\pm \mu_{ss}$ turn out to be unstable in favor of time dependent pulsations which amount to 200% amplitude modulation. Near the threshold the pulsing occurs at the frequency $2\mu_{ss}$. The pulsing is effectively equivalent to phase-locked coexistence of the two detuned solutions. Samples of numerical solutions of this model are shown in Fig. IV-9.

266

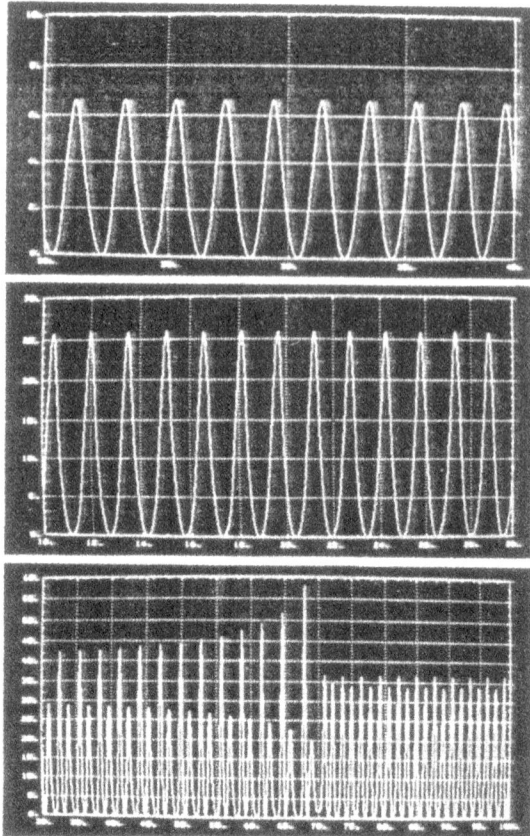

Figure IV-9. Numerical solutions of the intensity
versus time for increasing (a-c) excitation of the laser.

How is this model illustrated by the standing-wave pumped FIR laser? The standing wave
pump field interacts with two different velocity groups (±v) of the FIR gas with speeds
dependent on the detuning of the pump laser field from resonance. The excited medium then
consists of two homogenously broadened groups of atoms with a difference between their
resonant frequencies dependent on the detuning of the pump field.

Pulsing would be expected if the pump detuning is of order the homogenous linewidth and if the laser cavity is tuned near resonance for the FIR transition. This is exactly as observed in the experiments.[10]

Pulsations are shown in Fig. IV-10 and agree in many details with those predicted by this simple model.

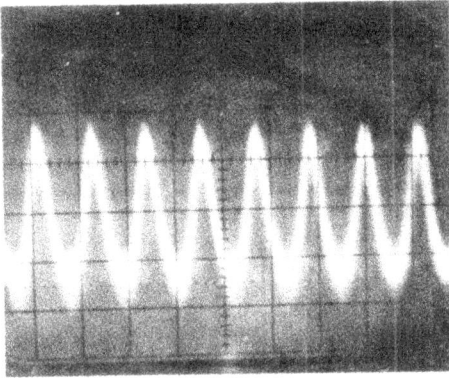

Figure IV-10. Oscilloscope display of intensity pulsations for standing wave pumped FIR HCOOH laser from Ref. 10.

A caution is in order about the modesplitting analysis. It works well here for the threshold of the detuned laser solutions because there is no laser intensity for excitations below the threshold and hence no saturation effects. In this case the gain and dispersion conditions factor and the graphical solution of the dispersion condition is exact. This is not usually the case though it was approximately true for the inhomogeneously broadened laser.

The analysis here only tells us of the presence of two symmetrically detuned steady state solutions. It does not tell us if they are stable (giving bistability) or unstable (in favor of "mode coexistence" type pulsations or other pulsing). These features are found by perturbative treatments (stability analysis) and numerical solutions of the model.

C. Laser with Saturable Absorber (LSA)

The operation of a laser with a homogeneously broadened gain medium and a homogeneously broadened absorber is another interesting case for single mode laser dynamics. Two special cases are relatively easy to understand. For a complete review see section 9 of

N.B. Abraham, P. Mandel, and L.M. Narducci, in Progress in Optics XXV, ed. E. Wolf (Elsevier, Amsterdam, 1988).

Fast absorber

If the absorber relaxation response is extremely fast then the absorber dynamics adiabatically follow the laser intensity and the effect of the absorber can be modelled by adding an intensity dependent loss term to the laser equations. For a laser described by rate equations this would be given by

$$\frac{dI}{dt} = -2\kappa\left(1 + \frac{\alpha}{1 + sI}\right)I - 2\kappa AIN \qquad \text{(IV-1a)}$$

$$\frac{dN}{dt} = -\gamma_\parallel(N - 1) - \gamma_\parallel IN \quad . \qquad \text{(IV-1b)}$$

The effect of α is to raise the laser threshold to $A_{thr} = 1 + \alpha$ and the effect of the saturability of the absorption is to open the possibility that the damped relaxation oscillations for ($\alpha = 0$) become undamped. This explanation of pulsations in ruby lasers was offered very early in the history of laser theories[11] and can be used to explain many cases of sustained pulsations in aged or deliberately damaged semiconductor lasers.[12] For materials with weakly damped relaxation oscillations the saturable absorber usually leads to the generation of large spikes in the output intensity with a spiking frequency lower than the frequency of the weak relaxation oscillations. This is called passive Q-switching.

Slow absorber

If the absorber responds more slowly than the gain medium, the absorber linewidth is correspondingly narrower than that of the gain medium. In this case one has gain and dispersion as shown schematically in Fig. IV-11.

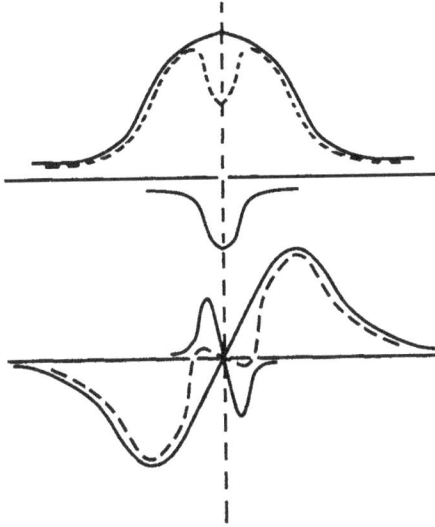

Figure IV-11. Schematic drawing of gain and dispersion for a slow absorber in a laser. Solid curves show amplifying and absorbing media separately. Dotted line gives the total effect.

For this case the mode splitting analysis again helps us to see the possible steady-state solutions. For suitable strengths and linewidths of the absorber and suitable cavity decay rates there are detuned laser field solutions for a resonantly tuned laser cavity with lower excitation thresholds than for the solutions with resonantly tuned laser fields. Possible steady state solutions are illustrated schematically in Fig. IV-12.

Dynamics of these solutions require more careful analysis. Unlike the similar case of the two-peaked gain medium the detuned solutions may be stable (bistability) or they may have the "coexisting mode" pulsing solution similar to that of the two-peaked gain case. Q-switched operation is commonly observed. Some recent experimental results can be found in Ref. 13a. Recent theoretical work by Chyba[13b] shows that with cavity detuning (or mismatch of the absorber and amplifier resonances) the apparent symmetry in the detuned solutions vanishes, giving more complex steady state and stability behavior.

270

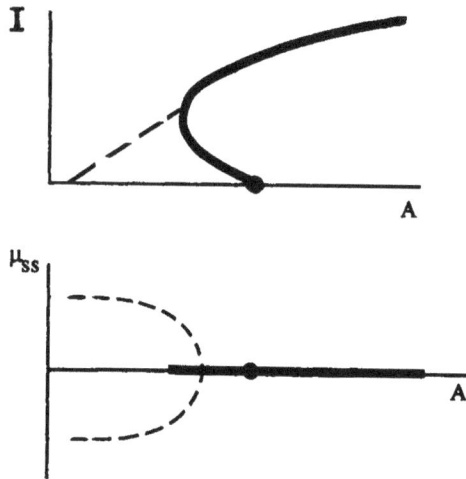

Figure IV-12. Example of possible steady state laser solutions for LSA.

Recent results[13a] in experimental studies have found period doublings, Q-switching and chaotic pulsing as illustrated in Fig. IV-13.

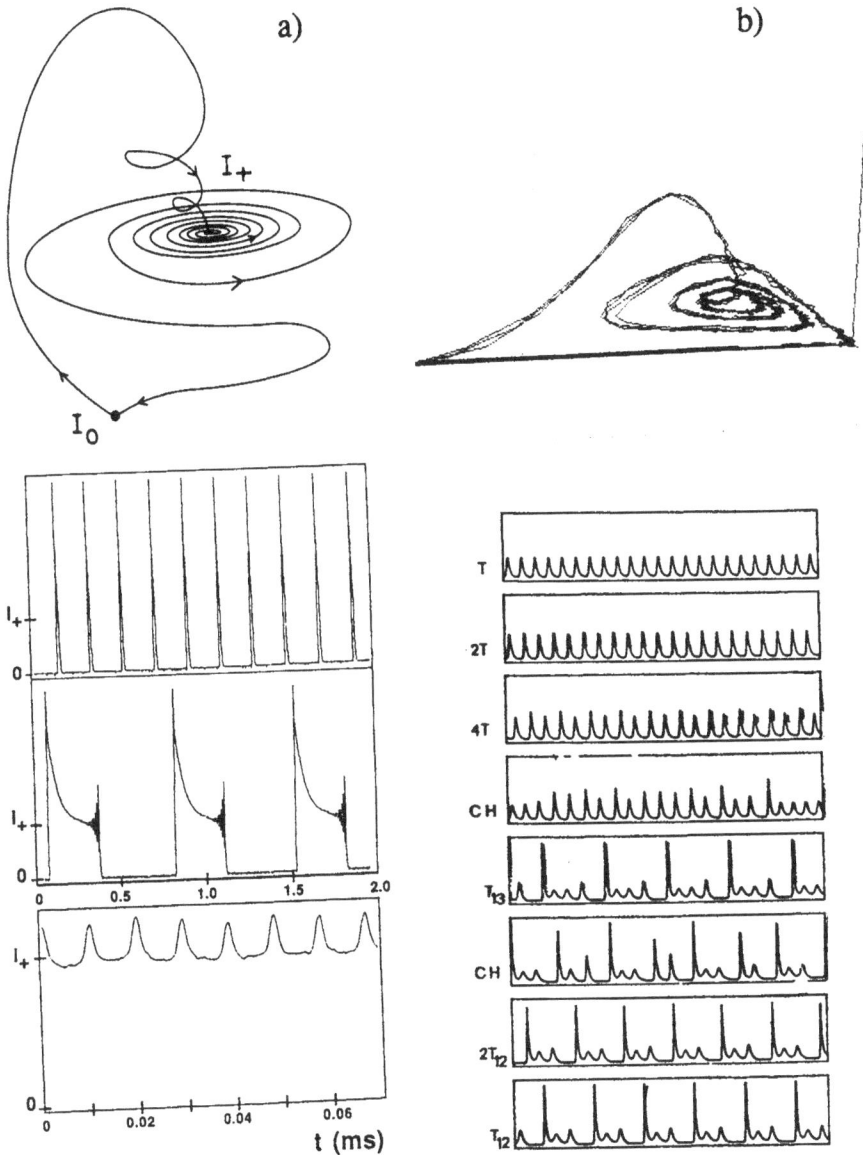

Figure IV-13a,b. Sample waveforms from a laser with saturable absorber.
a) after Arimondo *et al* . (Ref. 13a), b) courtesy of Glorieux and coworkers.

272

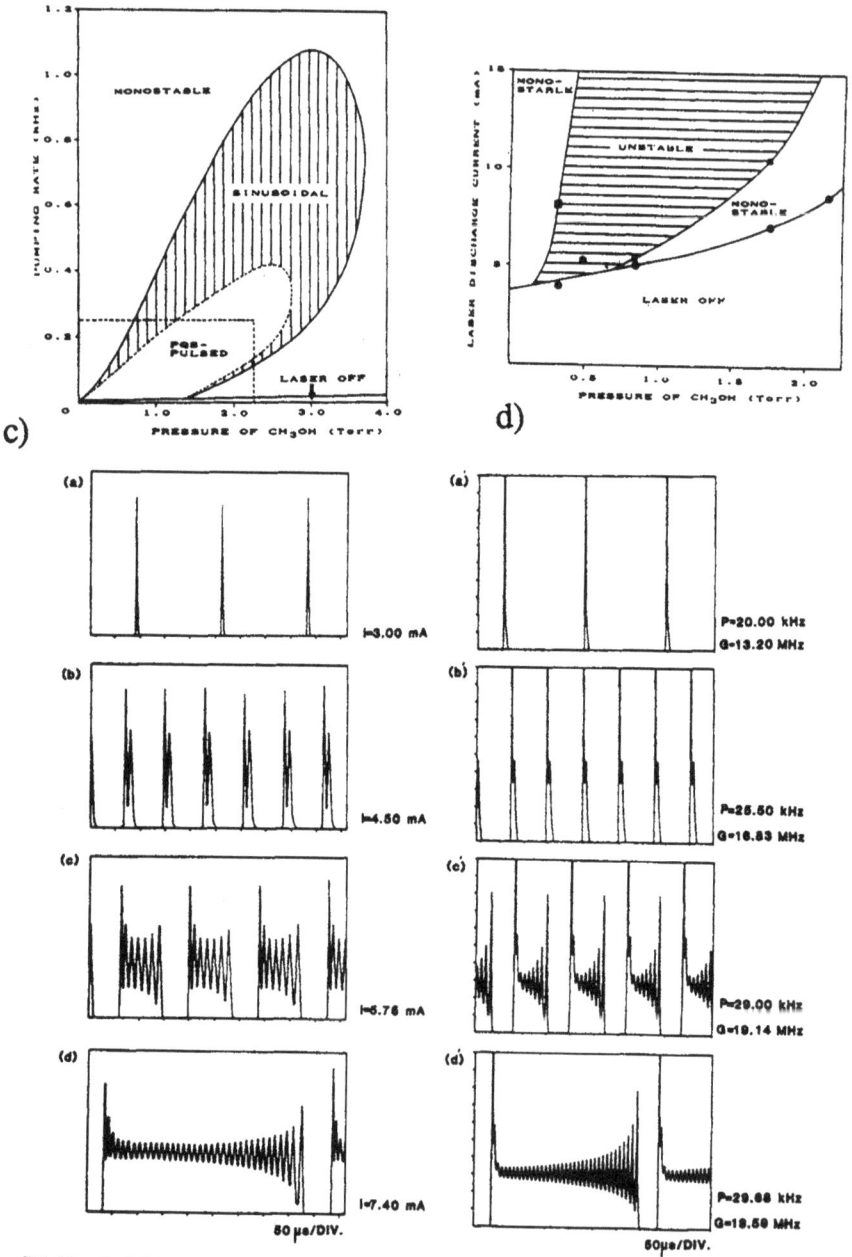

c)

d)

Figure IV-13c, d. LSA phase space diagram and examples of Q-switched pulses after Tanii *et al.*

D. Modulated Single-Mode Lasers

Many single mode lasers do not show spontaneous pulsations because their polarization relaxation rate exceeds the cavity decay rate ($\gamma_\perp > \kappa$). However, is is common for solid state, CO_2 and semiconductor lasers to have $\kappa >> \gamma_\perp$ and so these lasers exhibit damped relaxation oscillations at a frequency given approximately by

$$\sqrt{2\kappa\gamma_\parallel(A-1)} \, ,$$

where A is the laser threshold parameter (A is called 2C in Narducci's notes). Such damped oscillations are seen in a transient output of the laser if the excitation is switched on. This damped resonance is also found if one examines the noise spectrum of such lasers. Because the single mode operation is stable, perturbations by environmental or spontaneous emission noise are heavily damped, but the damping is weakest at the relaxation oscillation frequency resulting in a peak in the power spectrum as illustrated by McCumber[14] and Haken.[15]

When a parameter of a linear system is modulated, the output is modulated at the same frequency. Lasers show nonlinear dynamical response when they are periodically modulated. This effect is particularly strong if the modulation frequency matches the relaxation oscillation frequency, but resonances at many different rational ratios of the two frequencies have also been observed. Using this technique, Arecchi and coworkers[16] were able to drive a stable CO_2 laser into a variety of subharmonic periodic pulsing states and into chaotic pulsing and they were the first to identify the chaotic behavior as such. Similar results have been found for modulation of other lasers with damped relaxation oscillation.

The basic features of such laser phenomenology are indicated for a solid state laser in Fig. IV-14.

b)

a)

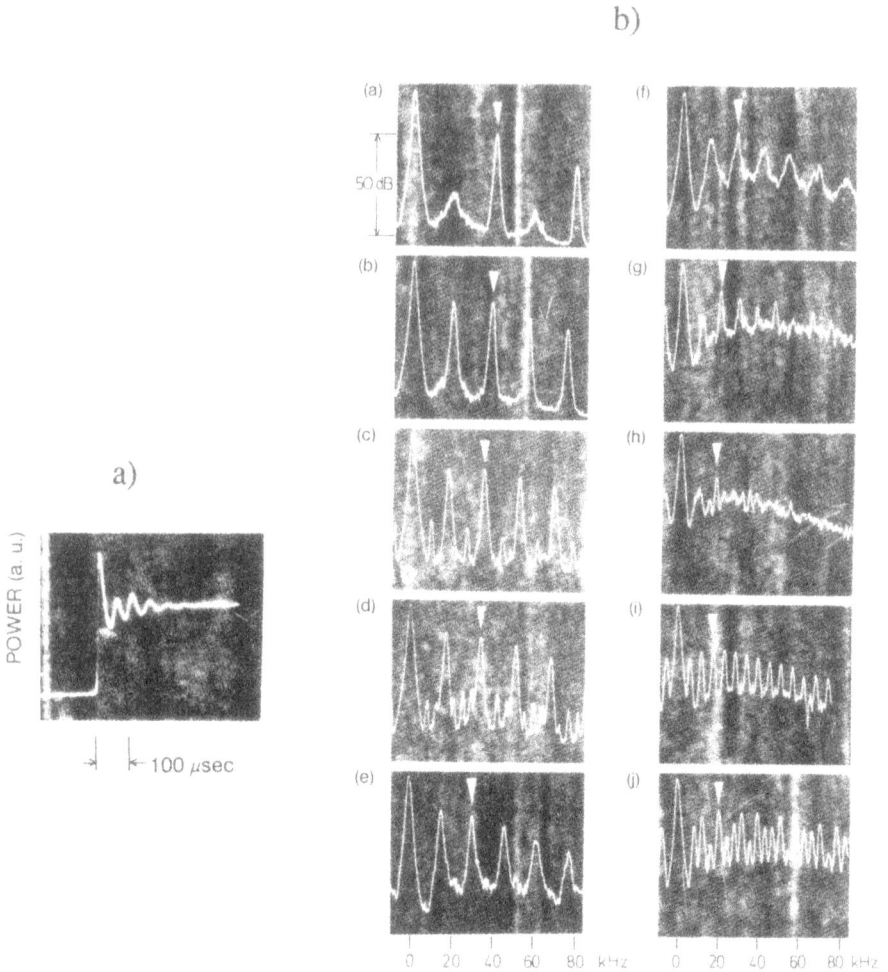

Figure IV-14. Results for a NdP_5O_{14} laser at 1.325μm with a periodically modulated pump (Ar+ laser) from ref. 17. A) Relaxation oscillations in response to a step change in the pump intensity. B) Spectrum of intensity pulsations for about 80% pump modulation and relaxation oscillation frequency at 19 kHz with modulation frequencies for a) - j) (in kHz) of 39.8, 37.9, 35.5, 35.3, 32.0, 29.0, 18.0, 17.5, 20.0, respectively.

These results show locked periodic and subharmonic oscillations and chaotic behavior when the driving frequency is approximately twice (a-f) or equal to (g-j) the relaxation oscillation frequency.

It is useful to note that modulation of the loss is much more effective in driving the nonlinear resonance than is modulation of the pump when ($\kappa >> \gamma_{\parallel}$).[19] Only a few percent modulation of the loss can achieve the same period-doublings as 80% modulation of the gain. This can be derived by linearizing the laser rate equations with respect to a perturbing modulation and writing them as a damped oscillator equation with a periodic forcing term.[18]

A detailed report of periodic and chaotic pulsations for loss modulation of a CO_2 laser is found in Ref. 19. Typical results of their studies are shown here. The boundaries for non-sinusoidal response of the medium and for period doublings and chaos are shown in Fig. IV-15.

276

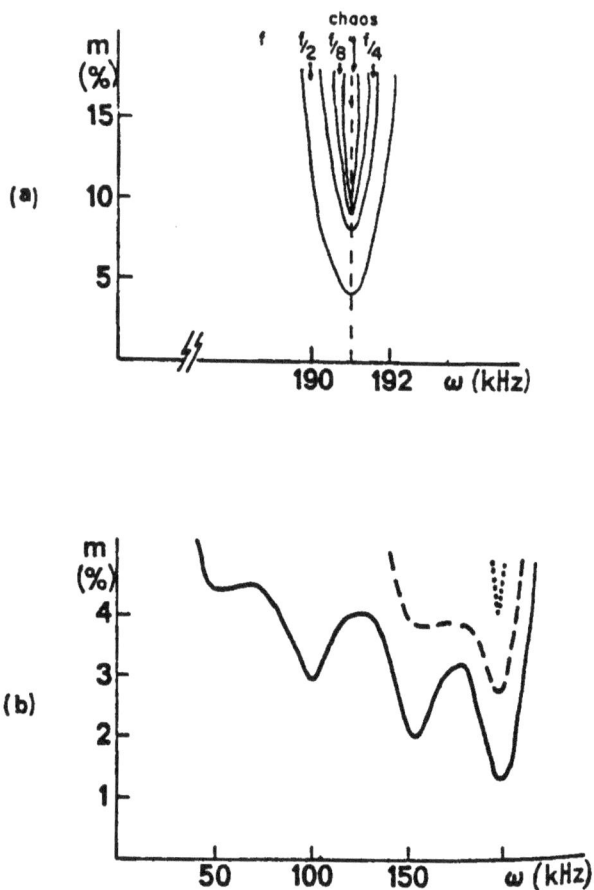

Figure IV-15. Boundaries for sinusoidal, period-doubled and chaotic response for the modulation depth versus driving frequency for a CO_2 loss modulated laser (from Ref. 19a).

Several sample pulsation waveforms and their spectra are shown in Fig. IV-16a.

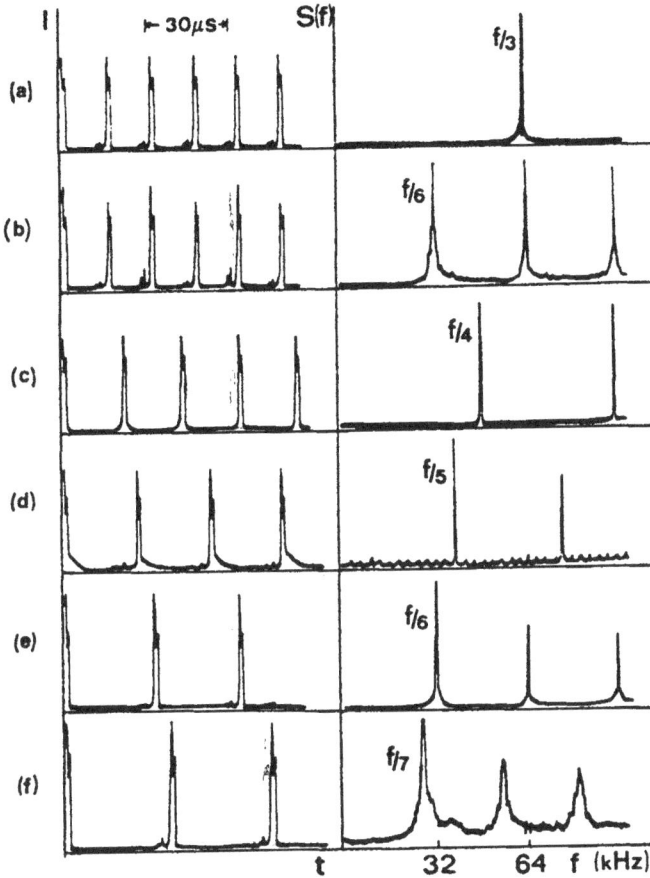

Figure IV-16a. Sample pulsation waveforms and spectra for a loss modulated CO_2 laser from Ref. 19a.

The chaotic attractors have been characterized in detail and have dimensions only slightly greater than 2.0. Another interesting way to characterize periodically forced systems is by stroboscopic measurements of one sample per forcing period at a fixed phase of the driving

278

oscillation. This gives a "Poincaré section" of the attractor. It is not easy to generate such plots for self-pulsing systems, but for those with a fixed external clock it can be very instructive as illustrated in Fig. IV-16b. A more complete theory of the modulated CO_2 laser has been developed recently (Ref. 19b) and several key results for coexisting attractors are shown in Fig. IV-17.

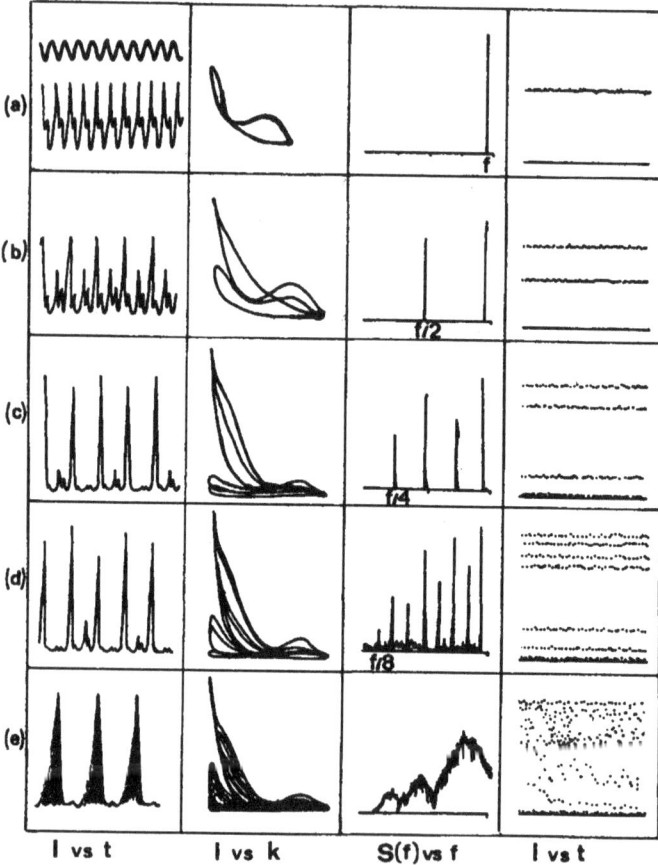

| I vs t | I vs k | S(f) vs f | I vs t |

Figure IV-16b. Poincaré section results for the loss modulated CO_2 laser after Ref. 19a.

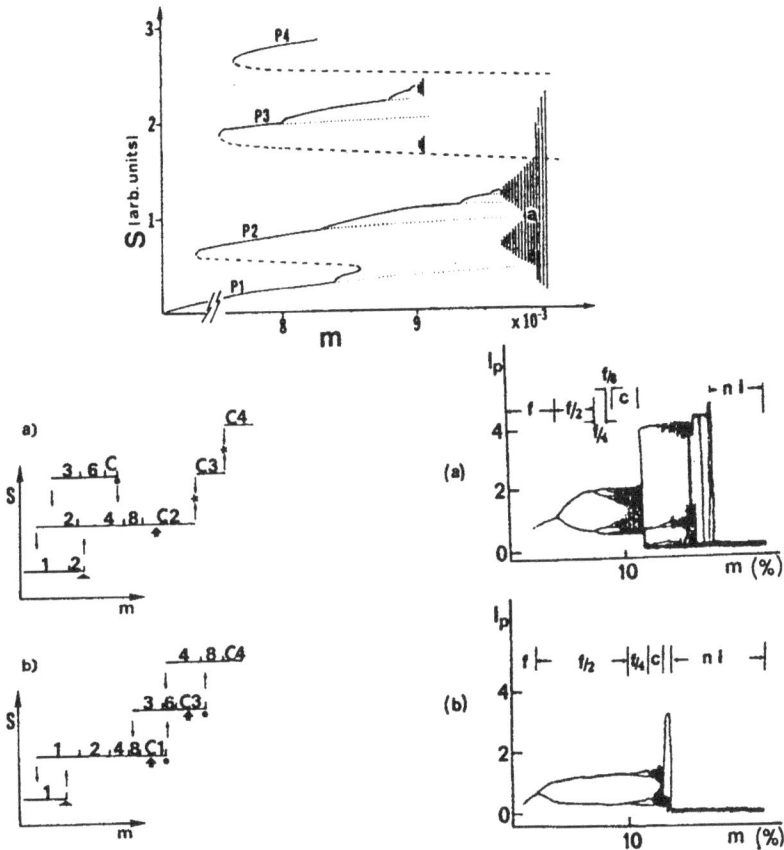

Figure IV-17. Bifurcation diagrams (theory and expt.) for a modulated CO_2 laser after Rev. 19b.

Other parameters can be modulated with similar results as illustrated by the striking resemblance to the figures just presented of experimental data for a CO_2 laser with modulation of the cavity length.[20] In this case one must be careful to use the field amplitude (rather than the intensity) in modelling the dynamics because cavity length modulation results in phase (and frequency) modulation as well as amplitude modulation. Length modulation is also sensitive to the specific value of the average cavity detuning. When the laser is tuned to the peak of its power output, modulation generates a response only at the first harmonic of the modulation frequency. These and other effects are considered in Refs. 20 and 21.

Modulation of semiconductor laser output will find many technological applications, particularly because of the use of such lasers in optical communications systems. The relaxation oscillation frequencies are generally in the range of 0.5 to 4.0 GHz and so the dynamical sensitivity to the modulation rate comes at values which are close to the desired bit rates in communications systems. The response of semiconductor lasers to modulation with the discovery of harmonic and subharmonic resonances and the discovery of period doubled and chaotic solutions can be found in part in references 22 and 23. Selected results are shown in Fig. IV-18.

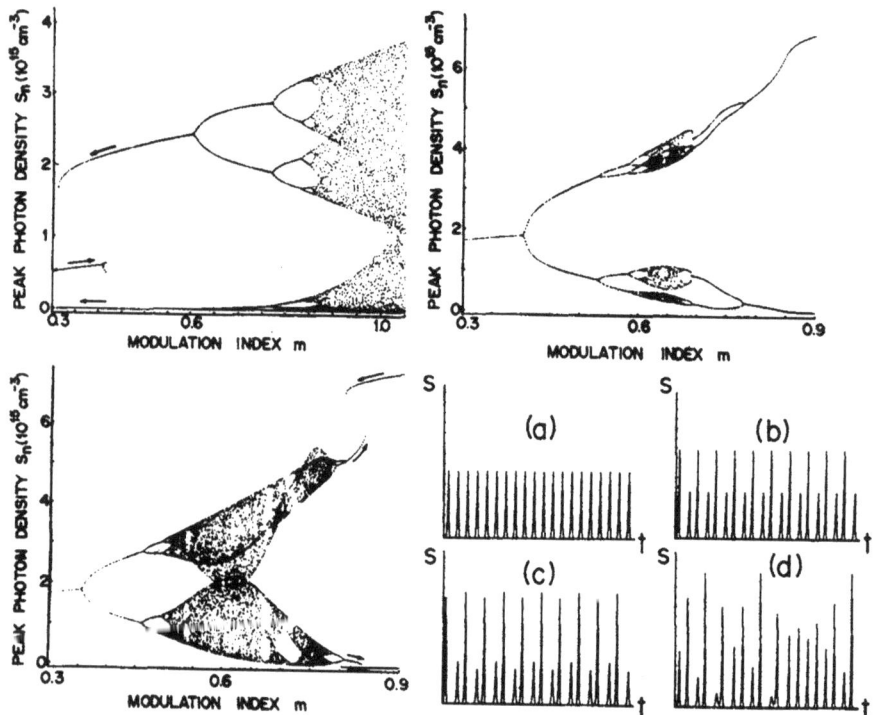

Figure IV-18. Selected bifurcation diagrams and time dependent results for a model of a modulated semiconductor laser from Ref. 23b.

E. Laser with an Injected Signal (LIS)

Another common way in which a laser is "driven" by external forcing is by injecting a signal from another laser. This is often done if one wants increased power at a particular frequency which is achieved if the injected laser is "slaved" (phase and frequency-locked) to the "master oscillator."

Several authors have carefully analyzed the dynamics of such systems with attention to the conditions under which the locking does not occur.[24,25] It would be trivial if there were no interaction between the slave laser and the injected signal as in that case one would observe only the beat note between the two lasers generated by the nonlinear mixing process of an intensity (square-law) detector. For sufficient detuning of the injected signal from the free-running frequency of the injected laser, a simple beat is all that is observed.

Studies of nonlinear dynamics focus on cases of interaction. Nonlinear mixing of the fields occurs in the laser medium of the slave laser. The population inversion is modulated at the beat frequency between the two lasers and the nonlinear susceptibility provides a four-wave mixing process that generates a mirror image frequency of the master oscillator as shown schematically in Fig. IV-19.

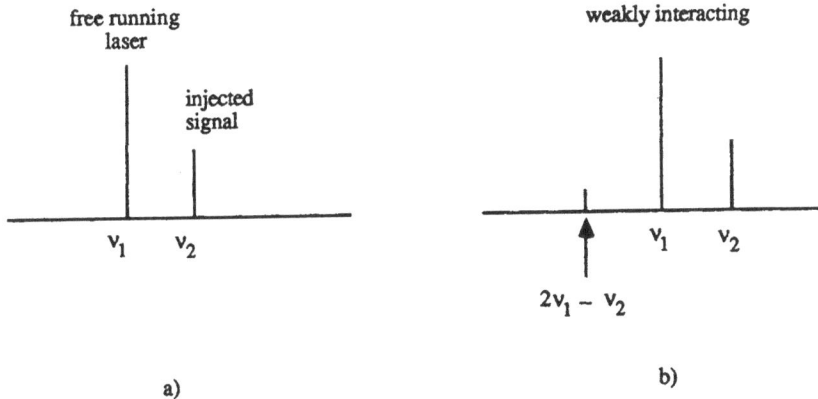

Fig. IV-19. Spectral illustrations of interactions in a laser with an injected signal: a) free running, b) weak nonlinear dynamical interaction.

Alternatively, the free-running laser may be sufficiently susceptible to the influence of the injected signal that it locks its frequency to that of the injected signal. Usually the free-running laser can only be pulled in frequency by approximately its cavity linewidth but the amount of pulling depends on the strength of the injected signal. Close to the locking regime strong frequency modulation and Q-switched pulsing may appear because the slave laser nearly locks to the master laser for a long time and then relatively abruptly switches its phase by 2π with a corresponding large pulse in the intensity output. This is illustrated by theoretical and experimental results for CO_2 laser from Refs. 24 and 26, respectively, as shown in Fig. IV-20.

a)

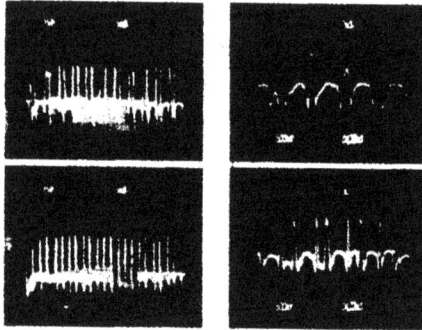

b)

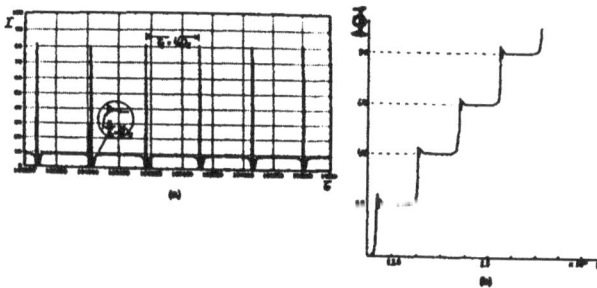

Figure IV-20. a) Experimental and b) Theoretical results for phase slippage and Q-switched like pulsing in a CO_2 laser with an injected signal.

Because of damped resonances in the free running laser, there is also a susceptibility to periodic pulsing when the detuning between the lasers matches a subharmonic or harmonic of the relaxation oscillation frequency.

Similar results have been observed for semiconductor lasers,[27] among others. However, LIS experiments to study nonlinear dynamics are extremely difficult because they require stability of the slave laser parameters to a few percent of the relaxation oscillation frequency and phase stability of the master oscillator to a few hundredths of a radian on the time scale of the relaxation oscillations.[28] The LIS dynamics are apparently somewhat less sensitive to noise in the pump laser amplitude.[28]

F. Bidirectional Ring Lasers

When one studies linear wave equations with boundary conditions there result eigenmodes and the state of the system can be written as a linear combination of the eigenmodes. We carry this bias into the study of nonlinear systems such as lasers. Often the problems are described in terms of eigenmodes and combinations of eigenmodes, where the degree of interaction of the modes may either lead to competition and suppression of all but one mode or to relatively "peaceful coexistence." In this section and in the following two sections, two examples of simple "multimode laser" behavior will be examined to illustrate the nature of solutions of certain nonlinear laser problems which can still be viewed helpfully as the nonlinear interaction of single mode solutions. This means that we can describe the problem in terms of slowly varying amplitudes of modal functions and the nonlinear interaction leads to the correlation and coupling of the variations of the amplitudes.

The simplest case of "multimode laser dynamics" is that of a symmetric ring laser with a small transverse aperture size and a short round-trip length so that only one eigenmode in each direction can interact with net gain from the amplifying medium. The two unidirectional solutions are degenerate in frequency and they interact because they compete for the gain of the medium. This problem also has technological significance because the frequency degeneracy can be broken if the plane of the laser ring rotates about an axis perpendicular to it. As the frequency splitting of the linear eigenmodes is proportional to the rotation frequency, the bidirectional ring laser has found application as a gyroscope. Various aspects of this problem have been reviewed in Refs. 29-31.

A bidirectional ring laser is shown schematically in Fig. IV-21, where E_1 and E_2 represent the slowly varying amplitudes of the outputs in the two directions.

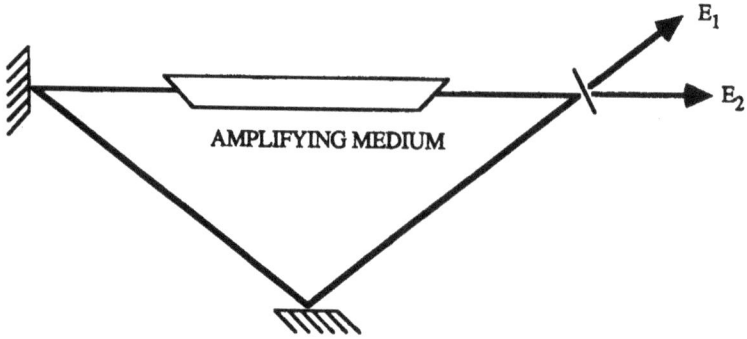

Figure IV-21. Bidirectional Ring Laser.

The most common model for this problem is used to describe lasers for which the material variables can be adiabatically eliminated and for which the gain saturation can be described by a third order non-linearity[32]

$$\frac{dE_1}{dt} = a_1E_1 - \left(|E_1|^2 + \beta\,|E_2|^2 \right) E_1 \qquad\qquad \text{(IV-2a)}$$

$$\frac{dE_2}{dt} = a_2E_2 - \left(|E_2|^2 + \beta\,|E_1|^2 \right) E_2 \qquad\qquad \text{(IV-2b)}$$

where a_1 and a_2 combine the linear gain and loss for each mode and β describes the degree of cross-saturation of the modes. The value of β is fixed by the particular laser material and the parameters with values ranging from $\beta = 2$ for homogeneously broadened media with a stationary standing wave grating burned into the population inversion to $\beta = 0$ for sufficiently detuned operation of Doppler-broadened media for which the two modes interact with completely different velocity classes.

Considering the symmetric case ($a_1 = a_2 = a$), the system of equations (IV-2) has four steady state solutions:

The "TRIVIAL SOLUTION":

$$I_1 = 0 \; ; \; I_2 = 0 \; , \tag{IV-3a}$$

two "UNIDIRECTIONAL SOLUTIONS":

$$I_1 = a \; ; \; I_2 = 0 \; , \tag{IV-3b}$$

$$\text{and } I_1 = 0 \; ; \; I_2 = a \; , \tag{IV-3c}$$

and a "BIDIRECTIONAL SOLUTION",

$$I_1 = \frac{a}{1+\beta} \; ; \; I_2 = \frac{a}{1+\beta} \; , \tag{IV-3d}$$

where $I_i = |E_i|^2$. Above the lasing threshold (a=0) the trivial solution is unstable. The unidirectional solutions are stable in favor of the bidirectional solution for $\beta > 1$ and the stability reverses for $\beta < 1$.[32]

No interesting dynamical behavior results for lasers described by this model unless there is some additional coupling of the forward and backward fields by backscattering from dust, imperfections, or surfaces in the beam line or by back reflection from external mirrors.

If the dynamics of the population inversion are retained, as is necessary for many solid state, CO_2 and semiconductor lasers, then dynamical instabilities are found without the need for backscattering or other couplings between the two fields.[30,31] The population inversion must be described by its longitudinal spatial average and the lowest modulation of the longitudinal spatial pattern by the saturating effect of the standing wave pattern of the counterpropagating fields. The model then becomes:

$$\frac{dE_1}{dt} = -\kappa_1 E_1 + \kappa_1(1+i\Delta)\,\tilde{A}\left(E_1 D_0 + E_2 D_1^*\right) \tag{IV-4a}$$

$$\frac{dE_2}{dt} = -\kappa_2 E_1 + \kappa_1(1+i\Delta)\,\tilde{A}\left(E_2 D_0 + E_1 D_1\right) \tag{IV-4b}$$

$$\frac{dD_0}{dt} = -\gamma_{\parallel}(D_0 - 1) -\tilde{\gamma}_{\parallel}\left(|E_1|^2 + |E_2|^2\right) - \tilde{\gamma}_{\parallel}\left(E_1 E_2^* D_1 + E_1^* E_2 D_1^*\right) \tag{IV-4c}$$

$$\frac{dD_1}{dt} = -\gamma_{\parallel}D_1 -\tilde{\gamma}_{\parallel}D_1\left(|E_1|^2 + |E_2|^2\right) - \tilde{\gamma}_{\parallel}E_1^* E_2 D_0 \; , \tag{IV-4d}$$

where D_0 and D_1 are the uniform population inversion and the complex amplitude of its spatial modulation at one half of the laser wavelength, A is the usual laser excitation parameter, and the tilde denotes division by $1 + \Delta^2$, where Δ is the cavity detuning from the material resonance

286

divided by the polarization decay rate (hence Δ is the detuning measured as a fraction of the homogeneous half linewidth of the medium).

Using this model[31] it can be shown that the bidirectional solution is always unstable and that the unidirectional solutions are both stable when $\Delta = 0$ if $\kappa_1 = \kappa_2$. Detuning of the laser leads to destabilization of the unidirectional laser solutions with a stability boundary as shown schematically in Fig. IV-22.

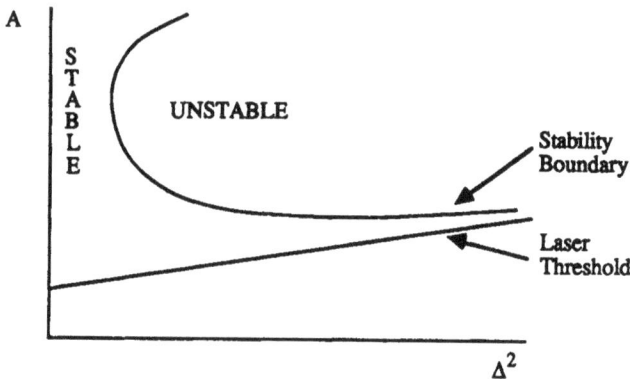

Figure IV-22. Schematic of the stability boundary for the unidirectional solutions of equations (IV-4).

Samples of the time dependent solutions that can be obtained are shown in Fig. IV-23 with examples of slow bidirectional switching, high frequency ringing at the relaxation oscillation frequency and highly irregular pulsations depending on the gain and detuning. In Fig. IV-24 we also show some examples of time dependent outputs observed experimentally for one direction of a bidirectional FIR laser.[33a] The FIR laser pulsations have many of the qualitative features of the model results. Similar pulsations were seen in a bidirectional CO_2 laser.[33b]

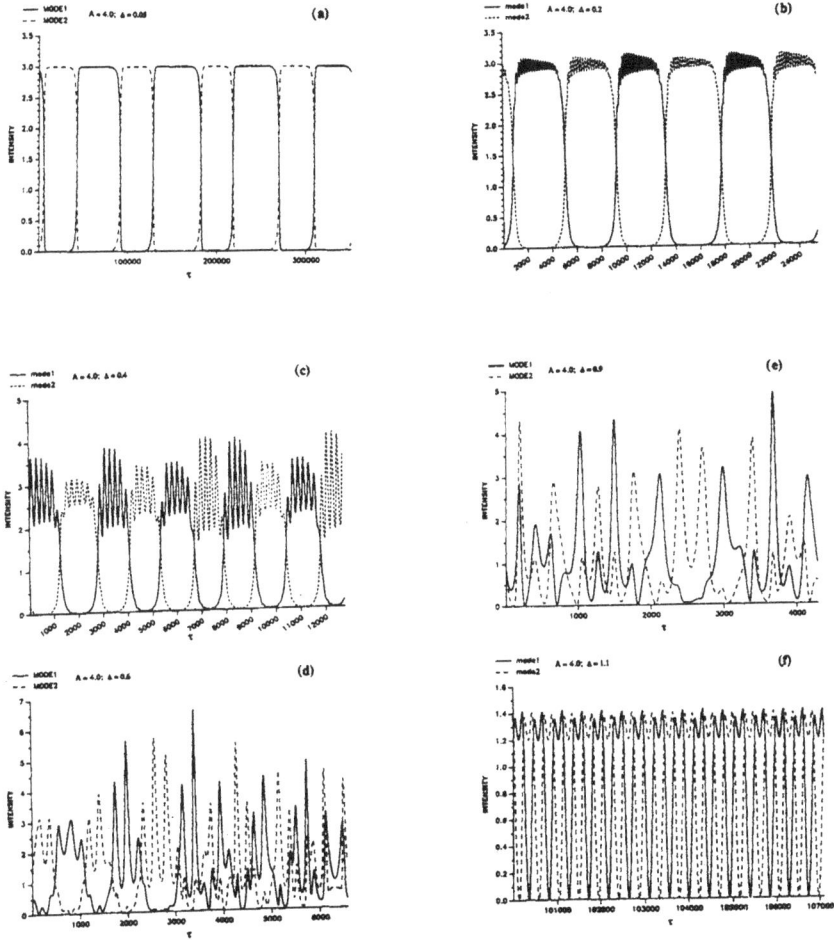

Figure IV-23. Samples of time dependent outputs
from numerical solutions of Ref. 31.

288

a)

b)

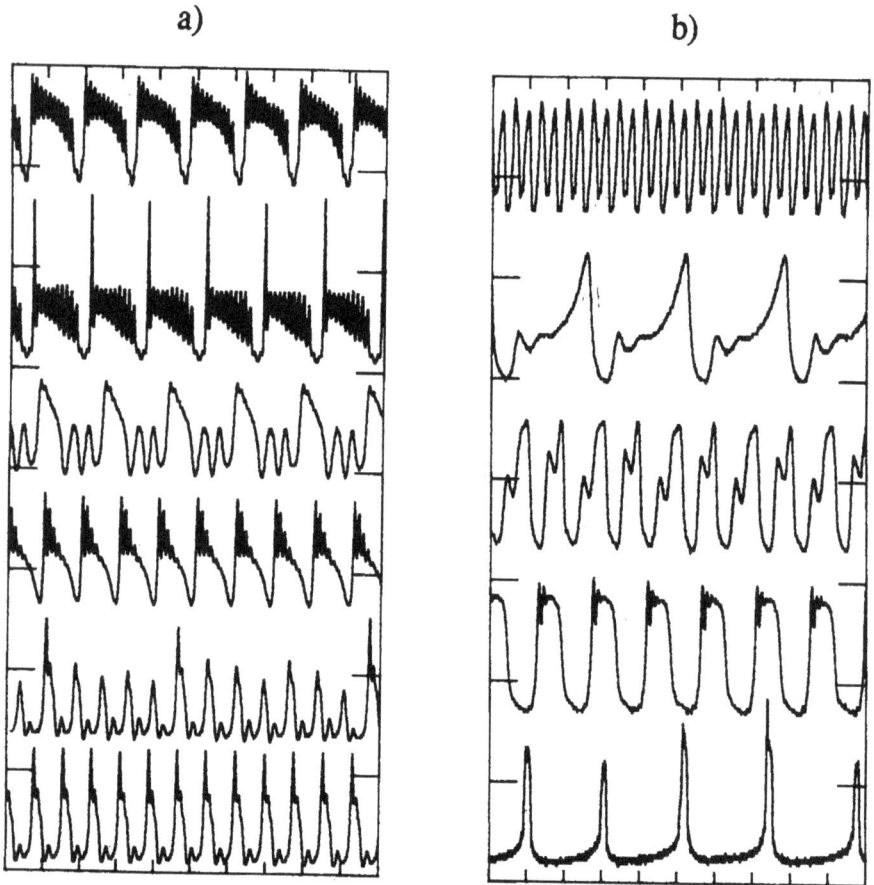

Figure IV-24. Samples of experimentally observed pulsations for output in one direction of a bidirectional FIR laser at a) 81.5 μm using NH_3 and b) 153 μm using $^{15}NH_3$ (courtesy of C.O. Weiss).

For much of the behavior of this laser we can speak in terms of the single mode solutions. The bidirectional solution is like a linear combination of the two single mode solutions, though the nonlinearity chooses a specific combination instead of allowing arbitrary amounts of the two modes. The dynamical behavior appears to be switching of the relative dominance of the two modes and can be referred to as dynamical mode competition. The details are quite subtle and include significant phase and frequency dynamics as well as amplitude pulsations as shown recently in Ref. 34.

G. The Risken-Nummedal Multimode Instability in Homogeneously Broadened Lasers

The multimode instability[35] in homogeneously broadened lasers has a strong connection with the single mode instability. Both use the coherent Rabi oscillations to provide gain at sideband frequencies. The two instabilities differ in that the single mode case uses the bad cavity condition to achieve a dispersive mode-splitting resonance at the sideband frequencies (thereby an instability of the same spatial mode) while the multimode case uses other cavity resonance frequencies matched to the sideband frequencies (to create an instability of modes with different spatial patterns). The threshold gain condition is the same in both cases, relying on the formula

$$A > A_{CO} = 5 + 3\left(\frac{\gamma_{\parallel}}{\gamma_{\perp}}\right) + 2\left[4 + 6\left(\frac{\gamma_{\parallel}}{\gamma_{\perp}}\right) + 2\left(\frac{\gamma_{\parallel}}{\gamma_{\perp}}\right)^2\right]^{1/2} \quad (IV-5)$$

where A_{CO} gives the condition that the gain of the sideband frequencies exceeds the loss at a particular symmetrically located pair of sidebands frequencies. The further condition of a suitable cavity loss (single mode instability) or a suitable cavity mode spacing (multimode instability) must also be met.

As many homogeneously broadened lasers with low losses ("good cavities") can reach 10, 20, or even 100 times above threshold, it is surprising that the predicted transitions from single mode to multimode instabilities at $A > A_c \sim 10$ have not been clearly observed.

The position of cavity mode frequencies to see the multimode effect is given by the following conditions. First we recall the definition of the cavity mode frequencies used earlier:

$$\tilde{\alpha}_m \equiv \frac{2\pi cm}{L\gamma_{\perp}} , \quad (IV-6)$$

where m is an integer, $\tilde{\alpha}_m = \frac{\omega_m - \omega_A}{\gamma_{\perp}}$, and the single mode steady state operates at $\omega_{CO} = \omega_A$.

290

At the critical threshold where gain from the coherent Rabi oscillation just exceeds the loss, we have

$$\tilde{\alpha} = \left\{ \frac{\gamma_{\parallel}}{\gamma_{\perp}} \left[3(A-1) - \gamma_{\parallel}/\gamma_{\perp} \right] \right\}^{1/2} \left(1 + 0(\tilde{\kappa})\right),$$

(IV-7)

where $\tilde{\alpha} = \dfrac{(\omega - \omega_A)}{\gamma_{\perp}}$, and $\tilde{\kappa} = \dfrac{\kappa}{\gamma}$.

This frequency can be evaluated in two common limits:

$$\gamma_{\parallel}/\gamma_{\perp} \approx 0 \qquad\qquad \tilde{\alpha} = \sim 3.3 \left(\frac{\gamma_{\parallel}}{\gamma_{\perp}}\right)^{\frac{1}{2}} \qquad \text{(IV-8a)}$$

$$\gamma_{\parallel}/\gamma_{\perp} \approx 1 \qquad\qquad \tilde{\alpha} = \sim 4.5 \left(\frac{\gamma_{\parallel}}{\gamma_{\perp}}\right)^{\frac{1}{2}} \qquad \text{(IV-8b)}$$

If the instability is to be observed at (or near) the threshold, either

$$\tilde{\alpha}_1 = \tilde{\alpha},$$
$$\tilde{\alpha}_m = \tilde{\alpha}, \text{ or equivalently,}$$
$$\tilde{\alpha}_1 \ll \tilde{\alpha}.$$

These set a condition on the minimum length, L_c, of such a laser. For $\tilde{\alpha}_1 = \tilde{\alpha}$ we obtain

$$L_c = \frac{4 \times 10^8 \text{ m/s}}{(\gamma_{\parallel}\gamma_{\perp})^{\frac{1}{2}}} \qquad \left(\frac{\gamma_{\parallel}}{\gamma_{\perp}} \approx 1\right) \qquad \text{(IV-9a)}$$

and

$$L_c = \frac{6 \times 10^8 \text{ m/s}}{(\gamma_{\parallel}\gamma_{\perp})^{\frac{1}{2}}} \qquad \left(\frac{\gamma_{\parallel}}{\gamma_{\perp}} \approx 0\right) \qquad \text{(IV-9b)}$$

For $\tilde{\alpha}_1 \ll \tilde{\alpha}$ we must have $L \gg L_c$.

Some typical examples can be calculated using the data tabulated earlier for different laser lines.

Laser	A_c	L_c	Typical L
CO_2 10 μm	9	600 m	1-2 m
Semiconductor ~1 μm resonators	9	2 cm	100-500 μm (external cavity may have ~ 5-30 cm)
Dye .6 μm	9	.4 cm	20-50 cm
Sodium .54 μm	15	4 m	50 cm
Nd:YAG 1.06 μm	9	6 m	50 cm

We see that the only cases with suitable modespacings are dye lasers (where one might expects complex effects because of the large number of intervening cavity modes) and semiconductor diodes with AR-coated surfaces and external resonators. One might use unusually large resonators for YAG or CO_2 to try to observe this effect.

Thus far, the only clear reports of instabilities similar to the predicted multimode instability are those of Hillman and Stroud and coworkers[36] and Kuznetsov and coworkers.[37] An example of the dye laser results is shown in Fig. IV-25 and results for a semiconductor laser in an external resonator (but with the semiconductor material proton-bombarded to create saturable absorption) are shown in Fig. IV-26.

Calculations from the simple two level atom theoretical models have not satisfactorily reproduced the dye laser result. More recent work using molecular band models (effectively inhomogeneous broadening) seems to provide an adequate explanation.[38]

With the nearest illustrations involving complicating factors systems (dye bands or semiconductor absorbers), one of the simplest multimode models for space-time instabilities remains without a clear experimental illustration.

$$\Delta\lambda = 0 \text{ Å}$$

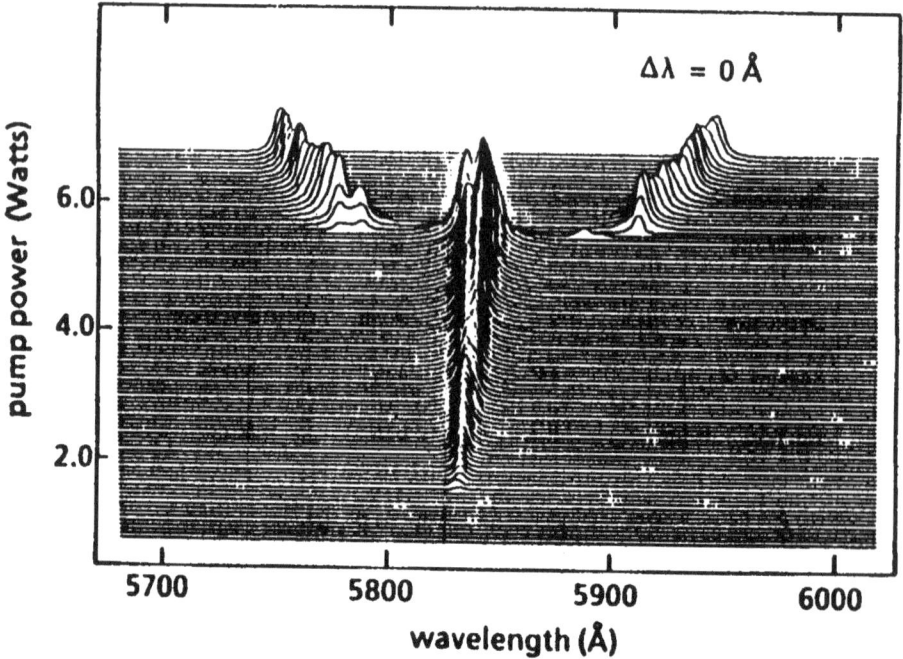

Figure IV-25. Spectral change in a dye laser with increased pumping after Ref 36c.

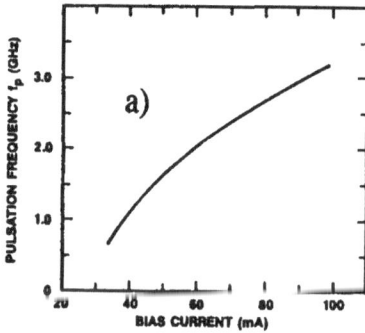

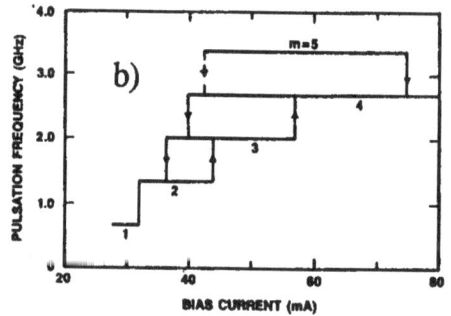

Figure IV-26. a) Change in the relaxation oscillation frequency versus excitation of a laser diode with saturable absorption caused by proton bombardment. b) Multistability of pulsing solutions of an AR-coated diode in an external resonator. Pulsing corresponds to multimode behavior and the mode spacing corresponds to the observed pulsing frequency.

H. Combination-tone generation in 3-mode lasers

Another example of "easy-to-understand" mode interactions is that which occurs when three or more modes are operating at different optical frequencies.

Suppose we begin with eigenmodes as shown in Fig. IV-27 and attempt to picture their dynamical interactions by perturbative reasoning.

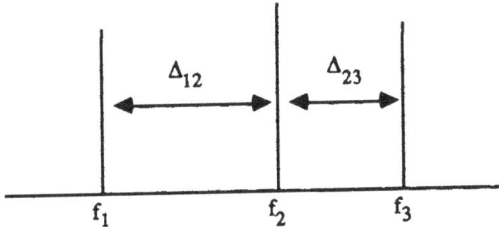

Figure IV-27. Optical frequencies of three eigenmodes.

Suppose that the intermodal spacings are not equal ($\Delta_{12} \neq \Delta_{23}$) as may occur if there is non-uniform mode pulling (as is true in inhomogeneously broadened lasers) or if the modes have different transverse profiles.

The third-order nonlinearity of the gain medium of the laser provides a term for the growth electric field which corresponds to

$$\frac{dE}{dt} = -\kappa E + \beta \, |E|^2 \, E + \ldots \qquad \text{(IV-10)}$$

and which generates new fields at frequencies $f_1 + f_3 - f_2$, $2f_2 - f_3$, and $2f_2 - f_1$. These are pictured on the spectral plot of the field in Fig. IV-28.

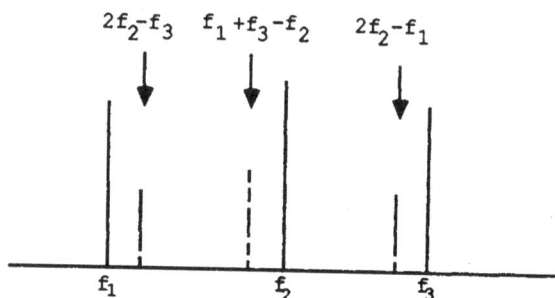

Figure IV-28. Three modes with third order combination frequencies.

Lamb (c.f. ref. 32) has suggested that these third-order combination tones act as injected signals pulling the mode frequencies into a locked condition. This forces $\Delta_{12} = \Delta_{23} = \Delta$ and leads to phase-locked mode-locking and intensity pulses at the frequency Δ. The details of mode locking are a complicated matter of dynamics and are discussed in Narducci's notes and in the references cited in Ref. 38.

A different form of dynamical behavior occurs if the combination tones do not lock the modes but remain as new fields created in the medium. Then the population inversion and the output intensity are modulated at not only Δ_{12}, Δ_{23} and Δ_{13} but also at the difference frequency between the combination tones and the mode frequencies $\delta = f_1 - (2f_2 - f_3) = \Delta_{12} - \Delta_{23}$. Periodic low frequency pulsations can thus be generated (if $\delta \ll \Delta_{ij}$) and these pulsations have been observed to show period doubling, quasiperiodic and chaotic behaviors.[39,40]

Low frequency pulsations in multi-transverse-mode HeNe lasers at 3.39 μm[39,40] have recently been interpreted in terms of nonlinear dynamics and chaos, but such effects were seen very early in multimode laser dynamics and there are many pictures of quasiperiodic, period-doubled and chaotic pulsations in the laser literature of the last 25 years as cited, in part, in Ref. 39b.

As with other modulated laser systems, one might also expect resonances between the low frequency combination-tone-generated modulation at δ with the relaxation oscillation frequency of the amplitudes of the individual modes and the population inversion, since the modulation of the population inversion at frequency δ impresses an amplitude modulation at that frequency on each mode.

Examples of the HeNe laser data are shown in Figs. IV-29 - IV-32. Fig. IV-29 shows a sequence of homodyne and heterodyne spectra indicating the onset of complex low frequency pulsations as Δ_{12} approaches Δ_{23} with variations in the detuning. Fig. IV-30 shows the hysteresis when there is variation of the laser excitation driving the laser into a mode-locked condition which then is stable for lower excitation if the excitation current is slowly reduced.

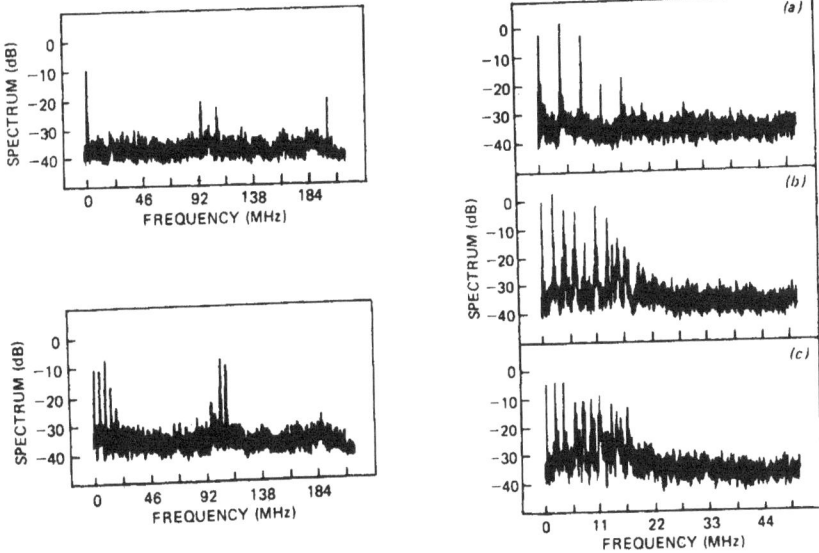

Figure IV-29. Homodyne and heterodyne spectra from Ref. 39b.

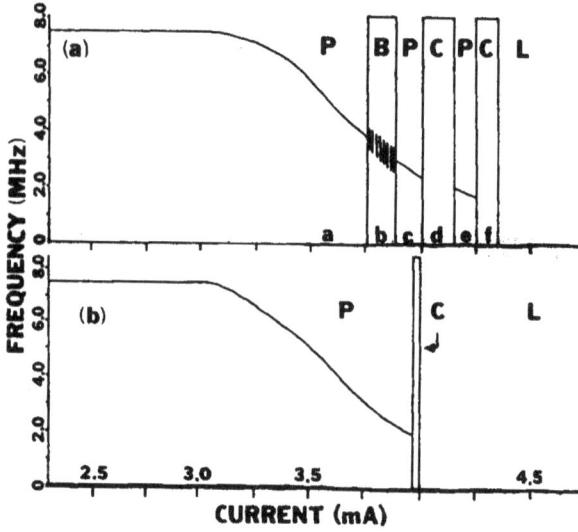

Figure IV-30. Variation of low-frequency intensity pulsation spectra with increasing and decreasing laser excitation from Ref. 39b. Regions are denoted as P (periodic), B (broadened peaks), C (chaotic) and L (mode locked - no low frequency pulsing).

Samples of results from Ref. 40 are shown in Figs. IV-31 and IV-32 where clear period-doubling, quasiperiodic, and intermittency routes to chaos are shown in such a three-mode laser.

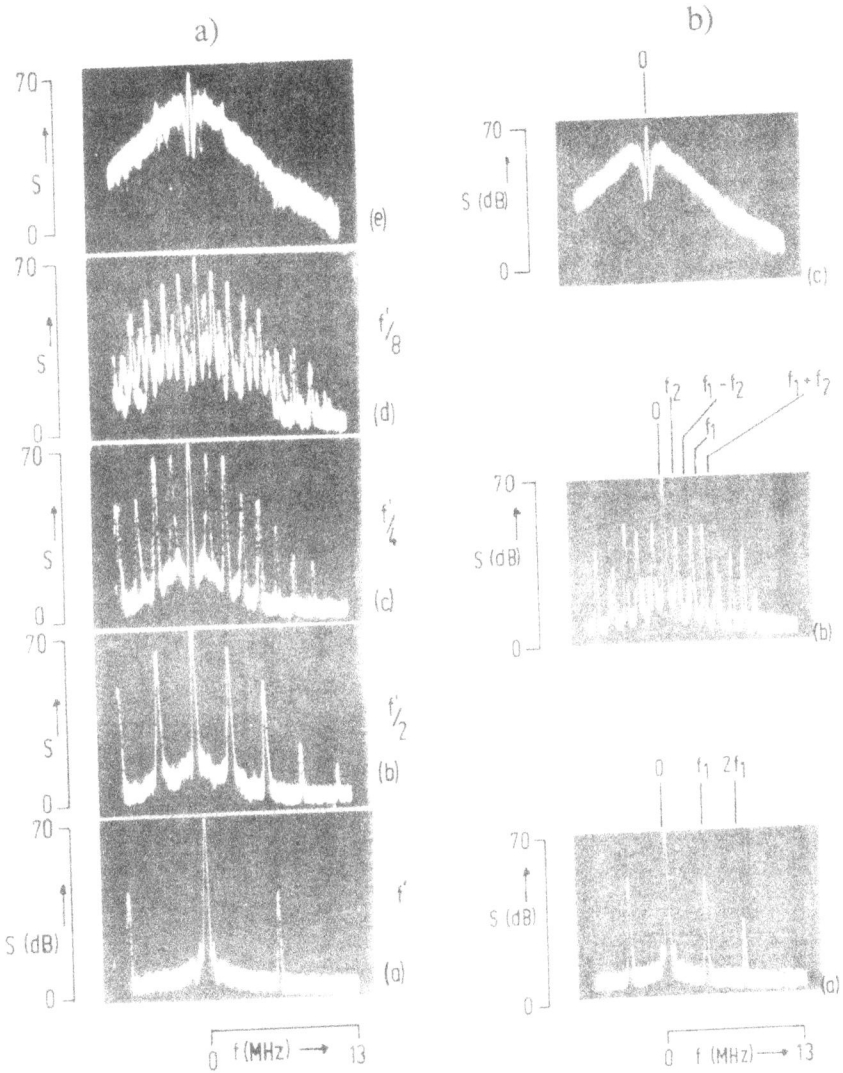

Figure IV-31. a) Period doubling and b) quasiperiodic routes to chaos
for tilting of one laser mirror in the experiments of Ref. 40.

a)

b)

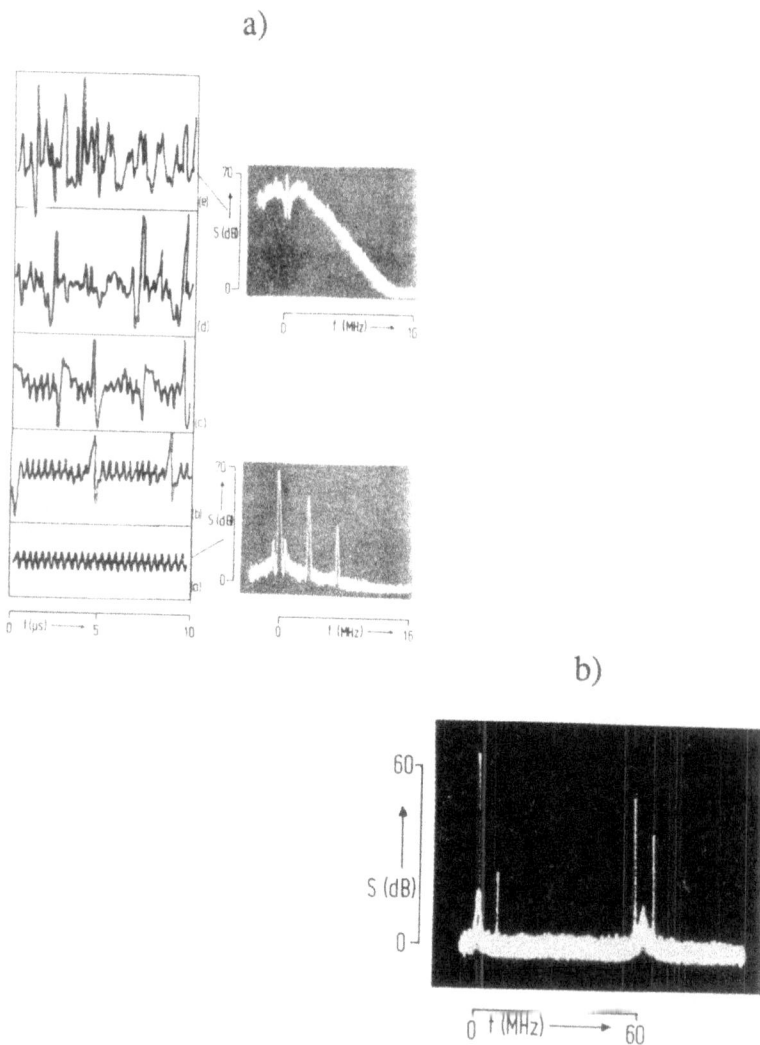

Figure IV-32. a) Intermittency route to chaos and b) intensity beat spectrum showing Δ_{12} and Δ_{23} from Ref. 40.

I. References

1. C.O. Weiss and W. Klische, Opt. Commun. **51**, 47 (1984).

2. M.A. Dupertuis, M.R. Siegrist and R.R.E. Salomaa, IEEE J. Quantum Electron. QE-23, 1217 (1987); J.C. Ryan and N.M. Lawandy, Opt. Commun. **64**, 54 (1987); R. Corbalán, F. LaGuarta, J. Pujol and R. Vilaseca, J. Opt. Soc. Am. B **5**, to be published May 1988; and private communication; P.A. Khandokhin, Ya. I. Khanin and I.V. Koryukin, Opt. Commun. **65**, 367 (1988).

3. W. Klische and C.O. Weiss, Phys. Rev. A **31** 4049 (1985).

4. C.O. Weiss, W. Klische, P.S. Ering and M. Cooper, Opt. Commun. **52**, 405 (1985).

5. C.O. Weiss and J. Brock, Phys. Rev. Lett. **57**, 2804 (1986).

6. T.Q. Wu, V. Schröder, F.I. Saad and W. Klische, Opt. Commun., **65** to be published (1988).

7. M.P. Sassi, N. Barbeau and C.O. Weiss, Appl. Phys. B **43**, 179 (1987).

8. E.N. Lorenz, J. Atmos. Sci. **20**, 130 (1963); C. Sparrow, *The Lorenz Equations: Bifurcations, Chaos and Strange Attractors* (Springer, Berlin, 1982).

9. V.S. Idiatulin and A.V. Uspenskii, Sov. Phys. - Rad. Eng. Electron. Phys. **18**, 422 (1973).

10. N.B. Abraham, D. Dangoisse, P. Gloriex and P. Mandel, J. Opt. Soc. Am. B **2**, 23 (1985); X.-G. Wu and P. Mandel, J. Opt. Soc. Am. B **3**, 724 (1986).

11. K. Shimoda, in *Optical Masers*, Microwave Research Symposia Series, Vol. XIII (Polytechnic Institute of Brooklyn, NY, 1963), page 95.

12. M. Kuznetsov, D.Z. Tsang, J.N. Walpole, Z.L. Liau and E.P. Ippen, in *Optical Instabilities*, eds. R.W. Boyd, M.G. Raymer and L.M. Narducci (Cambridge U. Press, 1986), page 281; M. Kuznetsov, IEEE J. Quantum Electron. QE-21, 587 (1985).

13. a) E. Arimondo, D. Dangoisse, C. Gabbanini, E. Menchi and F. Papoff, J. Opt. Soc. Am. B **4**, 892 (1987); D. Hennequin, F. deTomasi, B. Zambon, and E. Arimondo, Phys. Rev. A **37**, 2243 (1988); M. Tackikawa, K. Tanii, M. Kajita and T. Shimizu, Appl. Phys. B, **39**, 83 (1986); M. Tachikawa, K. Tanii and T. Shimizu, J. Opt. Soc. Am. B **4**, 387 (1987); K. Tanii, M. Tachikawa, M. Kajita and T. Shimizu, J. Opt. Soc. Am. B **5**, 24 (1988); see also J. Opt. Soc. Am. B **5**, May 1988; b) D.E. Chyba, N.B. Abraham, and A.M. Albano, Phys. Rev. A **35**, 2936 (1987) and D.E. Chyba, private communication.

14. D.E. McCumber, Phys. Rev. **141**, 306 (1966).

15. H. Haken, in *Handbuch der Physik*, vol. XXV/2c, ed. L. Genzel (Springer, Heidelberg, 1970).

16. F.T. Arecchi, R. Meucci, G.P. Puccioni and J.R. Tredicce, Phys. Rev. Lett. **49**, 1217 (1982).

17. W. Klische, H.R. Telle, and C.O. Weiss, Opt. Lett. **9**, 561 (1984).

18. J.R. Tredicce, N.B. Abraham, G.P. Puccioni and F.T. Arecchi, Opt. Commun. **55**, 131 (1985).

19. a) J.R. Tredicce, F.T. Arecchi, G.P. Puccioni, A. Poggi and W. Gadomski, Phys. Rev. A **34**, 2073 (1986); b) H.G. Solari, E. Eschenazi, R. Gilmore and J.R. Tredicce, Opt. Commun. **64**, 49 (1987).

20. T. Midavaine, D. Dangoisse, and P. Glorieux, Phys. Rev. Lett. **55**, 1989 (1985); D. Dangoisse, P. Glorieux and D. Hennequin, Phys. Rev. A **36**, 4775 (1987).

21. B.K. Goswami and D.J. Biswas, Phys. Rev. A **36**, 975 (1987).

22. Y.C. Chen, H.G. Winful and J.M. Liu, Appl. Phys. Lett. **47**, 208 (1985).

23. a) M. Tang and S. Wang, Appl. Phys. Lett. **48**, 900 (1986); b) C.H. Lee, T.H. Yoon and S.Y. Shin, Appl. Phys. Lett. **46**, 95 (1985).

24. J.R. Tredicce, F.T. Arecchi, G.L. Lippi and G.P. Puccioni, J. Opt. Soc. Am. B **2**, 173 (1985).

25. D.K. Bandy, L.M. Narducci, L.A. Lugiato, J. Opt. Soc. Am. B **2**, 148 (1985).

26. J.L. Boulnois, A. Van Lerberghe, P. Cottin, F.T. Arecchi and G.P. Puccioni, Opt. Commun. **58**, 124 (1986).

27. F. Mogensen, G. Jacobsen, and H. Olesen, Opt. Quantum Electron. **16**, 183 (1984).

28. M. Bambrilla, L.A. Lugiato, G. Strini and L.M. Narducci in *Coherence, Cooperation and Fluctuations*, eds. F. Haake, L.M. Narducci, D.F. Walls (Cambridge U. Press, 1986) p. 185.

29. *Physics of Optical Ring Gyros*, eds. S.F. Jacobs, M. Sargent III, M.O. Scully, J. Simpson, V. Sanders and J.E. Killpatrick (SPIE, Bellingham, 1984).

30. Ya. I. Khanin, in *Optical Instabilities*, eds. R.W. Boyd, M.G. Raymer and L.M. Narducci, (Cambridge U. Press, Cambridge, 1986), p. 212; P.A. Khandokhin and Ya. I. Khanin, J. Opt. Soc. Am. B **2**, 226 (1985).

31. H. Zeghlache, P. Mandel, N.B. Abraham, L.M. Hoffer, G.L. Lippi and T. Mello, Phys. Rev. A **37**, 470 (1988).

32. M. Sargent III, M.O. Scully and W.E. Lamb, Jr., *Laser Physics* (Academic Press, Reading, 1974); W.E. Lamb, Jr., Phys. Rev. **134**, A1429 (1964).

33. a) N.B. Abraham and C.O. Weiss, private communication; b) G.L. Lippi, J.R. Tredicce, N.B. Abraham and F.T. Arecchi, Opt. Commun. **53**, 129 (1985).

34. L.M. Hoffer, G.L. Lippi, N.B. Abraham and P. Mandel, Opt. Commun. **65**, to be published (1988).

35. H. Risken and K. Nummedal, J. Appl. Phys. **39**, 4662 (1988).

36. a) L.W. Hillman, J. Krasinski, R.W. Boyd and C.R. Stroud, Jr., Phys. Rev. Lett. **52**, 1605 (1984); b) L.W. Hillman, J. Krasinski, K. Koch and C.R. Stroud, Jr., J. Opt. Soc. Am. B **2**, 211 (1985); c) C.R. Stroud, Jr., K. Koch, S. Chakmajian, and L.W. Hillman, in *Optical Chaos*, eds. J. Chrostowski and N.B. Abraham, SPIE Proceedings vol. 667 (Bellingham, 1986), p. 47.

37. M. Kuznetsov, D.Z. Tsang, J.N. Walpole, Z.L. Liau, and E.P. Ippen, Appl. Phys. Lett. **51**, 895, (1987).

38. F. Hong and H. Haken, Opt. Commun. **64**, 454 (1987) and J. Opt. Soc. Am. B. **5** (to be published May 1988).

39. P.W. Smith, M. Duguay and E. Ippen, Prog. Quantum Electron. **3**, 105 (1974).

40. a) N.B. Abraham, T. Chyba, M. Coleman, R.S. Gioggia, N.J. Halas, L.M. Hoffer, S.-N. Liu, M. Maeda and J.C. Wesson, in *Laser Physics*, eds. J. Harvey and D.F. Walls (Springer, Berlin, 1983) p. 107; b) N.J. Halas, S.-N. Liu and N.B. Abraham, Phys. Rev. A **28**, 2915 (1983).

41. C.O. Weiss and H. King, Opt. Commun. **44**, 59 (1982); C.O. Weiss, A. Godone, A. Olafsson, Phys. Rev. A **28**, 892 (1983).

302

V. Postscript

It is now almost two and one-half years since we prepared lecture notes for the Jilin summer school in China. Since then the modern study of laser instabilities and chaos has rapidly expanded. Fortunately, recent meetings have led to conference proceedings and compendia of research papers which make the current literature relatively available in efficient ways to the interested student. Listed below are modern reviews and compilations of research papers which may be consulted as additional references.

A. Reviews

N.B. Abraham, L.A. Lugiato and L.M. Narducci, J. Opt. Soc. Am. B **2**, 7 (1985).

R.G. Harrison and D. Biswas, Prog. Quantum Electron. **11**, 127 (1985).

J.R. Ackerhalt, P.W. Milonni, and M.-L. Shih, Phys. Rep. **128**, 205 (1985).

P.W. Milonni, M.-L. Shih and J.R. Ackerhalt, *Chaos in Laser-Matter Interactions* (World Scientific, Singapore, 1987).

N.B. Abraham, P. Mandel and L.M. Narducci, in *Progress in Optics XXV*, ed. E. Wolf (Elsevier, Amsterdam, 1988).

C.O. Weiss and R. Vilaseca, *Laser Dynamics*, Series on Nonlinear--Phenomena, Methods and Applications, ed. H. Schuster (Physics-Verlag, Weinheim, to be published).

B. Compendia of Research Articles

N.B. Abraham, L.A. Lugiato, and L.M. Narducci, eds., special issue on "Instabilities in Active Optical Media," J. Opt. Soc. Am. B **2**, January 1985.

R.W. Boyd, M.G. Raymer and L.M. Narducci, eds. (Proceedings on International Workshop on Instabilities and Dynamics of Lasers and Nonlinear Optical Systems, June 1985) *Optical Instabilities*, (Cambridge U. Press, 1986).

J.R. Chrostowski and N.B. Abraham, eds., *Optical Chaos*, SPIE Proceedings Vol. 667, June 1986, (SPIE, Bellingham, 1986).

F.T. Arecchi and R.G. Harrison, eds., *Instabilities and Chaos in Quantum Optics*, (Springer Series in Synergetics, vol. 34, ed. H. Haken), (Springer, Berlin, 1987).

N.B. Abraham, F.T Arecchi and L.A. Lugiato, eds., *Instabilities and Chaos in Quantum Optics II* (Proceedings of a NATO ASI, July 1987) (Plenum, NY, 1988).

C.O. Weiss, Opt. and Quantum Electron. **20**, 1 (1988).

D.K. Bandy, A.N. Oraevskii and J.R. Tredicce, eds., special issue on "Dynamics in Lasers" J. Opt. Soc. Am. B **5**, May 1988.

www.ingramcontent.com/pod-product-compliance
Lightning Source LLC
Chambersburg PA
CBHW061626220326
41598CB00026BA/3888